Methods for Assessing Soil Quality

Related Society Publications

BIOREMEDIATION: Science and Applications

Defining Soil Quality for a Sustainable Environment

Methods of Soil Analysis

Pesticides in the Soil Environment: Processes, Impacts, and Modeling

Reactions and Movement of Organic Chemicals in Soils

Soil and Water Science: Key to Understanding Our Global Enviornment

Soil Testing and Plant Analysis, Third Edition

Sorption and Degradation of Pesticides and Organic Chemicals in Soil

For information on these titles, please contact the ASA, CSSA, and SSSA Headquarters Office; Attn.: Marketing; 677 South Segoe Road; Madison, WI 53711-1086. Telephone: (608) 273-8080. Fax: (608) 273-2021.

Methods for Assessing Soil Quality

Methods for Assessing Soil Quality builds on two previous publications, *Defining Soil Quality for a Sustainable Environment* (SSSA Spec. Publ. 35, 1994) and *Soil Health and Sustainability* (Doran et al., 1996). The papers in this publication evolved out of the work and collaboration of the North Regional Committee on Soil Organic Matter and Soil Quality (NCR-59) and the North Central Regional Committee on Impact of Accelerated Erosion on Soil Properties and Productivity (NC-174).

Editors
John W. Doran and Alice J. Jones

Editorial Committee
Richard P. Dick, Rattan Lal, Birl Lowery, Charles W. Rice, and Diane E. Stott

Contributing Administrative Advisors
Signe Betsinger, George E. Ham, and Jerry Klonglan

Editor-in-Chief SSSA
Jerry M. Bigham

Managing Editor
David M. Kral

Associate Editor
Marian K. Viney

SSSA Special Publication Number 49

**Soil Science Society of America, Inc.
Madison, Wisconsin, USA**

1996

Cover Design: John W. Doran
Cover Artwork: Sally Kettler

Copyright © 1996 by the Soil Science Society of America, Inc.

ALL RIGHTS RESERVED UNDER THE U.S. COPYRIGHT LAW OF 1978 (P. L. 94-553)

Any and all uses beyond the "fair use" provision of the law require written permission from the publishers and/or author(s); not applicable to contributions prepared by officers or employees of the U.S. Government as part of their official duties.

Reprinted in 1999.

Soil Science Society of America, Inc.
677 South Segoe Road, Madison, Wisconsin 53711 USA

Library of Congress Registration Number: 96-72332

Printed in the United States of America

CONTENTS

	Page
Foreword	ix
Preface	xi
Contributors	xv
Conversion Factors for SI and non-SI Units	xxi

    Introduction: Importance of Soil Quality to Health and Sustainable Land Management
        David F. Bezdicek, Robert I. Papendick, and Rattan Lal 1

1 Linkages between Soil Quality and Plant, Animal, and Human Health
        L. J. Cihacek, W. L. Anderson, and Phillip W. Barak 9

2 Quantitative Indicators of Soil Quality: A Minimum Data Set
        John W. Doran and Timothy B. Parkin 25

3 Farmer-Based Assessment of Soil Quality: A Soil Health Scorecard
        Douglas E. Romig, M. Jason Garlynd, and Robin F. Harris 39

4 A Conceptual Framework for Assessment and Management of Soil Quality and Health
        Robin F. Harris, Douglas L. Karlen, and David J. Mulla 61

5 On-Farm Assessment of Soil Quality and Health
        Marianne Sarrantonio, John W. Doran, Mark A. Liebig, and Jonathan J. Halvorson 83

6 Standardized Methods, Sampling, and Sample Pretreatment
        Richard P. Dick, David R. Thomas, and Jonathan J. Halvorson 107

7 Physical Tests for Monitoring Soil Quality
        M. A. (Charlie) Arshad, Birl Lowery, and Bob Grossman 123

8 Soil Water Parameters and Soil Quality
        Birl Lowery, M. A. (Charlie) Arshad, Rattan Lal, and William J. Hickey 143

9	Soil Organic Carbon and Nitrogen Lawrence J. Sikora and Diane E. Stott 157
10	Measurement and Use of pH and Electrical Conductivity for Soil Quality Analysis Jeffrey L. Smith and John W. Doran 169
11	Assessing Soil Nitrogen, Phosphorus, and Potassium for Crop Nutrition and Environmental Risk Deborah L. Allan and Randy Killorn 187
12	Role of Microbial Biomass Carbon and Nitrogen in Soil Quality Charles W. Rice, Thomas B. Moorman, and Mike Beare 203
13	Potentially Mineralizable Nitrogen as an Indicator of Biologically Active Soil Nitrogen Laurie E. Drinkwater, Cynthia A. Cambardella, Jean D. Reeder, and Charles W. Rice 217
14	Field and Laboratory Tests of Soil Respiration Timothy B. Parkin, John W. Doran and E. Franco-Vizcaíno 231
15	Soil Enzyme Activities and Biodiversity Measurements as Integrative Microbiological Indicators Richard P. Dick, Donald P. Breakwell, and Ronald F. Turco 247
16	Soil Invertebrates as Indicators of Soil Quality John M. Blair, Patrick J. Bohlen, and Diana W. Freckman 273
17	Tests for Risk Assessment of Root Infection by Plant Pathogens Ariena H. C. van Bruggen and Niklaus J. Grünwald 293
18	Assessing Organic Chemical Contaminants in Soil Thomas B. Moorman 311

SOIL QUALITY ASSESSMENT-PRELIMINARY CASE STUDIES:

19	Soil Quality in Central Michigan: Rotations with High and Low Diversity of Crops and Manure E. Franco-Vizcaíno 327
20	Impact of Farming Practices on Soil Quality in North Dakota John C. Gardner and Sharon A. Clancy 337

21 Use of Soil Quality Indicators to Evaluate Conservation
    Reserve Program Sites in Iowa
      Douglas L. Karlen, Timothy B. Parkin,
      and Neal S. Eash 345

22 Quantifying Soil Condition and Productivity in Nebraska
      Gail L. Olson, Betty F. McQuaid, Karen N. Easterling,
      and Joyce Mack Scheyer 357

23 Soil Quality Assessment Training for Environmental Educators
    of Grades 5 through 12
      Betty F. McQuaid 371

24 A Comparative Study of Soil Quality in Two Vineyards Differing
    in Soil Management Practices
      Stamatis Stamatiadis, A. Liopa-Tsakalidi, L. M. Maniati,
      P. Karageorgou, and E. Natioti 381

25 Soil Quality Information Sheets
      Gary B. Muckel and Maurice J. Mausbach 393

26 Measuring Sustainability of Agricultural Systems
    at the Farm Level
      A. A. Gomez, David E. Swete Kelly, J. K. Syers,
      and K. J. Coughlan 401

FOREWORD

Everyone is in favor of excellent water and air quality. Likewise no one would argue against having excellent soil quality. The concepts of water quality and air quality have been accepted by both the public and the scientific community. For example, if water is suitable for drinking or human consumption, it is considered to have acceptable water quality. For many, the concept of *soil quality* is newer and the standard against which soil quality is compared has not been as well defined. In fact, soil quality is evaluated against the intended use of the soil. It is for this reason that the concept of soil quality is still a bit controversial.

What is soil quality anyway? Soil quality, as referenced in the preface of this publication, states that it is "the capacity of the soil, within land use and ecosystem boundaries, to sustain biological productivity, maintain environmental quality, and promote plant, animal, and human health". An SSSA–ASA symposium in 1992 entitled "Defining Soil Quality for A Sustainable Environment" attempted to define more clearly the soil quality concept. The contents of that symposium were published in 1994 in SSSA Special Publication 35.

The information presented in this publication *Methods for Assessing Soil Quality* results from the continued pursuit by a number of dedicated researchers to develop methodologies to assess soil quality for a range of soils and their uses. The chapters in this publication evolved primarily from the research, debate and actions of two regional research committees, NCR-59 (Soil Organic Matter and Soil Quality) and NC-174 (Impact of Accelerated Erosion on Soil Properties and Productivity). Both committees, working from different viewpoints, use different methods to assess soil quality parameters on a wide range of soils in the Midwestern USA. A number of case studies, some from countries overseas, are included.

The contents of this special publication are progressive, at times provocative, and undoubtedly will stimulate continued debate on the soil quality concept.

D. Keith Cassel
President, SSSA

PREFACE

SOIL QUALITY AND HEALTH: INDICATORS OF SUSTAINABILITY

Soil quality are words being used today, not only across the USA but around the world, to describe the soil's ability to produce food and fiber and to function as an important interface with the environment. It is becoming part of the vocabulary of farmers and ranchers as well as environmentalists, politicians, and researchers. Increasing familiarity and use of the words *soil quality* reflect the growing awareness that soil is an essential component of the biosphere. Soil is required for significant production of food and fiber. It also makes a major contribution to maintaining and enhancing air and water quality at the local, regional, national, and global level.

By functioning as a *living* filter, through which water is cycled and chemicals are altered, soil influences environmental quality and the overall functioning of the biosphere. Soil quality can be broadly defined as the capacity of a soil to function, within land use and ecosystem boundaries, to sustain biological productivity, maintain environmental quality, and promote plant, animal, and human health (after Doran & Parkin, 1994; Karlen et al., 1997). The terms *soil quality* and *soil health* are often used interchangeably. Some people prefer the term *soil health* because it portrays soil as a living, dynamic organism that functions holistically rather than as an inanimate object. Others prefer the term *soil quality* and descriptors of its innate physical, chemical, and biological characteristics. In this book the terms *soil health* and *soil quality* are used synonymously; however, *soil quality* tends to be used more often because of the orientation of the authors.

Soil quality affects three essential facets of sustainable land management: *Productivity* of crops and livestock, *Environmental Quality* of natural resources, and *Health* of plants, animals, and humans. To successfully assess soil quality, today's researchers are challenged to develop research philosophies and approaches that facilitate holistic, system-oriented investigations. Diverse disciplines that span production, environmental quality, and health will be needed to implement such investigations and to generate technologies that can be adapted and used by land managers. Thus, it is not reasonable to expect that we can assess soil quality using only traditional reductionistic approaches that confine our interpretations to narrow scientific disciplines (Bouma, 1997). Nor is it reasonable to assume that such investigations, although interdisciplinary, can be fully successful without the involvement of agricultural producers and land managers as active partners in the research process. This point was emphasized at a recent international symposium on "Advances in Soil Quality for Land Management: Science, Practice, and Policy" held in Ballarat, Australia, in April 1996; the theme of this conference was "Soil Quality is in the Hands of the Land Manager."

The importance of soil quality assessment to sustainability and agricultural policy decision making was highlighted in the National Research Council report entitled *Soil and Water Quality: An Agenda for Agriculture*. This study concluded that, "Protecting soil quality, like protecting air and water quality, should be a fundamental goal of national environmental policy" (National Research Council, 1993). The need to develop methodology to characterize and define management factors controlling the degradation, maintenance, and rehabilitation of soil quality is gaining greater national and international recognition. During the last 5 yr, concern about deficiencies in the basic understanding of soil quality and lack of mechanistically-based soil quality methodology, particularly the soil biota, was a major focus of conferences and publications on sustainability and soil quality in the USA (Rodale, Pennsylvania; ASA and SSSA, Minnesota), Thailand, Hungary , the Netherlands, Australia, Canada, and elsewhere (see Acton & Gregorich, 1995; Greenland & Szabolcs, 1994).

In an effort to address the need for standardized methods and strategies for assessing soil quality and health, the NCR-59 Regional Technical Committee on 'Soil Organic Matter and Soil Quality' formed a subcommittee in September 1992. The committee's goal was to coalesce as much information as possible on the various soil physical, chemical, and biological properties considered to be essential in assessing soil quality and to provide standardized methods and protocols for soil quality assessment. Cross referencing of measurements from the literature is complicated by the fact that analytical results are methods dependent and protocols for measurement vary among investigators. Therefore, establishment of standardized methods and protocols was viewed essential to development of useful data bases and indices for soil quality. Disciplinary perspectives and applications for this approach were broadened in 1993, when the NC-174 Regional Technical Committee on 'Impact of Accelerated Erosion on Soil Properties and Productivity" joined NCR-59 as a partner in developing this book on soil quality methods and approaches.

There are two unique features of *Methods for Assessing Soil Quality* that separate it from other methods texts written by soil scientists. First is the inclusion of chapters that focus on the linkages of soil quality to the health of plants, animals, and humans, and farmer-based approaches to assessing soil quality. Second, are the concluding chapters highlighting preliminary case studies that discern land use and management impacts on soil quality; develop and synthesize possible soil quality indices for sustainability; and demonstrate educational tools and techniques to increase knowledge and understanding about soil quality and its role in the biosphere.

Methods for Assessing Soil Quality builds on two previous publications, *Defining Soil Quality for a Sustainable Environment* (SSSA Spec. Publ. 35; Doran et al., 1994) and *Soil Health and Sustainability* (Doran et al., 1996). This book bridges our understanding of the theory, methods, and applications of soil quality. Thus, research scientists, resource managers, consultants, farm owners and operators, and educators can quickly grasp and use portions of the information presented here. More important, a greater appreciation of the connectedness of soil quality across production, environmental, and societal entities can be gained by all readers.

Approaches presented in this book provide a unique illustration of how research techniques can be made transferable and relevant to agricultural producers and the general public. Authors include agricultural and environmental researchers, extension educators, Natural Resources Conservation Service staff, and other closely aligned scientists that have demonstrated an interest and involvement in assessment of soil quality and health. Many have already made considerable contributions in this area. Author experiences range from that of highly established professionals to those who are at the beginning of their professional careers. Selection of co-authors for each chapter has been made with the idea of achieving a balance of viewpoints, experience, and institutional representation.

The process involved with developing this book may prove to be as important as the final product. It provided a unique opportunity for professionals within the federal, state, and private sectors to work cooperatively on an issue of major concern to farming, agricultural sustainability and the general survival and well-being of people around the world.

It is our hope that this book will advance agricultural sustainability by providing standardized tools and approaches for assessing the effects of land management on soil quality and health. The integrative approach to assessing the effects of agricultural management on soil function and sustainability is intended to extend beyond the usual reductionistic constraints of the physical, chemical, and biological disciplines.

The editors extend thanks to the 59 contributors to this book and to the editorial committee: Richard P. Dick, Rattan Lal, Birl Lowery, Charles W. Rice, and Diane Stott, for their vigilance and persistence during review and revision of this book. Thanks also are extended to the administrative advisors for the NCR-59 and NC-174 regional committees, Signe Betsinger (Minnesota), George E. Ham (Kansas), and Jerry Klonglan (Iowa) for facilitating this joint project.

JOHN W. DORAN
USDA-ARS
University of Nebraska
Lincoln, Nebraska

ALICE J. JONES
University of Nebraska
Lincoln, Nebraska

REFERENCES

Acton, D.F., and L.J. Gregorich. 1995. The health of our soils: Toward sustainable agriculture in Canada. Agric. Agri-Food Can., Res. Branch Publ. 1906/E, Ottawa, Canada.

Bouma, J. 1997. Soil environmental quality: A European perspective. J. Environ. Qual. 26:26–31.

Doran, J.W., D.C. Coleman, D.F. Bezdicek, and B.A. Stewart (ed.). 1994. Defining soil quality for a sustainable environment. SSSA Spec. Publ. 35. SSSA, Madison, WI.

Doran, J.W., and T.B. Parkin. 1994. Defining and assessing soil quality. p. 3–21. *In* J.W. Doran et al. (ed.) Defining soil quality for a sustainable environment. SSSA Spec. Publ. 35. SSSA, Madison, WI.

Doran, J.W., M. Sarrantonio, and M.A. Liebig. 1996. Soil health and sustainability. Adv. Agron. 56:1–54.

Greenland, D.J., and I. Szabolcs. 1994. Soil resilience and sustainable land use. CAB Int., Wallingford, Oxon, England.

Karlen, D.L., M.J. Mausbach, J.W. Doran, R.G. Cline, R.F. Harris, and G.E. Schuman. 1997. Soil quality: A concept, definition, and framework for evaluation. Soil Sci. Soc. Am. J. 61:4–10.

National Research Council. 1993. Soil and water quality: An agenda for agriculture. Committee on Long-Range Soil and Water Conserv., Board on Agric., Natl. Res. Council, Natl. Academy Press, Washington, DC.

CONTRIBUTORS

Deborah L. Allan	Associate Professor of Soil Science, Department of Soil, Water, and Climate, 1991 Upper Buford Circle, University of Minnesota, St. Paul, MN 55108; telephone: 612-625-3158, fax: 612-625-2208, email: dallan@soils.umn.edu
W. L. Anderson	Professor of Soil Science, School of Agribusiness and Agriscience Middle Tennessee State University, School of Agribusiness and Agriscience, Murfreesboro, TN 37132; telephone: 615-898-2408, fax: 615-898-5169, e-mail: andersonw@acad1.mtsu.edu
M. A. (Charlie) Arshad	Research Scientist, Agriculture and Agri-Food Canada, Research Station Road, P.O. Box 29, Beaverlodge Alberta T0H 0C0 Canada; telephone: 403-354-5110, fax: 403-354-8171, email: arshadc@em.agr.ca
Phillip Barak	Assistant Professor of Soil Science, Department of Soil Science, University of Wisconsin, 1525 Observatory Drive, Madison, WI 53706; telephone: 608-263-5450, fax: 608-265-2595, email: barak@calshp.cals.wisc.edu
Mike Beare	Soil Scientist, New Zealand Institute of Crop and Food Research, Canterbury Agriculture and Science Centre, Private Bag 4704, Christchurch, New Zealand; telephone: 64-3-325-6400, fax: 64-3-325-2074, email: bearem@lincoln.cri.nz
David F. Bezdicek	Center for Sustainable Agriculture and Natural Resources, Department of Crop and Soil Sciences, P.O. Box 646420, Washington State University, Pullman, WA 99164-6240; telephone: 509-335-3644, fax: 509-335-8674, email: bezdicek@wsu.edu
John M. Blair	Assistant Professor of Biology, Division of Biology, Kansas State University, Manhattan, KS 66506; telephone: 913-532-7065, fax: 913-532-6653, email: jbiair@ksu.ksu.edu
Patrick J. Bohlen	Research Associate, Institute of Ecosystem Studies, Box AB, Millbrook, NY 12545; telephone: 914-677-7651, fax: 914-677-5976, email: cajv@vm.marist.edu
Donald P. Breakwell	Assistant Professor, Department of Life Sciences, Snow College, Ephraim, UT 84627
Cynthia A. Cambardella	USDA-ARS, National Soil Tilth Laboratory, 2150 Pammel Drive, Ames, IA 50011; telephone: 515-294-2921, fax: 515-294-8125, email: cindyc@nstl.gov
L. J. Cihacek	Associate Professor of Soil Science, Soil Science Department, North Dakota State University, Box 5638, Fargo, ND 58105; telephone: 701-231-8572, fax: 701-231-7861, email: cihacek@badlands.nodak.edu

CONTRIBUTORS

Sharon A. Clancy	Ecologist, North Dakota State University, Carrington Research Education Center, P.O. Box 219, Carrington, ND 58421; telephone: 701-652-2951, fax: 701-652-2055, email: recenter@ndsuext.nodak.edu
K. J. Coughlan	Program Coordinator, Land and Water Resources, ACIAR, Canberra, Australia
Richard P. Dick	Professor of Soil Science, 3017 Agriculture and Life Sciences, Oregon State University, Corvallis, OR 97331-7306; telephone: 541-737-5718, fax: 541-737-5725, email: dickr@ccs.orst.edu
John W. Doran	Soil Scientist, USDA-ARS, 116 Keim Hall, University of Nebraska, Lincoln, NE 68583-0915; telephone: 402-472-1510, fax: 402-472-0516, email: jdoran@unlinfo.unl.edu
Laurie E. Drinkwater	Rodale Research Center, 611 Siegfriedale Road, Kutztown, PA 19530, telephone: 610-683-1437, fax: 610-683-8548, email: ldrink@rodaleinst.org
Neal S. Eash	Assistant Professor, Plant and Soil Science Department, University of Tennessee, Knoxville, TN 37901
Karen N. Easterling	Statistician, Pharmaceutical Product Development, North Carolina State University, 1500 Perimeter Park Drive, Suite 300, Morrisville, NC 27560; telephone: 919-363-4343, email: easterkn@ppdi.com
E. Franco-Vizcaíno	Visiting Professor, Department Crop and Soil Science, Michigan State University, East Lansing, MI 48854-1325, telephone: 517-355-0223, fax: 517-353-5174, email: 228335mgr@msu.edu
Diana W. Freckman	Professor and Director, Natural Resource Ecology Laboratory, Colorado State University, Fort Collins, CO 80523; telephone: 970-491-1982, fax: 970-491-3945, email: dfreckman@lternet.edu
John Gardner	Director and Agronomist, North Dakota State University, Carrington Research Center, P.O. Box 219, Carrington, ND 58421; telephone: 701-652-2951, fax: 701-652-2055, email: jogardne@prairie.nodak.edu
M. Jason Garlynd	Department of Soil Science, 1525 Observatory Drive, University of Wisconsin, Madison, WI 53706-1299; telephone: 608-263-5691, fax: 608-265-2595
A. A. Gomez	SEARCA, Laguna 4031, Philippines; telephone: 632-818-1926, fax: 632-817-0598, email: aag@agri.searca.org
Bob Grossman	USDA-SCS, Room 152, Federal Building, Lincoln, NE 68508-5760; telephone: 402-437-5697, fax: 402-437-5336
Niklaus J. Grünwald	Research Assistant, Department of Plant Pathology, University of California, Davis, CA 95616; telephone: 916-752-7795, fax: 916-752-5674, email: njgrunwald@ucdavis.edu
Jonathan J. Halvorson	Research Assocaite, USDA-ARS, 215 Johnson Hall, Washington State University, Pullman, WA 99164-6421; telephone: 509-335-2263, fax: 509-335-3842, email: halvorjj@mail.wsu.edu

CONTRIBUTORS

Robin F. Harris	Professor of Soil Science and Chair, Department of Soil Science, 1525 Observatory Drive, University of Wisconsin, Madison, WI 53706-1299; telephone: 608-263-5691, fax: 608-265-2595, email: rfharris@facstaff.wisc.edu
William J. Hickey	Associate Professor, Department of Soil Science, University of Wisconsin, 1525 Observatory Drive, Madison WI 53706-1299; telephone: 608-262-9018, fax: 608-265-2595, email: wjhickey@facstaff.wisc.edu
Douglas L. Karlen	Research Soil Scientist, USDA-ARS, National Soil Tilth Laboratory, 2150 Pammel Drive, Ames, IA 50011, telephone: 515-294-3336, fax: 515-294-8125, email: dkarlen@nstl.gov
P. Karageorgou	Research Assistant, Goulandris Natural History Museum, 13 Levidou Street, 14562 Kifisia, Greece; telephone: 01-8087345, fax: 01-8080674
Randy Killorn	Professor of Soil Fertility, 2104 Agronomy Hall, Iowa State University, Ames, IA 50011; telephone: 515-294-1923, fax: 515-294-9985, email: rkillorn@iastate.edu
Rattan Lal	Professor of Soil Science, School of Natural Resources, Ohio State University, 2021 Coffey Road, Columbus, OH 43210-5445; telephone: 614-292-9069, fax: 614-292-4424 or 7432, email: lal.l@osu.edu
Mark A. Liebig	Graduate Research Assistant, Department of Agronomy, 197 Plant Science Hall, University of Nebraska, Lincoln NE 68583-0915; telephone: 402-472-9035, fax: 402-472-0516, email: mliebig@unlinfo.unl.edu
A. Liopa-Tsakalidi	Researcher, Foundation for Research and Technology, Institute of Chemical Engineering and High Temperature Chemical Processes, P.O. Box 1414, GR-26500 Patpas, Greece; telephone: 061-991527, fax: 061-991527
Birl Lowery	Professor, Department of Soil Science, 525 Observatory Drive, University of Wisconsin, Madison, WI 53706, telephone: 608-262-2752, fax: 608-265-2595, email: blowery@facstaff.wisc.edu
L.M. Maniati	Research Assistant, Goulandris Natural History Museum, 13 Levidou Street, 145 62 Kifissia, Greece; telephone: 01-8087345, fax: 01-8080674
Maurice J. Mausbach	Director, Soil Quality Institute, USDA-NRCS, 2150 Pammel Drive, Ames, IA 50011; telephone: 515-294-4592, fax: 515-294-8125, email: mausbach@nstl.gov
Betty F. McQuaid	Soil Ecologist, USDA-NRCS, Watershed Sciences Institute, 1509 Varsity Drive, Raleigh, NC 27606; telephone: 919-515-9482, fax: 919-515-3593, email: betty_mcquaid@ncsu.edu
Thomas B. Moorman	Microbiologist, USDA-ARS, National Soil Tilth Laboratory, 2150 Pammel Drive, Ames, IA 50011; telephone: 515-294-2308, fax: 515-294-8125, email: moorman@nstl.gov.
Gary B. Muckel	Soil Scientist, USDA-NRCS, l00 Centennial Mall North, Room 152, Lincoln, NE 68508-3866; telephone: 402-437-4148, fax: 402-437-5336, email: gmuckel@nssc.nssc.nrcs.usda.gov

David J. Mulla	Professor and W.E. Larson Chair for Soil and Water Resources, Department of Soil, Water and Climate, 564 Borlaug Hall, University of Minnesota, 1991 Upper Buford Circle, St. Paul, MN 55108; telephone: 612-625-6721, fax: 612-624-4223, email: dmulla@soils.umn.edu
E. Natioti	Research Assistant, Goulandis Natural History Museum, 13 Levidou Street, 145 62 Kifisia, Greece; telephone: 01-8087345, fax: 01-8080674
Gail L. Olson	Lockheed Martin Idaho Technologies, P.O. Box 1625, Idaho Falls, ID 83415-2107; telephone: 208-526-4069, fax: 208-526-0603, email: olsogl@inel.gov
Robert I. Papendick	USDA-ARS, 215 Johnson Hall, Washington State University, Pullman, WA 99164-6421; telephone: 509-335-1552, fax: 509-335-3842
Timothy B. Parkin	Soil Microbiologist, USDA-ARS, National Soil Tilth Laboratory, 2150 Pammel Drive, Ames, IA 50011; telephone: 515-294-6888, fax: 515-294-8125, email: parkin@nstl.gov
Jean D. Reeder	Soil Scientist, USDA-ARS, Crops Research Laboratory, 1701 Center Avenue, Fort Collins, CO 80526; telephone: 970-498-4236, fax: 970-482-2909, email: jdreeder@lamar.colostate.edu
Charles W. Rice	Associate Professor of Soil Microbiology, Department of Agronomy, Throckmorton Plant Sciences Center, Kansas State University, Manhattan, KS 66506-5501; telephone: 913-532-7217, fax: 913-532-6094, email: cwrice@ksu.edu
Douglas E. Romig	Department of Soil Science, University of Wisconsin, Madison, WI 53706; telephone: 608-265-4850, fax: 608-265-2595, email: dromig@emnrdsf.state.nm.us
Marianne Sarrantonio	Assistant Professor of Agroecology, 101 Eisenberg Hall, Slippery Rock University, Slippery Rock, PA 16057; telephone: 412-738-2972, fax: 412-738-2959, email: marianne.sarrantonio@sru.edu
Joyce Mack Scheyer	Soil Scientist, USDA-NRCS, National Soil Survey Center, Federal Building, Room 152, 100 Centennial Mall North, Lincoln, NE 68508-3866; telephone: 402-437-5698, fax: 402-437-5336, email: mack@nssc.nrcs.usda.gov
Lawrence J. Sikora	Soil Microbiologist, USDA-ARS, Soil Microbial Systems Laboratory, Building 318, BARC-East, Beltsville, MD 20705; telephone: 301-504-9384, fax: 301-504-8370, email: lsikora@asrr.arsusda.gov
Jeffrey L. Smith	Soil Biochemist, USDA-ARS, 215 Johnson Hall, Washington State University, Pullman, WA 99164-6421; telephone: 509-335-7648, fax: 509-335-3842, email: jlsmith@mail.wsu.edu
Stamatis Stamatiadis	Laboratory Director, Goulandris Natural History Museum, 13 Levidou Street, 145 62 Kifissia, Greece; telephone: 01-8087345, fax: 01-8080674, email: stam@greece_nature.ath.forthnet.gr
D.E. Stott	Research Soil Microbiologist, USDA-ARS, National Soil Erosion Research Laboratory, 1196 Soil Building, Purdue University, West Lafayette, IN 47907-1196; telephone: 317-494-6657, fax: 317-494-5948, email: stottd@soils.ecn.purdue.edu

CONTRIBUTORS

David E. Swete Kelly	Maroochy Horticultural Research Station, P.O. Box 5083, Nambour, Queensland 4560, Australia; telephone: 011-617-441-2211, fax: 011-617-441-2235, email: swetekd@dpi.qld.gov.au
J. K. Syers	Professor, Department of Agriculture and Environmental Science, The University of Newcastle Upon Tyne, Newcastle Upon Tyne, United Kingdom
David R. Thomas	Professor, Department of Statistics, 44 Kidder, Oregon State University, Corvallis, OR 97331; telephone: 541-737-1983, fax: 541-737-3489, email: thomas@stat.orst.edu
Ronald F. Turco	Department of Agronomy, 1150 Lily Hall, Purdue University, West Lafayette, IN 47907-1150; telephone: 317-494-8077, fax: 317-494-6508, e-mail rturco@dept.agry.purdue.edu
Ariena H.C. van Bruggen	Professor of Plant Pathology, Department of Plant Pathology, University of California, Davis, CA 95616; telephone: 916-752-5026, fax: 916-752-5674, email: ahvanbruggen@ucdavis.edu

Conversion Factors for SI and non-SI Units

Conversion Factors for SI and non-SI Units

To convert Column 1 into Column 2, multiply by	Column 1 SI Unit	Column 2 non-SI Units	To convert Column 2 into Column 1, multiply by
Length			
0.621	kilometer, km (10^3 m)	mile, mi	1.609
1.094	meter, m	yard, yd	0.914
3.28	meter, m	foot, ft	0.304
1.0	micrometer, µm (10^{-6} m)	micron, µ	1.0
3.94×10^{-2}	millimeter, mm (10^{-3} m)	inch, in	25.4
10	nanometer, nm (10^{-9} m)	Angstrom, Å	0.1
Area			
2.47	hectare, ha	acre	0.405
247	square kilometer, km² (10^3 m)²	acre	4.05×10^{-3}
0.386	square kilometer, km² (10^3 m)²	square mile, mi²	2.590
2.47×10^{-4}	square meter, m²	acre	4.05×10^3
10.76	square meter, m²	square foot, ft²	9.29×10^{-2}
1.55×10^{-3}	square millimeter, mm² (10^{-3} m)²	square inch, in²	645
Volume			
9.73×10^{-3}	cubic meter, m³	acre-inch	102.8
35.3	cubic meter, m³	cubic foot, ft³	2.83×10^{-2}
6.10×10^4	cubic meter, m³	cubic inch, in³	1.64×10^{-5}
2.84×10^{-2}	liter, L (10^{-3} m³)	bushel, bu	35.24
1.057	liter, L (10^{-3} m³)	quart (liquid), qt	0.946
3.53×10^{-2}	liter, L (10^{-3} m³)	cubic foot, ft³	28.3
0.265	liter, L (10^{-3} m³)	gallon	3.78
33.78	liter, L (10^{-3} m³)	ounce (fluid), oz	2.96×10^{-2}
2.11	liter, L (10^{-3} m³)	pint (fluid), pt	0.473

CONVERSION FACTORS FOR SI AND NON-SI UNITS

Mass

To convert Column 1 into Column 2, multiply by	Column 1 SI Unit	Column 2 non-SI Unit	To convert Column 2 into Column 1, multiply by
2.20×10^{-3}	gram, g (10^{-3} kg)	pound, lb	454
3.52×10^{-2}	gram, g (10^{-3} kg)	ounce (avdp), oz	28.4
2.205	kilogram, kg	pound, lb	0.454
0.01	kilogram, kg	quintal (metric), q	100
1.10×10^{-3}	kilogram, kg	ton (2000 lb), ton	907
1.102	megagram, Mg (tonne)	ton (U.S.), ton	0.907
1.102	tonne, t	ton (U.S.), ton	0.907

Yield and Rate

To convert Column 1 into Column 2, multiply by	Column 1 SI Unit	Column 2 non-SI Unit	To convert Column 2 into Column 1, multiply by
0.893	kilogram per hectare, kg ha^{-1}	pound per acre, lb acre^{-1}	1.12
7.77×10^{-2}	kilogram per cubic meter, kg m^{-3}	pound per bushel, lb bu^{-1}	12.87
1.49×10^{-2}	kilogram per hectare, kg ha^{-1}	bushel per acre, 60 lb	67.19
1.59×10^{-2}	kilogram per hectare, kg ha^{-1}	bushel per acre, 56 lb	62.71
1.86×10^{-2}	kilogram per hectare, kg ha^{-1}	bushel per acre, 48 lb	53.75
0.107	liter per hectare, L ha^{-1}	gallon per acre	9.35
893	tonnes per hectare, t ha^{-1}	pound per acre, lb acre^{-1}	1.12×10^{-3}
893	megagram per hectare, Mg ha^{-1}	pound per acre, lb acre^{-1}	1.12×10^{-3}
0.446	megagram per hectare, Mg ha^{-1}	ton (2000 lb) per acre, ton acre^{-1}	2.24
2.24	meter per second, m s^{-1}	mile per hour	0.447

Specific Surface

To convert Column 1 into Column 2, multiply by	Column 1 SI Unit	Column 2 non-SI Unit	To convert Column 2 into Column 1, multiply by
10	square meter per kilogram, m^2 kg^{-1}	square centimeter per gram, cm^2 g^{-1}	0.1
1000	square meter per kilogram, m^2 kg^{-1}	square millimeter per gram, mm^2 g^{-1}	0.001

Pressure

To convert Column 1 into Column 2, multiply by	Column 1 SI Unit	Column 2 non-SI Unit	To convert Column 2 into Column 1, multiply by
9.90	megapascal, MPa (10^6 Pa)	atmosphere	0.101
10	megapascal, MPa (10^6 Pa)	bar	0.1
1.00	megagram, per cubic meter, Mg m^{-3}	gram per cubic centimeter, g cm^{-3}	1.00
2.09×10^{-2}	pascal, Pa	pound per square foot, lb ft^{-2}	47.9
1.45×10^{-4}	pascal, Pa	pound per square inch, lb in^{-2}	6.90×10^3

(continued on next page)

Conversion Factors for SI and non-SI Units

To convert Column 1 into Column 2, multiply by	Column 1 SI Unit	Column 2 non-SI Units	To convert Column 2 into Column 1, multiply by
Temperature			
$1.00\ (K - 273)$	Kelvin, K	Celsius, °C	$1.00\ (°C + 273)$
$(9/5\ °C) + 32$	Celsius, °C	Fahrenheit, °F	$5/9\ (°F - 32)$
Energy, Work, Quantity of Heat			
9.52×10^{-4}	joule, J	British thermal unit, Btu	1.05×10^{3}
0.239	joule, J	calorie, cal	4.19
10^{7}	joule, J	erg	10^{-7}
0.735	joule, J	foot-pound	1.36
2.387×10^{-5}	joule per square meter, $J\ m^{-2}$	calorie per square centimeter (langley)	4.19×10^{4}
10^{5}	newton, N	dyne	10^{-5}
1.43×10^{-3}	watt per square meter, $W\ m^{-2}$	calorie per square centimeter minute (irradiance), $cal\ cm^{-2}\ min^{-1}$	698
Transpiration and Photosynthesis			
3.60×10^{-2}	milligram per square meter second, $mg\ m^{-2}\ s^{-1}$	gram per square decimeter hour, $g\ dm^{-2}\ h^{-1}$	27.8
5.56×10^{-3}	milligram (H_2O) per square meter second, $mg\ m^{-2}\ s^{-1}$	micromole (H_2O) per square centimeter second, $\mu mol\ cm^{-2}\ s^{-1}$	180
10^{-4}	milligram per square meter second, $mg\ m^{-2}\ s^{-1}$	milligram per square centimeter second, $mg\ cm^{-2}\ s^{-1}$	10^{4}
35.97	milligram per square meter second, $mg\ m^{-2}\ s^{-1}$	milligram per square decimeter hour, $mg\ dm^{-2}\ h^{-1}$	2.78×10^{-2}
Plane Angle			
57.3	radian, rad	degrees (angle), °	1.75×10^{-2}

CONVERSION FACTORS FOR SI AND NON-SI UNITS

Electrical Conductivity, Electricity, and Magnetism

To convert Column 1 into Column 2, multiply by	Column 1 SI Unit	Column 2 non-SI Unit	To convert Column 2 into Column 1, multiply by
10	siemen per meter, S m^{-1}	millimho per centimeter, mmho cm^{-1}	0.1
10^4	tesla, T	gauss, G	10^{-4}

Water Measurement

9.73 × 10^{-3}	cubic meter, m^3	acre-inches, acre-in	102.8
9.81 × 10^{-3}	cubic meter per hour, m^3 h^{-1}	cubic feet per second, ft^3 s^{-1}	101.9
4.40	cubic meter per hour, m^3 h^{-1}	U.S. gallons per minute, gal min^{-1}	0.227
8.11	hectare-meters, ha-m	acre-feet, acre-ft	0.123
97.28	hectare-meters, ha-m	acre-inches, acre-in	1.03 × 10^{-2}
8.1 × 10^{-2}	hectare-centimeters, ha-cm	acre-feet, acre-ft	12.33

Concentrations

1	centimole per kilogram, cmol kg^{-1}	milliequivalents per 100 grams, meq 100 g^{-1}	1
0.1	gram per kilogram, g kg^{-1}	percent, %	10
1	milligram per kilogram, mg kg^{-1}	parts per million, ppm	1

Radioactivity

2.7 × 10^{-11}	becquerel, Bq	curie, Ci	3.7 × 10^{10}
2.7 × 10^{-2}	becquerel per kilogram, Bq kg^{-1}	picocurie per gram, pCi g^{-1}	37
100	gray, Gy (absorbed dose)	rad, rd	0.01
100	sievert, Sv (equivalent dose)	rem (roentgen equivalent man)	0.01

Plant Nutrient Conversion

	Elemental	Oxide	
2.29	P	P$_2$O$_5$	0.437
1.20	K	K$_2$O	0.830
1.39	Ca	CaO	0.715
1.66	Mg	MgO	0.602

Introduction: Importance of Soil Quality to Health and Sustainable Land Management

David F. Bezdicek

Washington State University
Pullman, Washington

Robert I. Papendick

USDA-ARS and Washington State University
Pullman, Washington

Rattan Lal

Ohio State University
Columbus, Ohio

Soil, water, and air are three basic natural resources upon which most life depends. The balance between economic viability or destruction often depends on how we manage our soil resource base. For example, the soil provides nutrients for plant growth that are essential for animal and human nutrition. It provides the medium for the recycling and detoxification of organic materials and for the recycling of many nutrients and global gases. A healthy soil provides a link to plant, animal, and human health. History has repeatedly shown that mismanagement of the soil resource base can lead to poverty, malnutrition, and economic disaster.

Many nations have sought conservation policies to protect the soil resource base, to safeguard and preserve the food resource base, and to maintain air and water quality; however, soil resources continue to be degraded both nationally and globally through salinization, erosion, loss of tilth and biological activity, and build up of toxic compounds. Although national programs such as the National Resources Inventory monitor changes in erosion, a more comprehensive approach is needed that addresses changes in overall soil quality and health over time as a result of different land-use practices.

State and federal agencies receive many questions from the general public and from producers on the effect of farming practices, chemicals, and fertilizers on soil microorganisms, soil tilth, and environmental quality. We have acceptable measures of quality for water and air, but not for soil. Recently, there has been

---

Copyright © 1996 Soil Science Society of America, 677 S. Segoe Rd., Madison, WI 53711, USA. *Methods for Assessing Soil Quality*, SSSA Special Publication 49.

considerable national and international interest in soil quality and health as a key issue relating to agricultural sustainability (Acton & Gregorich, 1995; Papendick & Parr, 1992; National Research Council, 1993; Doran et al., 1994).

APPROACHES AND RELATED EFFECTS

Soil quality can be viewed in two different ways: (i) as an inherent attribute of soils that can be inferred from soil characteristics or indirect observations (e.g., erodibility or compactability); or (ii) as a capacity to perform certain productivity, environmental, and health functions.

Soil Quality Based on Inherent Soil Attributes

An *ideal* or native soil could be used as a reference to evaluate a soil in question assuming that the reference soil contained the ideal characteristics. This approach might work in tilled soils derived under native grassland using virgin sites, if available, as a reference. Soil quality characteristics from the reference soils could then be used to evaluate those soils in question. This approach would have to assume that native soils possessed ideal soil characteristics and that agricultural practices degrade soil; however, there are examples where sound agriculture practices can enhance soil quality through improvement of nutrient status, drainage, and physical characteristics. Using an ideal or native soil as a reference would not be possible for soils derived under forest vegetation and now under cultivation as the soil profile characteristics would have been changed considerably.

Desert soils of the West in their natural state were not ideal for agriculture. When first irrigated, infiltration rates were low and the soils tended to crust because of low organic matter and lack of structure; however, with good management, soil organic matter increased and concomitantly the quality characteristics improved such as water intake rates, water holding capacity, and fertility. Further, soil quality for a native dryland soil currently under rice culture could not be evaluated in reference to the native condition because the vegetation and environmental conditions have been changed markedly.

Soil Quality Based on Capacity to Perform

A more commonly accepted working definition of soil quality is based on its capacity to perform productivity, environmental, and health functions. Doran and Parkin (1994) have defined soil quality as "the capacity of the soil to function within ecosystem boundaries to sustain biological productivity, maintain environmental quality, and promote plant and animal health." Thus, the definition involves more than the capacity of soil to produce crops. The relation of soil quality to crop productivity is probably the best understood of the soil quality components.

Most soil scientists agree that soil organic matter is a key desirable component in all three performance categories. Yet, it alone is not an adequate indicator of soil quality or health. For example, the activity and proportion of organic matter pools and particulate organic matter (POM) may better reflect soil quality and complex crop rotations than total soil organic matter alone (Cambardella & Elliott, 1992; Wander et al., 1994). Soil organic matter and its influence on soil tilth does have a direct influence on crop productivity by maintaining a favorable environment for root health through a desirable balance of water and soil porosity. From a plant nutrient viewpoint, soil pH influences the availability of a number of nutrients to plants and may be a factor in the relation of soil quality to environmental quality and animal and plant health.

Less understood is the relation of soil quality to environmental quality. Although many soil quality parameters favorable to plant productivity can enhance environmental quality, e.g., soil organic matter, the presence of certain heavy metals and organics, and excessive nitrates may not impair plant productivity, but could impair environmental quality and animal and human health. The relation of soil quality to plant, animal, and human health is probably the least understood. Relationship between soil pH and the potential for certain plant diseases are known (Pegg, 1978; Smiley & Cook, 1973). Availability of Se and other nutrients can influence animal health and possibility human health through the consumption of forages and food crops (Westermann & Robbins, 1974).

There is much yet to be learned on the broad causal relationship between soil quality and plant, animal, and human health. Focusing on the three components of soil quality is a beginning step to integrate these relationships and bring into the equation the interaction of plant and animal scientists, nutritionists, producers, environmentalists, and the consuming public. From a sustainable soil management viewpoint, growers, scientists, policy makers, and the public need a tangible and systematic means of assessing and monitoring the quality of the soil resource base both in the short and long term.

Soil Quality and Soil Resilience

Soil degradation is a major concern of modern times because of its adverse impacts on local, regional, and global scales. Consequently, there have been several attempts to assess soil degradation at global scale (Oldeman, 1994; World Resources Institute, 1992). The available statistics on soil degradation, is rather vague and subjective because it is not based on land use or science-based inputs and management; however, these can be improved by: (i) relating the severity of soil degradation to productivity under improved or science-based management and inputs; (ii) determining the relationship of soil properties and processes to soil quality as determined by its life support processes and environmental regulatory functions; and (iii) knowing soil resilience and soil capacity to restore itself.

The term resilience, as applied to soil science, is a recent concept (Greenland & Szabolcs, 1994). It has been defined as: (i) tolerance against stress (Szabolcs, 1994); (ii) ability of a system to return after disturbance to a new dynamic equilibrium (Blum & Santelises, 1994); and (iii) ability of a soil to resist adverse changes under a given set of ecological and land use conditions, and to return to its original dynamic equilibrium after disturbance (Rozanov, 1994; Lal,

Table I–1. Common soil stresses and related degradative processes.

Stress	Principal degradative processes
1. Heavy load due to vehicular traffic.	Physical degradation, e.g., crusting, compaction, structural decline, and poor soil tilth.
2. High intensity rains and overland flow, high wind velocity.	Accelerated erosion by water and wind.
3. High evaporative demand and high salt concentration in the profile.	Drought, aridization or desertification, salinization or sodication.
4. Poor internal drainage, and slow surface drainage	Soil wetness and anaerobiosis.
5. Intensive cropping.	Chemical degradation, nutrient imbalance, soil organic matter depletion.
6. Intensive use of agrichemicals and monoculture.	Biological degradation, acidification, reduction in soil biodiversity.

1994a; Oldeman, 1994). The first definition implies knowledge of the specific stress(es) involved.

Common soil degradative processes in agricultural systems may be erosion, leaching, compaction, fertility depletion, loss of soil organic matter, and others (Table I–1). To be quantitative, we must know the response of soil to these stresses. Although some progress has been made in quantifying these responses, there is a lack of understanding of critical limits of soil properties and processes in relation to onset and severity of degradative processes. These critical limits differ among soils because they depend on intrinsic soil characteristics.

There are several factors affecting soil resilience (Lal, 1994b), which comprise intrinsic soil properties and endogenous factors. Intrinsic soil properties governing soil resilience are related to soil quality. Some important properties that affect soil quality and also determine soil resilience include soil structure, soil water, retention and transmission properties, cation-exchange capacity and exchangeable cations, soil organic matter content and its transformations, nutrient supplying capacity, and soil pH. Important among external factors are land use and management system. With these factors in mind soil resilience can be expressed as:

$$S_r = f(\text{soil quality, land use, management})_t \quad [1]$$

where S_r is soil resilience, and t is time. Soil quality is not independent of land use and management. Therefore, soil resilience is directly related to soil quality and factors affecting it (Lal, 1993):

$$S_q = f(W_c \times S_c \times R_d \times e_d \times N_c \times B_d)_t \quad [2]$$

where S_q is soil quality, W_c is water capacity or nonlimiting water range, S_c is structural index, R_d is rooting depth, e_d is charge density or a measure of cation-exchange capacity (CEC), N_c is nutrient supplying capacity, and B_d is soil biodiversity. In most soils, it may be possible to define soil quality in terms of one or two most critical soil properties, e.g., structural index, or nutrient capacity. Formation of a functional relationship from Eq. [2] may involve development of one or several pedotransfer functions (Larson & Pierce, 1991).

Table I–2. Relation between soil quality and water quality.

Soil properties–processes	Water quality characteristics
A. Direct effects	
1. Parent material	Salt concentration, softness–hardness
2. Organic matter content	Color
3. Soil structure and erodibility	Turbidity
4. Cation-exchange capacity	Dissolved load
5. Anaerobiosis	BOD and COD†
6. Texture	Suspended load
B. Indirect Effects	
1. Tillage methods	Sediment concentration and suspended load
2. Chemical inputs	Dissolved load, eutrophication
3. Farming system	Biomass
4. Drainage	Dissolved load

† BOD, biological oxygen demand; COD, chemical oxygen demand.

Soil Quality and Water Quality

Fresh water is a scarce resource, especially in arid and semiarid regions. Consequently, water quality is important to human and animal health, and it has been a major global concern in developing and developed countries (Lal & Stewart, 1994). The National Research Council (1993) report "Soil and Water Quality: an Agenda for Agriculture" espouses conservation and enhancement of soil quality as the first line of defense against air and water pollution. Soil quality affects water quality both directly and indirectly (Table I–2). Direct effects of soil quality on water quality are attributed to inherent soil characteristics, e.g., parent material, texture, and structure. Land use and soil management also affect water quality through the effects of soil quality. The magnitude of the effect of soil quality on water quality is greatly modified by management systems. Important management practices that affect water quality include tillage, application of fertilizers, pesticides and soil amendments, drainage, and farming system. Some practices that minimize dependence on off-farm or purchased inputs also can have adverse effects on water quality, e.g., improper use of farm yard manure, biosolids, and compost. Similarly, plowing under of a green manure crop, intensive tillage for seedbed preparation, and mechanical weed control can increase soil erosion.

There is a need to better define a relationship between soil quality and water quality. Some index may be used on some relevant soil properties with direct effect on water quality. Important among these properties are: (i) erodability that affects sediment load or turbidity; (ii) CEC and nutrient reserves that affect leaching intensity and dissolved load; and (iii) soil organic matter content that affects buffer capacity against leaching. An index defining water quality in relation to soil characteristics may involve pedotransfer functions of the type shown in Eq. [3]:

$$q_w = f(CEC, K, SOC, i_c) \qquad [3]$$

where q_w is water quality based on suspended–dissolved load or BOD–COD etc., CEC is the soil cation-exchange capacity, K is soil erodibility, SOC is soil organic carbon content, and i_c is infiltration rate.

Soil Quality and the Greenhouse Effect

World soils play an important role in regulating the concentrations of radiatively-active gases in the atmosphere. World soils constitute the largest terrestrial pool of C estimated at about 1550 Pg (Lal et al., 1995), and play a major role in the global C cycle. In addition to C, world soils also contain about 95 Tg of N. The atmospheric pools of C and N (in the form of CO_2, CH_4, and NO_x) are increasing at the cost of soil C and N pools. Soil quality plays an important role in regulating the gaseous fluxes from soil-related processes. For example, improving soil quality tends to sequester atmospheric C in soil and may decrease the release of N oxides from soil. Agricultural practices that affect gaseous fluxes include tillage, fertilizer application, cultivation of rice paddies, and crop rotation.

Inherent soil properties with direct effect on flux of radiatively-active gases are soil organic C, soil temperature, soil water, and aeration (Eq. [4]):

$$J_q = f(SOC, K_t, C_t, f_a) \qquad [4]$$

where J_q is gaseous flux of CO_2 or NO_x, SOC is soil organic C, K_t is thermal conductivity, C_t is soil thermal capacity, and f_a is aeration porosity. Once again, there is a strong need to develop appropriate indices and pedotransfer functions relating soil quality to the flux of radiatively-active gases from soil-related processes.

Needs and Uses of Soil Quality for Sustainable Land Management

Producers, researchers, and policy makers are interested in an integrative soil quality index to monitor changes in soil over time. Many farmers are concerned about the long-term effects of contemporary farming practices on the soil resource (Beus et al., 1990). They continually evaluate new products and practices through on-farm testing, but they lack a good indicator that is sensitive to changes in the soil system. Farmers using alternative cropping and tillage practices often report improvements in soil tilth and reduced disease problems. A number of miracle products and management recommendations carry claims regarding improved soil health or fertility that are difficult to substantiate or refute within our current soil testing framework. With the increased adaptation of precision farming techniques and yield monitors on crop harvesters, sensitive soil quality indicators may be used to identify reasons for poor yielding areas that cannot be explained by routine soil tests. Farmers need some index of soil quality with adequate sensitivity to help them evaluate the economic potential of new options and their impact on the soil resource.

Ideally, a soil quality index would be formulated by an integration of parameters that account for the soils's capacity to perform the productivity, environmental, and health functions. Obviously, there are numerous soil properties that change in response to changes in management practices and land use, some of which are highly sensitive, whereas others are more subtle. These properties also will vary spatially and temporally, and thus their usefulness will depend on the desired spatial (e.g., plot, field, watershed, or region) or temporal (e.g., diurnal, seasonal, or annual) parameters. Moreover, the sensitivity of different potential indicators of soil quality change may vary considerably for different spatial and temporal scales. Indicator criteria for developing a soil quality index should therefore include both its sensitivity and variability for the appropriate scale.

An overview of uses for a soil quality index has been presented by Granatstein and Bezdicek (1992). In the USA, the 1985 Farm Bill mandated soil and water conservation compliance for commodity program participants. The USDA-Natural Resources Conservation Service (USDA-NRCS) relies heavily on predictive soil loss equations and easily measurable parameters (e.g., surface crop residue or tillage practices) in developing conservation plans with farmers. Some growers who use alternative practices such as green manure, cover crops, and alternative tillage practices feel that their farming systems have improved soil quality and thus reduced erosion potential relative to other farming systems on the same soil type. They question the need for additional practices such as reduced tillage or strip-cropping to satisfy minimum surface crop residue requirements in their farm plans. Soil loss technology application by USDA-NRCS needs to account more for site specific field changes due to management. Some type of soil quality index might assist in resolving this issue. The use of soil quality parameters for evaluating the impact of the Conservation Reserve Program (CRP) may be helpful in setting new policies and in justifying the return in terms of soil improvement from the 10-yr public investment.

Soil quality assessment could be useful to researchers and policy makers in setting research priorities in reference to policy decisions and for measuring changes in the soil resource base. For example, considerable resources have been committed nationally to the development of no-till technology. Many of the soil quality tests currently used indicate that no-till enhances quality attributes (Doran, 1980; Dormaar & Lindwall, 1989); however, no-till systems tend to require larger amounts of fertilizers and pesticides, although this is region dependent. A soil quality assessment would be useful for identifying and evaluating promising approaches that include the attributes of no-till and conventional tillage systems.

A soil quality assessment also could improve economic assessments of agriculture through assigning values on practices to agricultural land based on environmental considerations (e.g., erosion, water and air pollution, or presence of contaminants) and resource-use efficiency in addition to proven yield. Agencies could use an evaluation to monitor the long-term soil resource base. For example, the inclusion of soil quality measurements in the National Resources Inventory (NRI) would expand the usefulness of the inventory beyond soil erosion. If relationships between soil quality and human health are established, consumer-minded markets might use soil quality criteria in food decisions linking the market force to land stewardship.

REFERENCES

Acton, D.F., and L.J. Gregorich. 1995. The health of our soils: Towards sustainable agriculture in Canada. Agric. Agri-food Canada, Ottawa, ON, Canada.

Beus, C., D.F. Bezdicek, J.E. Carlson, D.A. Dillman, D. Granatstein, B.C. Miller, D. Mulla, K. Painter, and D.L. Young. 1990. Prospects for sustainable agriculture in the Palouse: Farmer experiences and viewpoints. XB1016, Agric. Res. Ctr., Washington State Univ., Pullman, WA.

Blum, W.E.H., and A.A. Santelises. 1994. A concept of sustainability and resilience based on soil functions. p. 535–542. *In* D.J. Greenland and I. Szabolcs (ed.) Soil resilience and sustainable landuse. CAB Int., Wallingford, Oxon, England.

Cambardella, C.A., and E.T. Elliott. 1992. Particulate soil organic matter changes across a grassland cultivation sequence. Soil Sci. Soc. Am. J. 56:777–783.

Doran, J.W. 1980. Soil microbial and biochemical changes associated with reduced tillage. Soil Sci. Soc. Am. J. 44:765–771.

Doran, J.W., D.C. Coleman, D.F. Bezdicek, and B.A. Steward (ed.). 1994 Defining soil quality for a sustainable environment. SSSA Spec. Publ. 35. ASA and SSSA, Madison WI.

Doran, J.W., and T.B. Parkin. 1994. Defining and assessing soil quality. p. 3–21. *In* J.W. Doran et al. (ed.) Defining soil quality for a sustainable environment. SSSA Spec. Publ. 35. ASA and SSSA, Madison WI.

Dormaar, J.F., and C.W. Lindwall. 1989. Chemical differences in dark brown chernozemic Ap horizons under various conservation tillage systems. Can. J. Soil Sci. 69:481–488.

Granatstein, D., and D.F. Bezdicek. 1992. The need for a soil quality index: Local and regional perspectives. Am. J. Altern. Agric. 7:12–16.

Greenland, D.J., and I. Szabolcs. 1994. Soil resilience and sustainable landuse. CAB Int., Wallingford, Oxon, England.

Lal, R. 1993. Tillage effects on soil degradation, soil resilience, soil quality and sustainability. Soil Tillage Res. 27:108.

Lal, R. 1994a. Sustainable landuse systems and soil resilience. p. 41–68. *In* D.J. Greenland and I. Szabolcs (ed.) Soil resilience and sustainable landuse. CAB Int., Wallingford, Oxon, England.

Lal, R. 1994b. Landuse and soil resilience. p. 246–261. *In* Proc. 15th World Cong. of Soil Sci., Acapulco, Mexico. 10–16 July 1994.

Lal, R., J. Kimble, E. Levine, and C. Whitman. 1995. World soils and greenhouse effect: An overview. Soils and greenhouse effect. p. 1–7. *In* R. Lal et al. (ed.) Soils and global change. Lewis Publ., Boca Raton, FL.

Lal, R., and B.A. Stewart. 1994. Soil processes and water quality. p. 1–6. *In* R. Lal and B.A. Stewart (ed.) Soil processes and water quality. Lewis Publ., Boca Raton, FL.

Larson, W.E., and F.J. Pierce 1991. Conservation and enhancement of soil quality. *In* Evaluation for sustainable land management in the developing world. IBSRAM Proc. 12. Vol. 2. Int. Board on Soil Res. and Manage., Bangkok, Thailand.

National Research Council. 1993. Soil and water quality: An agenda for agriculture. Natl. Academy Press, Washington, DC.

Oldeman, L.R. 1994. The global extent of soil degradation. p. 99–118. *In* D.J. Greenland and I. Szabolcs (ed.) Soil resilience and sustainable landuse. CAB Int., Wallingford, Oxon, England.

Papendick, R.I., and J.F. Parr. 1992. Soil quality: The key to sustainable agriculture. Am. J. Altern. Agric. 7:2–3.

Pegg, K.G. 1978. Soil application of elemental sulfur as a control of *Phytophthora cinnamomi* root and heart rot of pineapple. Aust. J. Exp. Agric. Anim. Husb. 17:859–861.

Rosanov, B.G. 1994. Stressed soil systems and soil resilience in drylands. p. 238–245. *In* Proc. 15th World Cong of Soil Sci., Acapulco, Mexico. 10–16 July 1994.

Smiley, R.W., and R.J. Cook. 1973. Relationship of take-all of wheat and rhizosphere pH in soils fertilized with ammonium vs. nitrate-nitrogen. Phytopathology 63:882–890.

Szabolcs, I. 1994. The concept of soil resilience. p. 33–40. *In* D.J. Greenland and I. Szabolcs (ed.) Soil resilience and sustainable landuse. CAB Int., Wallingford, Oxon, England.

Wander, M.M., S.J. Traina, B.R. Stinner, and S.E. Peters. 1994. Organic and conventional management effects on biologically-active soil organic matter pools. Soil Sci. Soc. Am. J. 58:1130–1139.

Westermann, D.T., and C.W. Robbins. 1974. Effect of SO_4–S fertilization on Se concentrations of alfalfa. Agron. J. 66:207–208.

World Resources Institute. 1992. Towards sustainable development. A guide to the global environment, WRI, Washington, DC.

1 Linkages between Soil Quality and Plant, Animal, and Human Health

L. J. Cihacek

North Dakota State University
Fargo, North Dakota

W. L. Anderson

Middle Tennessee State University
Murfreesboro, Tennessee

Phillip W. Barak

University of Wisconsin
Madison, Wisconsin

Photosynthetic plants form the basis for life on earth and the food chain that support animal life. Animals are supported by plants and provide meat and milk with humans being at the top of the food chain. Most nutrient elements enter plants as dissolved mineral salts in water taken up from the soil. Human consumption of plant and animal products, however, influences the biological, chemical, and physical properties of soils through changes in physical and nutrient composition, structure, water relationships, and erosion. Disturbing the food production cycle with excesses or deficiencies of nutrients will affect each of the plant, animal, and human components of the food chain. The soil is the primary supplier of water and nutrients for plants, which provide nutrients to animals and humans. Maintaining soil quality for plants will help to ensure normal growth and development for plants, animals, and humans.

Karlen et al. (1997) have defined soil quality as "the fitness of a specific kind of soil, to function within its capacity and within natural or managed ecosystem boundaries, to sustain plant and animal productivity, maintain or enhance water and air quality, and support human health and habitation". This definition is similar to those proposed by Acton and Gregorich (1995), Doran and Parkin (1994) and Larson and Pierce (1991) and provides a concept that is useable to quantify soil quality in spite of "natural differences" among soil orders and even between the same soil series found in different locations.

Copyright © 1996 Soil Science Society of America, 677 S. Segoe Rd., Madison, WI 53711, USA.
Methods for Assessing Soil Quality, SSSA Special Publication 49.

The purpose of this chapter is to present some positive and negative relationships and linkages between soil quality with respect to supplying essential nutrients and the balanced nutrition and health of plants, animals, and humans.

IMPORTANCE OF SOIL FACTORS IN MAINTAINING PLANT, ANIMAL, AND HUMAN HEALTH

Plants growing in soil absorb essential as well as nonessential plant elements. Nonessential plant elements, however, may be essential for normal growth and development of animals and humans. Examples of nonessential plant elements that are essential to animal nutrition are I, Cr, Se, and Ni.

A quality soil is one in which both the essential and nonessential plant elements are present in adequate amounts for animal and human health. Liebig (1855) recognized soil quality relative to supplying adequate plant nutrients. He noted that "in a soil rich in mineral nutrients, the yield of a field cannot be measured by adding more of the same substance." Recognition of a linkage between soils and animal nutrition is not a new concept. In 1878, President Welch of the Iowa State Agricultural College, proposed a study of soils in relation to animal and human nutrition (Troeh et al., 1991). Later, the 1898 USDA-Bureau of Animal Industry report contained a section on the relationship of animal bone disease and noncalcareous soils (Troeh et al., 1991).

Doran et al. (1996) outlined three avenues where soils interact with and affect the health of animals. Avenues may be direct or indirect where plants act as intermediaries for animals and humans. The first avenue is direct poisoning of plants and/or animals by chemicals in the soil. This may be due to an anthropomorphic event or geological process. Anthropomorphic examples include industrial spills, improper use of chemicals, or land disposal of waste. Many instances of soil contamination with heavy metals such as Cd, Pb, Hg, Ni, Zn, Cr, or Cu, organic or petroleum products such as trichloroethylene or diesel fuel, or pesticides such as DDT, aldrin or triazines have been reported in the scientific literature or popular press. These substances are significant in that they can cause toxic effects by directly entering the food chain by consumption of plants or water containing elevated concentrations or by concentration in the food chain by animals such as DDT in birds of prey. Many pollutants such as Hg or some pesticides can be stored in organs or fatty tissue in humans and become highly toxic because they persist in the body and continue to accumulate until they cause physiological and neurological disorders and cancer.

The second avenue is the ability of soil to filter contaminants from water. Soil acts as a binding agent to remove many harmful elements and chemicals from water as it passes through the soil. Within the soil profile, some chemicals, especially positively charged ones, can be adsorbed or precipitated out before entering the groundwater. Organic components are, in general, removed from water in the upper profile where the soil organic matter is high. Soils also have the ability to adsorb many pollutant gases (Smith et al., 1973; Bremner & Banwart, 1976; Cihacek & Bremner, 1988, 1990) and much of the storage of contaminants takes place in the organic matter rich surface soil. Soil, however, also can detach due to impact and raindrop water action to contaminate surface water bodies and streams and by wind action to generate dust-laden air that can move for long distances. Erodibility factors depend on particle-size distribution and soil organic matter contents. Transported soil sediments and adsorbed chemicals can physically and chemically produce offsite contamination of surface soil, surface water, and air.

The final avenue in which soil impacts plant, animal, or human health is by providing a medium that contains deficient, adequate, or excessive quantities of essential and/or nonessential nutrients to plants that can alter the quantity and quality of the plant growth. This effect may be due to the presence of inorganic or organic nutrient elements or chemicals in the soil or biological agents that alter a plant's physiological functioning. Here, intensive human consumption of plant and animal products can remove significant quantities of nutrients that can alter the physical and chemical conditions of the soil and alter the quantity and quality of plant growth, too.

A comparison of the function of elements in plants, animals and humans is shown in Table 1–1. Differences in function of an element can be noted between plants, animals, and humans, but similarities also exist. For example, while Ca is essential for teeth and bones of only animals and humans, Ca also is essential for cell permeability in plants, animals, and humans. Likewise, N, P, and S are all essential for proteins in plants, animals, and humans. Allaway (1975, 1984) reviewed interactions between soil and mineral nutrients and their effects on animals and humans.

Although the above are aspects of soil that most professional soil scientists use to define soil quality, another aspect of soil quality needs to be considered. The concept of soil quality also considers the intrinsic value of soil that focuses on the unique and irreplaceable characteristics of the resource, apart from its importance to crop growth or ecosystem function. Assigning intrinsic value to soil is not widely explored by professional soil scientists or included in economic models of resources. Intrinsic values associated with natural resources including the soil are held in various forms by naturalists and people who see a special relationship with the earth or are concerned in holistic natural resource management. Integrating information from all aspects of soil science is central to the concept of soil quality.

LINKAGE CASE EXAMPLES

In the following sections the discussion will focus on a few selected soils-supplied element interactions relative to their relationship to plant, animal, and human health.

Selenium Relationships in Plant, Animal, and Human Health

Selenium is found throughout the earth's crust (Mayland et al., 1989). Selenium toxicity frequently occurs in the Great Plains and Rocky Mountain states.

Table 1–1. Major element functions in plants, animals, and humans.

Elements	Major plant functions†	Major animal functions‡	Major human functions§
N	Proteins Nucleoprotein, DNA, RNA Chlorophyll	Proteins Major constituent of organs and soft structures	Proteins Major constituent of organs and soft structures
P	Energy storage and transfer Structural component nucleoprotein	Stuctual parts of bone and teeth Energy storage Component of DNA and RNA PO_4 important in acid-base balance	Structural parts of bone and teeth Energy storage Component of DNA and RNA O_4 important in acid-base balance
K	Not structural Enzyme activation Osmotic effects–water relations Translocation of assimilate CO_2 N uptake and protein synthesis Starch synthesis Resistance to certain plant diseases	Osmotic effect–water relation Acid-base balance Muscle activity Nerve function	Osmotic effect–water relation Acid-base balance Muscle activity Nerve function
Ca	Structure and permeability of cell membranes and walls	Bone and teeth formation Blood coagulation Nerve function Muscle activity Cell membrane permeability Milk production Egg shell formation	Bone and teeth formation Blood coagulation Nerve function Muscle activity Cell membrane permeability Milk production
Mg	Chlorophyll Ribosomes Enzyme activation for polypeptides and carbohydrates	Bone and teeth formation Enzyme activator primarily in glycolytic system	Bone and teeth formation Enzyme activator primarily in glycolytic system
S	Oil synthesis Synthesis of S amino acids Components of Coenzyme A, biotin, thiamine, chlorophyll Ferredoxins in CO_2 and NO_3 reduction and chlorophyll synthesis	Nerve function Muscle activity S-containing amino acids Component of biotin and thiamine	Nerve function Muscle activity S-containing amino acids Component of biotin and thiamine
Fe	Porphyrin, and nonheme molecules Electron transport reactions	Oxygen transport: hemoglobin Energy metabolism: cytochromes	O_2 transport: hemoglobin Energy metabolism: cytochromes

LINKAGES

	Plant	Animal
	Enzymes for NO$_3$, N$_2$ reduction and chlorophyll synthesis	Muscles: myoglobin Same as animal Conversion of betacarotene to vitamin A Detoxification of drugs Production of antibodies
Mn	Photosynthesis - O$_2$ reduction Electron transport reactions Nodule formation	Synthetic and metabolitic enzymes Bone formation Insulin production
Cu	Photosynthesis - O$_2$ evolution Electron transport reactions Enzymes for RNA and IAA	Synthetic and metabolitic enzymes Bone formation Growth and reproduction Red blood cell formation Cofactor in enzyme functions for protein metabolism Cofactor in several oxidation-reduction enzyme systems Hemoglobin synthesis Maintenance of myelin of nerves Bone formation Hair pigmentation Cofactor in enzyme functions for oxidative reactions

Zn	Auxin metabolism
	  a. Tryptaphan synthetase
	  b. Tryptamine metabolism
	Many enzyme systems
		Component or cofactor of several enzyme systems
		Bone and feather development
		Cofactor in enzyme functions for growth
		Cofactor in enzyme functions for replication
		Sexual maturation
		Fertility and reproduction
		Night vision
		Immune responses
		Taste and appetite

Mo	Enzyme: nitrate reductase	Purine metabolism Stimulates microbial activity in rumen Enzyme cofactor
Cl	Osmotic effects–water relations Acid-base balance Disease control	Osmotic effects/water relations Acid-base balance Hydrochloric acid in digestion Osmotic effects–water relations Acid-base balance Hydrochloric acid in digestion
Co	Need by legumes Growth of rhizobia	Component of vitamin B12 Needed by rumen bacteria for growth and B12 synthesis Component of vitamin B12
Va	Substitute for Mo	Nonessential to animals Generally nonessential to humans¶
Na	Essential for halaphytic plants Osmotic effects–water relations	Osmotic effects–water relations Acid-base balance Cell membrane permeability Muscle activity Osmotic effects–water relations Acid-base balance Cell membrane permeability Muscle activity

(continued on next page)

Table 1-1. Continued.

Elements	Major plant functions	Major animal functions	Major human functions
Si	Essential–nonessential Accumulate in roots Drought resistance Cell wall structure No biochemical role	Nonessential to animals	Generally nonessential to humans#
B	Not structural Growth and development of new cells Translocation of sugars, starches, N and P Synthesis of amino acids and proteins Nodule formation in legumes	Nonessential to animals	Nonessential to animals
I	Nonessential to plants	Part of thyroid hormone, regulates metabolism	Part of thyroid hormone, regulates metabolism
F	Nonessential to plants	Component of bone and teeth	Component of bones and teeth
Se	Nonessential to plants	Cofactor of antioxidant enzymes Maintain muscle condition	Cofactor of antioxidant enzymes Involved in Vitamin E absorption
Cu	Nonessential to plants	Sugar metabolism	Cofactor in insulin Involved in carbohydrate and lipid metabolism
Sn	Nonessential to plants	Generally believed to be nonessential to animals††	Generally believed to be nonessential to humans
Ni	May or may not be essential to plants	Non-ssential to animals	Generally believed to be nonessential to humans‡‡

† From Epstein, 1972; Chapman, 1965, 1973; Tisdale et al., 1985.
‡ From Crampton & Harris, 1969; Maynard & Loosli, 1969; Jurgens, 1993.
§ From Davis & Sherer, 1994; Hamilton et al., 1988.
¶ May affect growth and development (Hamilton et al., 1988).
# May affect bone calcification (Hamilton et al., 1988).
†† May be essential to animals and humans (Hamilton et al., 1988).
‡‡ May be essential for development and function of many body tissues (Hamilton et al., 1988).

Livestock poisoning can result from grazing on Se accumulator plants or drinking from Se enriched drainage water.

Selenium deficiency and toxicity are human public health problems in the People's Republic of China (Hamilton et al., 1988). People in northeastern, Pacific Northwest, and extreme southeastern U.S. regions, eastern Finland, and part of New Zealand are known to be prone to Se deficiency (Davis & Sherer, 1994).

Certain plant species tend to sequester Se when grown in high Se soils. Rosenfeld and Beath (1964) defined three categories of accumulator plants. The first category are called primary indicators and accumulate several thousand milligrams of Se per kilogram of plant material. Plants in this group include the genus *Astralagus, Machaeranthera, Haplopappus,* and *Stanleya*. The second category called secondary selenium absorbers accumulate a few hundred milligrams of Se per kilogram of plant material. Plants in this group include the genus *Aster, Atriplex, Castelleja, Grindelia, Gutierrezia, Machaeranthera,* and *Mentzelia*. The third group accumulate <50 mg of Se per kilogram of plant material. Plants in this category include the *Trifolium repens, Hilaria belangeri,* and *Bouteloua* sp. (Beeson & Matrone, 1976) as well as many species of grains, grasses and weeds. *Crucuferae* (mustard, cabbage, broccoli, and cauliflower) plants also accumulate Se.

Animals consuming plants high in Se show toxicity symptoms including garlic breath, lethargy, excess salivation, vomiting, dyspnea, muscle tremors, and respiratory distress, which can be fatal in advanced cases. Human symptoms of Se toxicity appear as increased dental caries, loss of hair and nails, skin lesions, hepatomegaly, polyneuritis, and gastrointestinal disturbance (Combs & Combs, 1986).

Although Se is toxic at high concentrations, low Se or Se deficiencies is often a concern in geographic areas where plant uptake is <0.1 mg kg^{-1}. Mikkelsen et al. (1989) reviewed the literature on soil and plant factors affecting Se uptake and Jackson (1988) has reviewed evidence of Se impacts on human health. The use of fertilizers and amendments affects Se availability. Liming increases Se availability while sulfate additions depress Se uptake. Phosphorus fertilizer additions give mixed results.

Nitrate–Nitrogen Relationship in Plant, Animal, and Human Health

Plant available N as nitrate enters the soil either through application of inorganic fertilizer materials or the decomposition of organic materials added to or incorporated in the soil. Nitrate is an essential source of N for plants to form proteins that then become sources of protein for animals and humans. Nitrate, however, can leach through the soil into subsurface waters or run off into surface waters.

Central to soil quality is the issue of maintaining balance to insure efficient crop production and minimize environmental contamination. We need adequate available N for plant growth but not so much that it will leach into groundwater.

Nitrate in drinking water or food can be reduced to nitrite and nitrosamines. Bacteria, in the stomach of 0- to 3-mo old infants and young ruminant animals, can reduce nitrate to nitrite. Children can develop cyanosis (*blue baby*). When

70% of the hemoglobin is oxidized to methemoglobin, death can occur. Nitrate, consumed by older animals, is rapidly absorbed and unused nitrate excreted (Hergert, 1986). Nitrate poisoning is not likely in older mammals unless they are sick and/or elderly. Health hazards from consuming nitrate from water may be small compared with other sources of nitrate in the diet. The USEPA has determined that 10 mg L^{-1} is the maximum safe level of nitrate in drinking water. If a human adult consumes eight glasses of water containing nitrate at this level per day, then 18.5 mg nitrate per day would be ingested. But an adult can ingest substantially more nitrate at the salad bar of many restaurants. Corré and Breimer (1979) reported on 29 studies of lettuce (*Lactuca sativa* L.) with nitrate ranging from 145 to 12 826 mg kg^{-1} in the fresh product with a mean of 2590 mg kg^{-1}. A 200 g serving of fresh lettuce, typical of that available in fast food restaurants, would contain 518 mg nitrate or 28 times the amount ingested daily in drinking water.

Soil quality can alter nitrate availability to plants and groundwater and may act as a source or sink for nitrate. Immobilizing nitrate by adding organic material can reduce nitrate leaching and lower plant nitrate concentration (Larson et al., 1983). The challenge we face in defining soil quality for animal crop production is to synchronize available N levels in soil so that they are nonlimiting for plant growth when the crop is present but minimal during noncropped periods of time when they are subject to loss by leaching.

Potassium and Magnesium in Plant, Animal, and Human Health

Potassium and Mg play physiologically significant roles in plants, animals, and humans. Potassium has a role in water relationships and osmotic pressure in cells. It facilitates transmission of nerve impulses and muscle activity in animals and humans. Magnesium is essential for chlorophyll production in plants and enzyme functions in plants, animals, and humans. Where soil K is high or high rates of K fertilizer are applied and soil Mg is low, Mg deficiencies occur in plants. Cattle (*Bos taurus*) or sheep (*Ovis aries*) grazing or feeding on forages with high K and low Mg can develop hypomagnesia.

In plants, Mg deficiency is characterized by interveinal yellowing (chlorosis) progressing to leaf discoloration, necrosis, and defoliation (Chapman, 1965, 1973). In cattle and sheep, hypomagnesia symptoms are restlessness, nervousness, loss of appetite, and neuromuscular hyperexcitability with increased pulse and respiratory rates and possibility of convulsions and death within 2 or 3 d (Russell, 1944). It is often referred to as grass tetany because of the muscle tremors, staggered gait and general muscle spasms associated by this disorder in animals grazing in lush pastures. Associated with this disorder is Mg loss from bone. Humans with diets low in Mg also are susceptible to tetany and neuromuscular irritability along with chronic nephritis as well as other disorders (Aikawa, 1963, 1971; Wacher, 1980). Magnesium loss from bone also accompanies a Mg deficiency in humans.

Magnesium deficiency in both animals and humans can be traced back to a dietary Mg deficiency. Voisin (1963) suggested that farming methods have resulted in an imbalance in soil and herbage, which then constitutes the diet of both animals and humans. As a result, it appears that Mg deficiency is influenced as much or more by local soil management as by geographical considerations (McDowell, 1985).

Sulfur Relationships in Plant, Animal, and Human Health

Sulfur has a significant role in affecting plant quality (Tisdale, 1977). Sulfur increases the vitamin A, chlorophyll, and protein content and quality, and decreases nonprotein N, nitrates, and N/S ratio. Effects of S on protein quality is significant in its effect on the synthesis of methionine and cystine. Both are S containing amino acids that are essential to the synthesis of proteins with high biological activity.

Soil provides plant available S in the form of sulfate. Its availability in soils is dependent on the mineralogy and organic matter content of soil. Since soil sulfates are very soluble, S can leach through the soil and filter into surface and subsurface waters in a manner much like nitrate. Since much larger quantities of N fertilizer are used in crop production and only small quantities of S are used as fertilizer, the leaching of S is generally of minimal concern from an environmental standpoint; however, there are instances in which S is emitted to the atmosphere causing acid rain and subsequently lowering the pH of water in streams and lakes as well as soils causing a decline in aquatic life and some plant species. The greatest immediate concern about S is that it is a common component of salts in soils and waters. Salts change osmotic potential relationships for water uptake by plants causing plants to die and soils to lose their productive quality. Salts also may cause decline in aquatic animal species. Excessive salts in drinking water may cause health problems in humans and animals.

Ruminant animals can use inorganic S because rumen microbes can synthesize S-containing amino acids (Goodrich & Garrett, 1986). Although methionine and cystine are essential amino acids for nonruminant animals and humans, these must be provided by dietary sources. Sulfur needs for nonruminant animals and humans can be satisfied by consumption of plant materials well supplied with S. This requires that the soil be of a quality that can adequately supply S to growing plants (Rendig, 1986).

Calcium and Phosphorus Relationships in Plant, Animal, and Human Health

Calcium in soils and plants was first studied in France in 1873 by Fliche and Grandeau (Beeson & Matrone, 1976). Since then numerous studies have examined the rate of Ca in plant nutrition. The Ca content of the soil varies significantly from one soil to another as well as in soils from one region to another. Regional variations are usually due to differences in climate and vegetation.

The Annual Report for 1898 from the Bureau of Animal Industry in the U.S. Department of Agriculture noted the occurrence of bone diseases in animals and were given names such as *loin disease, sweeny, creeping disease, down-in-the-back,* or *bone chewing*. Most of these problems appeared to occur on fenced rangeland. Animals fed alfalfa for at least part of the year did not exhibit these symptoms.

Phosphorus is a constituent of soil that usually is found as the mineral apatite. Adequate P in soils is essential to healthy plant growth and production.

Both Ca and P are structural components of bone. As early as 1862, Hoppe-Sayler noted the similarity between the composition of bone and mineral apatite (Dallemagne & Richelle, 1973). Deficiencies in either Ca or P in animal or human diets result in various bone diseases. Calcium deficiencies in humans also can result in premature births and low-birth weights, development of kidney stones, hypertension, and increases incidences of colon cancer.

Phosphorus deficiencies in animals results in loss of appetite, development of stiff joints, fragile bones and increased susceptibility to infections. In humans, P deficiencies affect a variety of muscular, respiratory, hematologic, neurologic, endocrine, cardiac, and renal functions as well as bone disease (Allen & Wood, 1994).

More than one-half of the Ca in human diets in the USA comes from dairy products while the remainder comes from green vegetables and Ca fortified foods. Phosphorus in diets comes from meat, milk, grains, and potatoes.

It is evident, therefore, that plants are important sources of Ca and P for animals and that humans depend on plants and animals and dairy products for dietary requirements.

Iodine Relationships in Plant, Animal, and Human Health

Although I is not known to be essential for plants, it is a highly essential element for good health of animals and humans.

Iodine can be taken up by plants but usually the I concentrations in plants is very low (Martin, 1965). In soil levels above 2.5 mg kg^{-1}, I can be toxic to plants. Most inland soils, however, have very low levels of I due to leaching from soil. Most of the I resides in the ocean and away from ready availability to animals and humans.

Iodine deficiencies in animals and humans are expressed similarly. In animals, iodine deficiency results in reproductive failure and thyroid insufficiency. In humans, it is expressed as increased incidences of stillbirths, abortions, and congenital abnormalities in fetuses and newborns while in older children and adults, goiter, cretinism, and hypothyroidism are common (Clugston & Helzel, 1994).

Endemic goiter has been common in inland populations of animals and humans. Treatment is usually accomplished by adding iodized salt to the diet.

LOCAL VERSUS GLOBAL PERSPECTIVE

Plants supplied with quality factors (support, heat, water, nutrients, free of toxic substances, or no competition) will produce dry matter with adequate nutrients for animals and humans. Anthropogenic manipulation of the soil environment has a profound effect on the nutrient quality of plants. There are two philosophical approaches to plant production: (i) using inorganic nutrient sources as primary nutrient sources; or (ii) using only organic nutrient sources. Either approach will effect plant nutrient content and also can influence plant uptake of contaminants that pose a threat to human and animal health such as heavy metals (Schramm et al., 1994). A common view is that if the soils are maintained at a high quality (or fertility) level, the health of animals and humans living on these soils will be at its best. This may be only partially true.

Historically in the USA, forages from pasture on noncalcareous soils are low in Ca and P unless limed and fertilized. Grazing animals will develop osteomalacia (Troeh et al., 1991). Forage from pasture on poorly drained soils, organic soils (Histosols) and mineral soils may contain concentrations of Mo and Cu that are toxic to cattle and sheep (Troeh et al., 1991). Animal and human nutritional problems like these that are related to specific geographical areas have been identified on soil maps (Troeh et al., 1991).

Animals often ingest soil (also known as geophagy) in order to obtain minerals not readily available from forage plants. Jones and Hanson (1985) discussed animal geophagy in terms of naturally occurring sites or licks where ungulates gather to consume soil and minerals in order to obtain nutrients and salt not readily available in their plant diets. They note that licks were predominant in the eastern USA where grazing lands were dissected and localized due to the terrain. In these areas, bison grazed on relatively short, dense bluegrass from lick to lick forming well defined trails. In the West, however, licks were less common due to the vast rolling expanses of grassland across which the bison moved in broad fronts. They propose that licks were less common because the animals consumed significant quantities of soil while they were grazing. They also note that even in recent times, bison at the National Bison Range of Montana eat soil in preference in salt blocks.

In some cultures, humans also are known to consume soil as a part of their diet. This consumption of soil usually prevalent under poor socioeconomic conditions is called pica. Hambidge et al. (1986) have reviewed on literature the role of pica on inducing Zn or correcting Zn deficiency in humans. Bothwell et al. (1979) note the roles of pica in inducing Fe deficiency in humans. Pica has been recognized in many parts of the world including the USA, Iran, Turkey, South America, India, and Africa.

In less developed parts of the world, locally produced crops have a major impact on the health of the local populace. Recently, Braun (1995) reviewed the effects of Zn deficiency in winter wheat (*Triticum aestivum* L.) on human health in Turkey. Humans living in areas of Zn deficient soils consuming wheat and wheat products have shown symptoms and diseases related to or caused by Zn deficiency including (Cadvar et al., 1983, 1988, 1991):

1. Growth retardation
2. Delayed or no sexual maturation
3. Increased births of anencephaletic babies (born with an open spine)
4. Lowered Zn serum content in mothers of anencephaletic babies
5. Negative effects on immune system
6. Hodgkin's disease

7. Chronic Zn-deficiency in patients with Mediterranean Anemia (Thalassania major) that is related to a genetic disorder.

A current concern is with regard to Cd in food crops. Cadmium is a heavy metal that accumulates in the liver and kidneys. Long-term toxic effects are on the kidney and renal cortex with other effects on pulmonary and cardiovascular systems.

Cadmium occurs naturally in small amounts in nature. But depending on sources of parent material, many soils may contain notable levels of Cd (Holmgren et al., 1993; Garrett, 1994). Phosphorus fertilizers are widely used throughout crop producing regions of the world. Depending on the source of the phosphate rock from which the P fertilizer is manufactured, many P fertilizers may contain significant amounts of Cd (Landner et al., 1995). This has led to efforts of regulating Cd in P fertilizers especially in European countries in order to control the Cd content of crops. These concerns of Cd in food crops has resulted in careful scrutiny of wheat, flax (*Linum usitatissimum* L.), and sunflower (*Helianthus annus* L.) seed by European countries and limitations on importing these grain crops from the USA and Canada.

Grant and Bailey (1995) and Grant et al., (1995), however, have shown that varietal variation in Cd uptake is related to genetic control. This shows that Cd uptake can be changed through breeding for low accumulating cultivars. Today, the USA is a global import market for fruits, vegetables, dairy, and meat products. Some of these products are produced on soils far from the point of consumption with no information on soil or product quality factors. Cadmium concerns illustrate how something that occurs in one area of the world can affect health and medical welfare in another part of the world.

Beeson and Matrone (1976) reviewed the relationship between soils and animal and human nutrition. They noted that on a local scale, increasing soil fertility to promote greater yields or efforts to reach maximum production may result in lower than normal concentration of essential (Ca and Zn) and nonessential (I, Co, and Se) elements in plants. It is difficult to demonstrate that a minimum quantity of an essential nutrient in a normal plant is deficient with respect to human needs; however, in most cases where the plants appear to grow normally and produce normal fruit and seed without deficiency symptoms, low levels of selected nutrients appear to be satisfactory in a normal human diet.

CONCLUSIONS

Both direct and indirect linkages occur between soil well supplied with essential plant nutrients and the health and quality of plants grown on these soils. This applies to both forage and food crops. These crops, in turn, provide their consumers—both animal and human—with essential factors that keep these organisms healthy and resistant to disease. If essential factors in the soil are absent or inadequate, then the health of plants, animals, and humans living off of this soil is negatively impacted.

In 1884, the USDA Bureau of Animal Industry was created to assist growers whose livestock were plagued with disease so that overseas markets were refusing to buy U.S. livestock. Many of these diseases could be related to nutrient deficiencies. At the present time, these diseases still exist. Maintaining a soil quality that provides sufficient nutrients to growing plants is the best way to keep the incidence of nutritionally induced disease at a very low level.

Today, consumers of plant and animal products are concerned about the quality of these products in order to maintain good health and quality of life. Maintaining nutritional quality in plants and animals in the food chain by maintaining soil quality is the most ecologically friendly approach to maintaining human health.

ACKNOWLEDGMENT

Thanks to J. Beckett and R. Garrigus (Animal Science), A. Halterlein (Plant Science), and T. Johnston (Food Science) at Middle Tennessee State University; J. Doran at the University of Nebraska and USDA-ARS; and D. Franzen, E. Deibert (Soil Science), and D. Meyer (Plant Sciences) at North Dakota State University; R. Lal at Ohio State University; and B. Lowrey at the University of Wisconsin for their helpful comments during the development of this manuscript.

REFERENCES

Acton, D.F., and L.J. Gregorich. 1995. Understanding soil health. p. 5–10. *In* D.F. Acton and L.J. Gregorich (ed.) The health of our soils: Toward a sustainable agriculture in Canada. Centre for Land Biol. Resour. Res., Research Branch, Agric. and Agri-Food, Ottawa, ON.

Aikawa, J.K. 1963. The role of magnesium in biologic processes. Charles C. Thomas Co., Springfield, IL.

Aikawa, J.K. 1971. The relationship of magnesium to disease in domestic animals and in humans. Charles C. Thomas Co., Springfield, IL.

Allaway, W.H. 1975. The effect of soils and fertilizers on human and animal nutrition. USDA Agric. Inf. Bull. 378. USDA, Washington, DC.

Allaway, W.H. 1984. Plants as sources of nutrients for people: An overview. p. 1–7. *In* R.M. Welch and W.H. Gabelman. (ed.) Crops as sources of nutrients in humans. ASA Spec. Pub. 48. ASA, Madison, WI.

Allen, L.H., and R.J. Wood. 1994. Calcium and phosphorus. p. 144–163. *In* M.E. Shils et al. (ed.) Vol. 1. Modern nutrition in health and disease. Lea & Febiger, Philadelphia, PA.

Beeson, K.C., and G. Matrone. 1976. The soil factor in nutrition: Animal and human. Marcel Dekker, New York.

Bothwell, T.H., R.W. Charlton, J.D. Cook, and C.A. Finch. 1979. Iron metabolism in man. Blackwell Scientific Publ., Oxford, England.

Braun, H.J. 1995. Zinc deficiency and wheat breeding: CIMMYT's experience in Turkey. p. 66–67. *In* S. Rajaram and G. Hettel (ed.) Wheat breeding at CIMMYT. Commemorating 50 years of research in Mexico for global wheat improvement. Wheat Special Rep. 29. CIMMYT, Mexico. D.F.

Bremner, J.M., and W.L. Banwart. 1976. Sorption of sulfur gases by soils. Soil Biol. Biochem. 8:79–83.

Cadvar, A.O., A. Arcasoy, S. Cin. E. Babacan, and S. Gozdasogler. 1983. Iron and zinc deficiency, iron and zinc absorption studies and response to treatment with zinc in geographic cases. p. 71–96. *In* A.S. Prasad et al. (ed.) Zinc deficiency in human subjects. Alan R. Liss, New York.

Cadvar, A.O., M. Bahceci, N. Akar, F.W. Dincer, and J. Eaten. 1991. Material bias zinc concentration in neutral tube defects in Turkey. Biol. Trace Elem. Res. 30(1):81–86.

Cadvar, A.O., M. Bahceci, N. Akar, J. Eaten, G. Babacan, A. Arcasay, and H. Yavuz. 1988. Zinc status in pregnancy and occurrence of anencephaly in Turkey. J. Trace Elem. Electrolytes Health Dis. 2:9–14.

Chapman, H.D. 1965. Diagnostic criteria for plants and soils. Quality Printing Co., Abilene, TX.

Chapman, H.D. 1973. Diagnostic criteria for plants and soils. H.D. Chapman, Riverside, CA.

Cihacek, L.J., and J.M. Bremner. 1988. Capacity of soils for sorption of sulfur dioxide. Commun. Soil Sci. Plant Anal. 19(16):1945–1964.

Cihacek, L.J., and J.M. Bremner. 1990. Capacity of soils for sorption of hydrogen sulfide. Commun. Soil Sci. Plant Anal. 21(5 and 6):351–363.

Clugston, G.A., and B.S. Helzel. 1994. Iodine. In M.E. Shils et al. (ed.). Vol. 1. Modern nutrition in health and disease. Lea & Febiger. Philadelphia, PA.

Combs, G.F., Jr., and S.B. Combs. 1986. The role of selenium in nutrition. Academic Press, Orlando, FL.

Corré, W.J., and T. Breimer. 1979. Nitrate and nitrite in vegetables. Centre for Agric. Publ. and Documentation, Wagenigen, the Netherlands.

Crampton, E.W., and L.E. Harris. 1969. Applied animal nutrition. W.H. Freeman & Co., San Francisco, CA.

Dallemagne, M.J., and L.J. Richelle. 1973. Inorganic chemistry of bone. p. 23–42. In I. Zephin (ed.) Biological mineralization. John Wiley & Sons, New York.

Davis, J., and K. Sherer. 1994. Applied nutrition and diet therapy for nurses. 2nd ed. W.B. Saunders Co., Philadelphia, PA.

Doran, J.W., and T.B. Parken. 1994. Defining and assessing soil quality. p. 3–21. In J.W. Doran et al. (ed.) Defining soil quality for a sustainable environment. SSSA Spec. Publ. 35. ASA, CSSA, and SSSA. Madison, WI.

Doran, J.W., M. Sarrantonio, and M.A. Leibig. 1996. Soil health and sustainability. p. 1–54. In D.L. Sparks (ed). Advances in agronomy. Vol. 56. Academic Press, New York.

Epstein, E. 1972. Mineral nutrition of plants: principles and perspectives. John Wiley & Sons, New York.

Garrett, R.G. 1994. The distribution of cadmium in A horizon soils in the prairies of Canada and adjoining United States. p. 73–82. In Current research 1994-B. Geol. Surv. of Canada.

Goodrich, R.D., and J.E. Garrett. 1986. Sulfur in livestock nutrition. p. 617–633. In M.A. Tabatabai (ed.) Sulfur in agriculture. Agron. Monogr. 27. ASA, CSSA, and SSSA, Madison, WI.

Grant, C.A., and L.D. Bailey. 1995. Cadmium research update. p. 126–132. In Proc. Western Nutrient Management Conf., Salt Lake City, UT. 9–10 Mar. 1995.

Grant, C.A., L.D. Bailey, M. Chandbury, K.R. Biona, and G.J. Racz. 1995. Fertilizer effects on cadmium concentration in crops. In Proc. Manitoba Soc. of Soil Sci. Annual Meeting. Univ. of Manitoba, Winnipeg, MB.

Hambidge, K.M., C.E. Casey, and N.F. Krebs. 1986. Zinc. p. 1–137. In W. Mertz (ed.). Trace elements in human and animal nutrition. 5th ed. Vol. 2. Academic Press, Orlando, FL.

Hamilton, E.M.N., E.N. Whitney, and F.S. Sizer. 1988. Nutrition: Concepts and controversies. 4th ed. West Publ. Co., St. Paul, MN.

Hergert, W.W. 1986. Consequences of nitrate in groundwater. Solutions (30)5:24–31.

Holmgren, G.G.S., M.W. Meyer, R.L. Chaney, and R.B. Daniels. 1993. Cadmium, lead, zinc, copper, and nickel in the agricultural soils of the United States of America. J. Environ. Qual. 22:335–348.

Jackson, M.L. 1988. Selenium: Geochemical distribution and associations with human heart and cancer death rates and longevity in China and the United States. Biol. Trace Elem. Res. 15:13–21.

Jones, R.L., and H.C. Hanson. 1985. Mineral licks, geophagy, and biogeochemistry of North American ungulates. Iowa State Univ. Press, Ames.

Jurgens, M.H. 1993. Animal feeding and nutrition. 7th ed. Kendall/Hunt Publ., Dubuque, IA.

Karlen, D.L., M.J. Mausbach, J.W. Doran, R.G. Clive, R.F. Harris, and G.E. Schuman. 1997. Soil quality: A concept, definition and framework for evaluation. Soil Sci. Soc. Am. J. 61:4–10.

Landner, L., J. Folke, M.O. Öberg, H. Mikaelson, and M. Aringberg-Laanatza. 1995. Cadmium in fertilizers. Consultants report prepared for OECD Cadmium Workshop, Sweden. 16–20 Oct. 1995.

Larson, W.E., and F.J. Pierce. 1991. Conservation and enhancement of soil quality. In Evaluation for sustainable land management in the developing world. Vol. 2. IBSRAM Proc. 12(2). Int. Board for Soil Res. and Manage., Bangkok, Thailand.

Larson, D., N. Spitz, E. Termine, P. Riband, H. Lafont, and J. Hariton. 1983. Effect of organic and mineral nitrogen fertilization on yield and nutritive value of butterhead lettuce. Qual. Plant Foods Hum. Nutr. 34:97–108.

Leibig, J. von. 1855. Principles of agricultural chemistry, with special reference to the late researches made in England. Walton & Maberly, London.

Martin, J.P. 1965. Iodine. p. 200–202. *In* H.D. Chapman (ed.). Diagnostic criteria for plants and soils. Quality Printing Co., Abilene, TX.

Mayland, H.F., L.F. James, K.E. Panter, and J.L. Sanderegger. 1989. Selenium in seleniferous environments. p. 15–50. *In* L.W. Jacobs (ed.) Selenium in agriculture and the environment. SSSA Spec. Publ. 23. ASA and SSSA, Madison, WI.

Maynard, L.A., and J.K. Loosli. 1969. Animal nutrition. 6th ed. McGraw-Hill, New York.

McDowell, L.R. 1985. Nutrition of grazing ruminants in warm climates. Academic Press, Orlando, FL.

Mikkelsen, R.L., A.L. Page, and F.T. Bringham. 1989. Factors affecting selenium accumulation by agricultural crops. p. 65–94. *In* L.W. Jacobs (ed.) Selenium in agriculture and the environment. SSSA Spec. Publ. 23. ASA and SSSA, Madison, WI.

Rendig, V.V. 1986. Sulfur and crop quality. p. 635–652. *In* M.A. Tabatabai (ed.) Sulfur in agriculture. Agron. Monogr. 27. ASA, CSSA, and SSSA, Madison, WI.

Rosenfeld, I., and O.A. Beath. 1964. Selenium, geobotany, biochemistry, toxicity and nutrition. Academic Press, New York.

Russell, F.C. 1944. Minerals in pasture: Deficiencies and excesses in relation to animal health. Tech. Comm. No. 15. Imperial Bur. of Animal Nutr., Aberdeen, Scotland.

Schramm et al., 1994.

Smith, K.A., J.M. Bremner, and M.A. Tabatabai. 1973. Sorption of gaseous atmospheric pollutants by soils. Soil Sci. 116:313–319.

Tisdale, S.L. 1977. Sulphur in forage quality and ruminant nutrition. Tech. Bull. 22. The Sulphur Inst., Washington, DC.

Tisdale, S.L., W.L. Nelson, and J.D. Beaton. 1985. Soil fertility and fertilizers. 4th ed. Macmillan, New York.

Troeh, F.R., J.A. Hobbs, and R.L. Donahue. 1991. Soil and water conservation. 2nd ed. Prentice-Hall, Englewood Cliffs, NJ.

Voisin, A. 1963. Grass tetany. Charles C. Thomas, Springfield, IL.

Wacher, W.E.C. 1980. Magnesium and man. Howard Univ. Press, Cambridge, MA.

2 Quantitative Indicators of Soil Quality: A Minimum Data Set

John W. Doran

USDA-ARS and University of Nebraska
Lincoln, Nebraska

Timothy B. Parkin

USDA-ARS and National Soil Tilth Laboratory
Ames, Iowa

Interest in evaluating soil quality has been stimulated by increasing awareness that *soil* is a critically important component of the earth's biosphere (Glanz, 1995). Soil functions in the production of food and fiber and also in the maintenance of the environment through acting as a filter and environmental buffer for water, air, nutrients, and chemicals. The quality and health of soils determine agricultural sustainability (Acton & Gregorich, 1995), environmental quality (Pierzynski et al., 1994), and, as a consequence of both—plant, animal, and human health (Haberern, 1992). Past management of nature to meet the food and fiber needs of increasing populations has taxed the resiliency of natural processes to maintain global balances of energy and matter (Doran et al., 1996). Within the last decade, inventories of the soil's productive capacity indicate severe degradation on well more than 10% of the earth's vegetated land as a result of soil erosion, atmospheric pollution, excessive tillage, over-grazing, land clearing, salinization, and desertification (Lal, 1994; Sanders, 1992). Findings from a project of the United Nations Environment Program on Global Assessment of Soil Degradation indicate that almost 40% of agricultural land has been adversely affected by human-induced soil degradation, and that more than 6% is degraded to such a degree that restoration of its original productive capacity is only possible through major capital investments (Oldeman, 1994). The quality of surface and subsurface water has been jeopardized in many parts of the world by intensive land management practices and the consequent imbalance in C, N, and water cycling in soil. At present, agriculture is considered the most widespread contributor to nonpoint source water pollution in the USA (CAST, 1992a; National Research Council, 1989). The present threat of global climate change and ozone depletion, through elevated levels of atmospheric gases and altered hydrological cycles, necessitates a better understanding of the effects of land management on soil processes. Soil management practices such as tillage, cropping patterns, and pesticide and fertilizer use are known to influence water quality. These manage-

---

Copyright © 1996 Soil Science Society of America, 677 S. Segoe Rd., Madison, WI 53711, USA.
Methods for Assessing Soil Quality, SSSA Special Publication 49.

ment practices also influence atmospheric quality through changes in the soil's capacity to produce, consume, or store important atmospheric gases such as carbon dioxide, nitrogen oxides, and methane (CAST, 1992b; Mosier et al., 1991; Rolston et al., 1993).

Developing *sustainable* agricultural management systems is complicated by the need to consider their utility to humans, their efficient use of resources, and their ability to maintain a balance with the environment that is favorable both to humans and most other species (Harwood, 1990). We are challenged to develop management systems that balance the needs and priorities for production of food and fiber with those for a safe and clean environment. In the USA, the importance of soil quality in maintaining this balance was iterated in a recent National Academy of Science publication, "Protecting soil quality, like protecting air and water quality, should be a fundamental goal of national environmental policy" (National Research Council, 1993). The same report recommended that U.S. Department of Agriculture (USDA) and the U.S. Environmental Protection Agency (USEPA) initiate an integrated effort to develop quantifiable standards and cost-effective monitoring methods that can be used to evaluate the effects of farming systems management on soil quality. Defining indicators of soil quality, however, is complicated by the need to consider the multiple functions of soil in maintaining productivity and environmental well-being and to integrate the physical, chemical, and biological soil attributes that define those functions (Papendick & Parr, 1992; Rodale Institute, 1991).

QUANTITATIVE INDICATORS OF SOIL QUALITY

Much like air or water, the *quality* of soil has a profound effect on the health and productivity of a given ecosystem and the environments related to it; however, unlike air or water for which we have quality standards, the definition and quantification of soil quality is complicated by the fact that it is not directly consumed by humans and animals as are air and water. Soil quality is often thought of as an abstract characteristic of soils that cannot be defined because it depends on external factors such as land use and soil management practices, ecosystem and environmental interactions, socioeconomic and political priorities, and so on. Perceptions of what constitutes a good soil vary depending on individual priorities for soil function and intended land use; however, to manage and maintain our soils in an acceptable state for future generations, *soil quality* must be defined, and the definition must be broad enough to encompass the many functions of soil. These considerations led Doran and Parkin (1994) to define *soil quality* as: "The capacity of a soil to function, within ecosystem and land-use boundaries, to sustain biological productivity, maintain environmental quality, and promote plant and animal health."

Quantitative assessment of soil quality is invaluable in determining the sustainability of land management systems. A framework for evaluation or an index of soil quality is needed to identify problem production areas, make realistic estimates of food production, monitor changes in sustainability and environmental quality as related to agricultural management, and to assist government agencies in formulating and evaluating sustainable agricultural and land-use policies (Acton, 1993; Granatstein & Bezdicek, 1992). Effective identification of appropriate indicators for soil health assessment depends on the ability of any approach to consider the multiple components of soil function, in particular, productivity and environmental well-being. Identification of indicators and assessment approaches is further complicated by the multiplicity of physical, chemical, and biological factors that control biogeochemical processes and their variation in intensity over time and space. Practical assessment of soil quality and health, however, requires consideration of the multiple functions of soil and their variations in time and space (Larson & Pierce, 1991).

INDICATORS OF SOIL QUALITY AND HEALTH: A MINIMUM DATA SET

The rapid acceleration of technological growth associated with industrial and postindustrial societies poses a risk to the health of natural ecosystems that are slow to change. Within the context of ecosystem health, Constanza et al. (1992) concluded that an ecological system is healthy if it is active, maintains its organization and autonomy over time, and is resilient to stress. They proposed a long-term strategy for the assessment and improvement of ecosystem health, based on the model used in the practice of human and animal medicine. The assessment of human health in medicine follows a six step sequence: (i) identify symptoms; (ii) identify and measure vital signs; (iii) make a provisional diagnosis (iv) conduct tests to verify the diagnosis; (v) make a prognosis; and (vi) prescribe a treatment.

Assessing soil quality and health can be likened to a medical examination of humans in which certain measurements are taken of the quality of certain parameters as basic indicators of system function (Larson & Pierce, 1991). In a medical exam, the physician takes measurements of body system functions such as temperature, blood pressure, pulse rate, and perhaps certain blood or urine chemistries. The physician also will take note of visible, outward signs of health status. If these basic indicators are outside specific ranges, more diagnostic tests can be conducted to help identify the cause of the problem and find a solution. For example, excessively high blood pressure may indicate a potential for system failure (death) through stroke or cardiac arrest. Because one of the causes of high blood pressure may be improper diet, lack of exercise, or high stress level, the physician may request a secondary blood chemistry test for cholesterol, electrolytes, etc. Assessment of stress level as a causative factor for high blood pressure is less straightforward and generally involves implementing some change in lifestyle followed by periodic monitoring of blood pressure to assess change. This is a good example of using a basic indicator both to identify a problem and to monitor the effects of management on the health of a system.

Applying this human health analogy to soil quality and health is fairly straightforward. Larson and Pierce (1991) proposed that a minimum data set (MDS) of soil parameters be adopted for assessing the health of world soils, and that standardized methodologies and procedures be established to assess changes in the quality of those factors. A set of basic indicators of soil quality and, therefore, health has not previously been defined, largely due to difficulty in defining soil quality and health, the wide range across which soil indicators vary in magnitude and importance, and disagreement among scientists and soil and land managers over which basic indicators should be measured.

Acton and Padbury (1993) defined soil quality attributes as measurable soil properties that influence the capacity of soil to perform crop production or environmental functions. Soil attributes are useful in defining soil quality criteria and serve as indicators of change in quality. Attributes that are most sensitive to management are most desirable as indicators and some such as soil depth, soil organic matter, and electrical conductivity are often affected by soil degradation processes (Arshad & Coen, 1992).

To be practical for use by practitioners, extension workers, conservationists, scientists, and policy makers the set of basic soil quality–health indicators should be useful across a range of ecological and socioeconomic situations. Indicators should:

1. Correlate well with ecosystem processes (this also increases their utility in process oriented modeling);
2. Integrate soil physical, chemical, and biological properties and processes and serve as basic inputs needed for estimation of soil properties or functions which are more difficult to measure directly.
3. Be relatively easy to use under field conditions and be assessable by both specialists and producers.
4. Be sensitive to variations in management and climate. The indicators should be sensitive enough to reflect the influence of management and climate on long-term changes in soil quality but not be so sensitive as to be influenced by short-term weather patterns.
5. Be components of existing soil data bases where possible.

The need for basic soil quality and health indicators is reflected in the question commonly posed by producers, researchers, and conservationists: "What measurements should I make or what can I observe that will help me evaluate the effects of management on soil function now and in the future?" *Too often scientists confine their interests and efforts to the discipline with which they are most familiar.* Microbiologists often limit their studies to soil microbial populations, having little or no regard for soil physical or chemical characteristics that define the limits of activity for microorganisms, plants, and other life forms. The proper approach in defining soil quality and health indicators must be holistic, not reductionistic. The indicators chosen also must be measurable by as many people as possible, especially managers of the land, and not limited to a select cadre of research scientists. Indicators should describe the major ecological processes in soil and ensure that measurements made reflect conditions as they exist in the field under a given management system. They should relate to major ecosystem functions such as C and N cycling (Visser & Parkinson, 1992) and be driving variables for process oriented models that emulate ecosystem function. Some indicators, such as soil bulk density, must be measured in the field so that laboratory analyses for soil organic matter and nutrient content can be better related to actual field conditions at time of sampling. Soil bulk density also is required for calculation of soil properties such as water-filled pore space (WFPS), which serves as an excellent integrator of soil physical, chemical, and biological soil properties and aeration dependent microbial processes important to C and N cycling in soil (Doran et al., 1990). A diagramatic representation of the relationship between soil WFPS and microbial activity is given by Parkin et al. (1996, this publication) in Fig. 14–2. Many basic soil properties are useful in estimating other soil properties or attributes that are difficult or too expensive to measure directly. A listing of these basic indicators and input variables and the soil attributes they can be used to estimate are given in Table 2–1.

Starting with the MDS proposed by Larson and Pierce (1991), Doran and Parkin (1994) developed a list of basic soil properties that meet many of the aforementioned requirements of indicators for screening soil quality and health. This initial list of soil quality indicators as reviewed and revised by the North Central Region 59 Technical Committee on Soil Organic Matter and the Soil Quality Working Group of the U.S. Department of Agriculture, Agricultural Research Service, is presented in Table 2–2. This recommended minimum data set of soil quality indicators forms the primary context for many of the methods discussed in other chapters of this book.

The appropriate use of soil quality indicators depends largely on how well these indicators are understood with respect to the ecosystem of which they are part. Thus, interpretation of the relevance of soil biological indicators apart from soil physical and chemical attributes and their ecological relevance is of little value and, with respect to assessment of soil quality or health, can actually be misleading. Data presented describing soil quality and financial performance of biodynamic and conventional farming management systems in New Zealand, are

Table 2–1. A limited listing of soil attributes or properties that can be estimated from basic input variables using pedotransfer functions or simple models.

Soil attribute or property	Basic input variables†	Reference
Cation-exchange capacity‡	Organic C + clay type and content	Larson & Pierce, 1994
Water retention characteristic (AWHC)	% sand, silt, clay, + organic C + BD†	Gupta & Larson, 1979
Hydraulic conductivity	Soil texture	Larson & Pierce, 1994
Aerobic and anaerobic microbial activity	WFPS† as calculated from BD and water content	Linn & Doran, 1984 Doran et al., 1990
C and N cycling	Soil respiration (soil temperature + WFPS)	Parkin et al., 1996
Plant/microbial activity or pollution potential	Soil pH + EC†	Smith & Doran, 1996
Soil productivity	BD, AWHC†, pH, EC, and aeration	Larson & Pierce, 1994
Rooting depth	BD, AWHC, pH	Larson & Pierce, 1994
Leaching potential	Soil texture, pH, organic C (hydraulic conductivity, CEC, depth)	Shea et al., 1992

† AWHC, available water holding capacity; BD, soil bulk density; EC, soil electrical conductivity; WFPS, water-filled pore space.

‡ Cation-exchange capacity ($cmol_c$ kg^{-1}) can be estimated by:

$$[(\% C/.58) \times 200] + [\% clay \times (average\ exchange\ capacity\ of\ clay\ types)]$$

where Montmorillonite = 100, Illite = 30, and Kaolinite = 8 $cmol_c$ kg^{-1} (meq 100 g^{-1})

Table 2–2. Proposed minimum data set of physical, chemical, and biological indicators for screening the condition, quality, and health of soil (after Doran & Parkin, 1994; Larson & Pierce, 1994).

Indicators of soil condition	Relationship to soil condition and function; rationale as a priority measurement	Ecologically relevant values or units; comparisons for evaluation
Physical		
Texture	Retention and transport of water and chemicals; modeling use, soil erosion, and variability estimate	% sand, silt, & clay; less eroded sites or landscape positions
Depth of soil, topsoil, and rooting	Estimate of productivity potential and erosion; normalizes landscape and geographic variability	cm or m; non cultivated sites or varying landscape positions
Infiltration and soil bulk density (SBD)	Potential for leaching, productivity, and erosivity; SBD needed to adjust analyses to volumetric basis	Minutes/2.5 cm of water and g/cm^3 row and/or landscape positions
Water holding capacity (water retention characteristic)	Related to water retention, transport, and erosivity; available H_2O: Calculate from SBD, texture, and OM	% (cm^3/cm^3), cm of available $H_2O/30$ cm; precipitation intensity
Chemical		
Soil organic matter (OM) (total organic C and N)	Defines soil fertility, stability, and erosion extent; use in process models and for site normalization	kg C or N/ha-30 cm; noncultivated or native control
pH	Defines biological and chemical activity thresholds; essential to process modeling	Compared with upper and lower limits for plant and microbial activity
Electrical conductivity	Defines plant and microbial activity thresholds; presently lacking in most process models	dS/m^1; compared with upper and lower limits for plant and microbial activity
Extractable N, P, and K	Plant available nutrients and potential for N loss; productivity and environmental quality indicators	kg/ha-30 cm; seasonal sufficiency levels for crop growth
Biological		
Microbial biomass C and N	Microbial catalytic potential and repository for C and N; modeling; Early warning of management effects on OM	kg N or C/ha-30 cm; relative to total C and N or CO_2 produced
Potentially mineralizable N (anaerobic incubation)	Soil productivity and N supplying potential; Process modeling; (surrogate indicator of biomass)	kg N/ha-30 cm/d; relative to total C or total N contents
Soil respiration, water content, and temperature	kg C/ha/d; relative microbial biomass activity, C loss vs. inputs and total C pool	Microbial activity measure (in some cases plants) process modeling; estimate of biomass activity

useful in illustrating this concern (Reganold et al., 1993; Table 2–3). Our analyses, however, are not intended as criticisms of this published work as the authors should be commended for their vision in choice of physical, chemical, and biological indicators of soil quality. One point of discussion, is the importance of expressing the results of soil quality tests on a volumetric rather than a gravimetric basis and in units for which ecological relevance can be readily ascertained. As illustrated in Table 2–3, the magnitude of differences in soil C, total N, respiration, and mineralizable N between management systems for samples expressed by weight of soil are 8 to 10% greater than where expressed on a volume basis using soil bulk density estimates. In cultivated systems soil bulk density can vary considerably across the soil surface due to mechanical compaction and throughout the growing season due to reconsolidation of soil after tillage. Soil bulk density also is directly proportional to the mass of any soil component for a given depth of soil sampled. Where samples are taken in the field under management conditions of varying soil densities, comparisons made using gravimetric analyses will err by the difference in soil density at time of sampling. The observed differences due to management in the New Zealand study were statistically significant; however, since results were expressed on a gravimetric basis, they may not be valid nor ecologically relevant. In cases such as this, where values for soil bulk density at time of sampling are not available, the use of soil indicator ratios (in this case mineralizable N to C) can reduce errors of interpretation associated with use of results expressed on a weight basis. Reganold and Palmer (1995) recommend calculating soil measurements on a volume basis per unit of topsoil or solum depth for most accurate assessment of management effects on soil quality.

Table 2–3. Reported and ecologically relevant mean values of aggregated soil quality data for 0- to 20-cm layer of 16 biodynamic (Bio.) and conventional (Conv.) farms in New Zealand (after Reganold et al., 1993)

Soil property	Biodynamic farms	Conventional farms	Ratio Bio./Conv.
Reported units & values			
0–5 cm bulk density, Mg m^{-3}	1.07	1.15	0.93*
Topsoil thickness, cm	22.8	20.6	1.11*
C, %	4.84	4.27	1.13*
Total N, mg kg^{-1})	4840	4260	1.14*
Mineralizable N, mg kg^{-1}	140.0	105.9	1.32*
Respiration, μL O$_2$ h^{-1}g^{-1}	73.7	55.4	1.33*
Ratio: Mineralizable N to C, mg g^{-1}	2.99	2.59	1.15*
Extractable P, mg kg^{-1}	45.7	66.2	0.69*
pH	6.10	6.29	0.97*
Ecologically relevant units & values			
0–20 cm bulk density†, g cm^{-3}	1.2	1.3	0.92
C, Mg C ha^{-1}	116.2	111.0	1.05
Total N, kg N ha^{-1}	11616	11076	1.05
Mineralizable N, kg N ha^{-1}14 d^{-1}	336	275	1.22
Respiration in laboratory, kg C ha^{-1} d^{-1}	2275	1850	1.23
Ratio: Mineralizable N to C	2.89	2.48	1.17*
Extractable P (excess)‡, kg P ha^{-1}	110 (50)‡	172 (112)	0.63*
pH units above 6.0 lower limit	0.1	0.3	0.33

* Values differ significantly at 0.01 probability level.
† Estimated, since data was only given for 0- to 5-cm depth.
‡ Threshold value for environmentally sound soil P level set at 60 kg P ha^{-1}.

Ellert and Bettany (1995) also illustrated the importance of accounting for differences in soil bulk density when estimating the storage of organic matter and nutrients in soil under different management schemes. They preposed sampling to different depths such that an equivalent mass of soil was compared for varying management situations; use of equivalent sampling depths, however, requires measurement of soil bulk density.

Choice of units of expression for soil quality indicators also can have an important bearing on determining the ecological relevance of measured values. In the New Zealand study, respiration of laboratory incubated soils from biodynamic farms averaged 73.7 mL O_2 $h^{-1}g^{-1}$, significantly greater (33%) than that from conventional farms. One interpretation of these results could be that the soils of the biodynamic farms are healthier since respiration was greater; however, if one assumes that for aerobic respiration a mole of O_2 is consumed for each mole of carbon dioxide produced, and the results are adjusted for soil density and expressed as kilogram of C released per hectare per day, a different picture emerges. The quantities of C released in 1 d from both the biodynamic and conventional farms are incredibly high and represent 2.0 and 1.7%, respectively, of the total C pools of these surface soils. While the values for soil respiration from disturbed soils incubated in the laboratory only represent a potential for release of readily metabolizable C (labile C), the results clearly demonstrate that more may not be better and that high rates of respiration may be ecologically detrimental as they represent potentials for depletion of soil organic C with accelerated enrichment of the atmosphere with carbon dioxide. When expressed in ecologically relevant units, it becomes obvious that the respiration rates observed in this study are of limited use in evaluating the status of soil quality and health between these different farming management systems when used as the only indicator.

Expression of soil quality indicators in ecologically relevant units, as shown in Table 2–2, facilitates establishing limits on interpretation thresholds that are at the same level of scale at which soils are managed. Ecologically relevant data from the New Zealand study (Table 2–3) will be used to illustrate this point. Levels of mineralizable N above that needed for crop production for biodynamic farms and extractable P levels above crop needs for conventional farms could represent a lower level of soil quality and health as a result of greater potential for environmental contamination through leaching, runoff, or volatilization losses. Specific upper limits for environmentally sound levels of soil P and N exist and are determined by local climatic, topographic, soil, and management situations (Sharpley et al., 1996). Again, an example that with respect to soil quality and health, more is not necessarily better and ecologically relevant units are needed for proper evaluation. Soil pH is another example of a soil quality attribute that must be referenced to a definable standard for upper and lower limits that are defined by the cropping system or biological processes of greatest ecological relevance. The above discussion serves to highlight the difficulty we have in interpreting the results of laboratory incubations and the need for in-field measurements of respiration and N cycling.

Indicators of soil quality and health are commonly used to make comparative assessments between agricultural management practices to determine their sustainability; however, the utility of comparative assessments of soil quality are limited because they provide little information about the processes creating the measured condition or performance factors associated with respective management systems (Larson & Pierce, 1994). Also, the mere analysis of soils, no matter how comprehensive or sophisticated does not provide a measure of soil quality or health unless the parameters are calibrated against designated soil functions (Janzen et al., 1992).

Quantitative Assessments

Quantitative assessments of soil quality and health will require consideration of the many functions that soils perform, their variations in time and space, and opportunities for modification or change. Criteria are needed to evaluate the impact of various practices on the quality of air, soil, water, and food resources. Soil quality and health can not be defined in terms of a single number, such as the 10 mg L^{-1} NO_3–N standard applied for drinking water, although such quantitative standards will be valuable to overall assessment. Assessments must consider specific soil functions being evaluated in their land use and societal contexts. Threshold values for key indicators must be established with the knowledge that these will vary depending upon land use, the specific soil function of greatest concern, and the ecosystem or landscape within which the assessment is being made. For example, soil organic matter concentration is frequently cited as a major indicator of soil quality. Threshold values established for highly weathered Ultisols in the southeastern USA indicate that surface soil organic matter levels of 2% (1.2% organic C) would be very good, while the same value for Mollisols developed under grass in the Great Plains, which commonly have higher organic matter levels, would represent a degraded condition limiting soil productivity. As pointed out by Janzen et al. (1992), the relationship between soil quality indicators and various soil functions does not always comply to a simple relationship increasing linearly with magnitude of the indicator, as is commonly thought. Simply put, bigger is not necessarily better.

Soil quality and health assessments will have to be initiated within the context of societal goals for a specific landscape or ecosystem. Examples include establishing goals such as enhancing water quality, soil productivity, biodiversity, or recreational opportunities. When specific goals have been established or are known, then critical soil functions needed to achieve those goals can be agreed upon, and the criteria for assessing progress toward achieving those goals can be set. Periodic assessments of soil quality and health with known indicators, thresholds, and other criteria for evaluation will then make it possible to quantify soil quality and health.

To accomplish such goals, several approaches for assessing soil quality have been proposed (Acton & Padbury, 1993; Doran & Parkin, 1994; Karlen et al., 1994; Larson & Pierce, 1994). A common attribute among all these approaches is that soil quality is assessed with respect to specific soil functions. Larson and Pierce (1994) proposed a dynamic assessment approach in which the dynamics, or change in soil quality, of a management system is used as a measure of its sustainability. They proposed use of a minimum data set of temporally variable soil properties to monitor changes in soil quality over time. They also proposed use of pedotransfer functions (Bouma, 1989) to estimate soil attributes which are too costly to measure and to interrelate soil characteristics in evaluation of soil quality (Table 2–1). Simple computer models are used to describe how changes in soil quality indicators impact important functions of soil, such as productivity. An important part of this approach is the use of statistical quality control procedures to assess the performance of a given management system rather than its evaluation by comparison to other systems. This dynamic approach for assessing soil quality permits identification of critical parameters and facilitates corrective actions for sustainable management.

Karlen and Stott (1994) presented a framework for evaluating site-specific changes in soil quality. In this approach they define a high quality soil as one that: (i) accommodates water entry; (ii) retains and supplies water to plants; (iii) resists degradation; and (iv) supports plant growth. They described a procedure by which soil quality indicators that quantify these functions are identified, assigned a priority or weight that reflects its relative importance, and are scored using a systems engineering approach for a particular soil attribute such as resistance to water erosion. Karlen et al. (1994) also demonstrated the utility of this approach in discriminating changes in soil quality between long-term crop residue and tillage management practices.

Doran and Parkin (1994) described a performance-based index of soil quality that could be used to provide an evaluation of soil function with regard to the major issues of: (i) sustainable production; (ii) environmental quality; and (iii) human and animal health. They proposed a soil quality index consisting of six elements:

$$SQ = f\ (SQE1, SQE2, SQE3, SQE4, SQE5, SQE6)$$

where Soil Quality Elements are:
SQE1 = food and fiber production
SQE2 = erosivity
SQE3 = groundwater quality
SQE4 = surface water quality
SQE5 = air quality; and
SQE6 = food quality.

One advantage of this approach is that soil functions can be assessed based on specific performance criteria established for each element, for a given ecosystem. For example, yield goals for crop production (SQE1); limits for erosion losses (SQE2); concentration limits for chemicals leaching from the rooting zone (SQE3); nutrient, chemical and sediment loading limits to adjacent surface water systems (SQE4); production and uptake rates for gases that contribute to ozone destruction or the greenhouse effect (SQE5); and nutritional composition and chemical residue of food (SQE6). This list of elements are restricted primarily to agricultural situations but other elements such as wildlife habitat quality could be easily added to expand the applications of this approach.

This approach would result in soil quality indices computed in a manner analogous to the soil tilth index proposed by Singh et al. (1990). Weighting factors are assigned to each soil quality element, with relative weights of each coefficient being determined by geographical considerations, societal concerns, and economic constraints. For example in a given region, food production may be the primary concern, and elements such as air quality may be of secondary importance. If such were the case, SQE1 would be weighted more heavily that SQE5. Thus this framework has an inherent flexibility in that the precise functional relationship for a given region, or a given field, is determined by the intended use of that area or site, as dictated by geographical and climatic constraints as well as socioeconomic concerns.

Assessment of soil quality and health is not limited to areas used for crop production, although this is the major emphasis of this book. Forests and forest soils are important to the global C balance as related to C sequestration and atmospheric levels of carbon dioxide. Soil organic matter and soil porosity, as estimated from soil bulk density, have recently been proposed among international groups as major soil quality indicators in forest soils (Richard Cline, 1995, personal communication). Criteria for evaluating rangeland health have recently been suggested in a National Research Council (1994) report that describes new methods to help classify, inventory, and monitor rangelands. Rangeland health is defined as the degree to which the integrity of the soil and the ecological processes of rangeland ecosystems are sustained. Assessment of rangeland health are based on the evaluation of three criteria: degree of soil stability and watershed function, integrity of nutrient cycles and energy flows, and presence of functioning recovery mechanisms.

SUMMARY

The minimum data set presented here provides a list of indicators deemed necessary for assessment of soil quality but does not provide a framework by which measurement of soil quality indicators can be interrelated to assess soil quality. This is discussed in detail by Harris et al. (1996, this publication); however, the process of identification and measurement of the basic physical, chemical, and biological components comprising the soil ecosystem facilitates appreciation by the researcher, consultant, or land manager of the broad effects of agricultural and land management on soil function and soil quality. Also, it can serve to identify specific soil attributes that are most important or need more detailed study within the unique constraints of soil, climatic, tillage, and cropping management systems, etc. and the social, economic and environmental concerns that may be unique to a certain geographical region. The specific use of soil quality indicators for on-farm assessment of soil quality is presented Sarrantonio et al. (1996, this publication).

REFERENCES

Acton, D.F. 1993. A program to assess and monitor soil quality in Canada: Soil quality evaluation program summary (interim). Centre for Land and Biol. Resour. Res. Contrib. no. 93-49. Research Branch, Agriculture Canada, Ottawa.

Acton, D.F., and L.J. Gregorich, 1995. The health of our soils: Toward sustainable agriculture in Canada. Agric. and Agri-food Canada, Ottawa, ON.

Acton, D.F., and G.A. Padbury. 1993. A conceptual framework for soil quality assessment and monitoring. p. 2-1–2-7. *In* D.F. Acton (ed.) A program to assess and monitor soil quality in Canada: Soil quality evaluation program summary (interim). Centre for Land and Biol. Rsour. Res. Contrib. no. 93-49. Res. Branch, Agriculture Canada, Ottawa.

Arshad, M.A., and G.M. Coen 1992. Characterization of soil quality: Physical and chemical criteria. Am. J. Altern. Agric. 7:12–16.

Bouma, J. 1989. Using soil survey data for quantitative land evaluation. Adv. Soil Sci. 9:177–213.

CAST. 1992a. Water quality: Agriculture's role. Task Force Rep. 120. Council for Agric. Sci. and Technol., Ames, IA.

CAST. 1992b. Preparing U.S. agriculture for global climate change. Task Force Rep. 119. Council for Agric. Sci. and Technol., Ames, IA.

Constanza, R., B.G. Norton, and B.D. Haskell. 1992. Ecosystem health: New goals for environmental management. Island Press, Washington, DC.

Doran, J.W., L.N. Mielke, and J.F. Power. 1990. Microbial activity as regulated by soil water-filled pore space. p. 94–99. *In* Trans. of the 14th Int. Congr. of Soil Science, Kyoto, Japan, August 1990. ISSS, Vienna, Austria.

Doran, J.W., and T.B. Parkin. 1994. Defining and assessing soil quality. p. 3–21 *In* J.W. Doran et al. (ed.) Defining soil quality for a sustainable environment. SSSA Spec. Publ. 35. SSSA, Madison, WI.

Doran, J.W., M. Sarrantonio, and M. Liebig. 1996. Soil health and sustainability. p. 1–54. *In* D. L. Sparks (ed.) Advances in agronomy. Vol. 56. Academic Press, San Diego, CA.

Ellert, B.H., and J.R. Bettany. 1995. Calculation of organic matter and nutrients stored in soils under contrasting management regimes. Can. J. Soil Sci. 75:529–538.

Glanz, J.T. 1995. Saving our soil: Solutions for sustaining earth's vital resource. Johnson Books, Boulder, CO.

Granatstein, D., and D.F. Bezdicek. 1992. The need for a soil quality index: Local and regional perspectives. Am. J. Altern. Agric. 17:12–16.

Gupta, S.C., and W.E. Larson. 1979. Estimating soil water retention characteristics from particle size distribution, organic matter percent, and bulk density. Water Resour. Res. 15:1633–1635.

Haberern, J. 1992. Viewpoint: A soil health index. J. Soil Water Conserv. 47:6.

Harwood, R.R. 1990. A history of sustainable agriculture. p. 3–19. *In* C.A. Edwards et al. (ed.) Sustainable agricultural systems. Soil Water Conser. Soc., Ankeny, IA.

Harris, R.F., D.L. Karlen, and D.J. Mulla. 1996. A conceptual framework for assessment and management of soil quality and health. p. 61–82. *In* J.W. Doran and A.J. Jones (ed.) Methods for assessing soil quality. SSSA Spec. Publ. 49. SSSA, Madison, WI.

Janzen, H.H., F.J. Larney, and B.M. Olson. 1992. Soil quality factors of problem soils in Alberta. p. 17–28. *In* Proc. Alberta Soil Science Workshop, Lethbridge, Alberta, Canada. Univ. of Alberta, Edmonton.

Karlen, D.L., and D.E. Stott. 1994. A framework for evaluating physical and chemical indicators of soil quality. p. 53–72. *In* J.W. Doran et al. (ed.) Defining soil quality for a sustainable environment. SSSA Spec. Publ. 35. SSSA, Madison, WI.

Karlen, D.L., N.C. Wollenhaupt, D.C., Erbach, E.C. Berry, J.B. Swan, N.S. Eash, and J.L. Jordahl. 1994. Crop residue effects on soil quality following 10-years of no-till corn. Soil Tillage Res. 31:149–167.

Lal, R. 1994. Sustainable land use systems and soil resilience. p. 41–67 *In* D.J. Greenland and I. Szabolcs (ed.) Soil resilience and sustainable land use. Proc. of a Symposium, Budapest, Hungary. 28 Sept. – 2 Oct. 1992. CAB Int, Wallingford, England.

Larson, W.E., and F.J. Pierce. 1991. Conservation and enhancement of soil quality. p. 175–203. *In* Int. Workshop on Evaluation for sustainable land management in the developing world, Chiang Rai, Thailand. 15–21 Sept. Int. Board for Soil Res. and Manage., Bangkok, Thailand.

Larson, W.E., and F.J. Pierce, 1994. The dynamics of soil quality as a measure of sustainable management. p. 37–51. *In* J.W. Doran et al. (ed.) Defining soil quality for a sustainable environment. SSSA Spec. Publ. 35. SSSA, Madison, WI.

Linn, D.M., and J.W. Doran, 1984. Effect of water-filled pore space on carbon dioxide and nitrous oxide production in tilled and nontilled soils. Soil Sci. Soc. Am. J. 48:1267–1272.

Mosier, A.R., D. Schimel, D. Valentine, K. Bronson, and W. Parton. 1991. Methane and nitrous oxide fluxes in native, fertilized and cultivated grasslands. Nature (London) 350:330–332.

National Research Council. 1989. Alternative agriculture. Committee on the role of alternative farming methods in modern production agriculture. Board on Agric., Natl. Res. Council. Natl. Academy Press, Washington, DC.

National Research Council. 1993. Soil and water quality: An agenda for agriculture. Committee on long-range soil and water conservation. Board on Agric., Natl. Res. Council. Natl. Academy Press, Washington, DC.

National Research Council. 1994. Rangeland health: New methods to classify, inventory, and monitor rangelands. Natl. Academy Press, Washington, DC.

Oldeman, L.R. 1994. The global extent of soil degradation. p. 99–118. *In* D.J. Greenland and I. Szabolcs (ed.) Soil resilience and sustainable land use. CAB Int., Wallingford, Oxon, England.

Papendick, R.I., and J.F. Parr. 1992. Soil quality. The key to a sustainable agriculture. Am. J. Altern. Agric. 7:2–3.

Parkin, T.B., J.W. Doran, and E. Franco-Vizcaino. 1996. Field and laboratory tests of soil respiration. p. 231–245. *In* J.W. Doran and A.J. Jones (ed.) Methods for assessing soil quality. SSSA Spec. Publ. 49. SSSA, Madison, WI.

Pierzynski, G.M., J.T. Sims, and G.F. Vance. 1994. Soils and environmental quality. Lewis Publ., CRC Press, Boca Raton, FL.

Reganold, J.P., and A.S. Palmer. 1995. Significance of gravimetric versus volumetric measurements of soil quality under biodynamic, conventional, and continuous grass management. J. Soil Water Conserv. 50:298–305.

Reganold, J.P., A.S. Palmer, J.C. Lockhardt, and A.N. Macgregor. 1993. Soil quality and financial performance of biodynamic and conventional farms in New Zealand. Science (Washington, DC) 260:344–349.

Rodale Institute. 1991. Conference report and abstracts for the Int. Conf. on the Assessment and monitoring of soil quality, Allentown, PA. 11–13 July 1991. Rodale Press, Emmaus, PA.

Rolston, D.E., L.A. Harper, A.R. Mosier, and J.M. Duxbury. 1993. Agricultural ecosystem effects on trace gases and global climate change. ASA Spec. Publ. 55. ASA, Madison, WI.

Sanders, D.W. 1992. International activities in assessing and monitoring soil degradation. Am. J. Altern. Agric. 7:17–24.

Sarrantonio, M., J.W. Doran, M.A. Liebig, and J.J. Halvorson, 1996. On-farm assessment of soil health. p. 83–105. *In* J.W. Doran and A.J. Jones (ed.) Methods for assessing soil quality. SSSA Spec. Publ. 49. SSSA, Madison, WI.

Sharpley, A., T.C. Daniel, J.T. Sims, and T.H. Pote. 1996. Determining environmentally sound soil phosphorus levels. J. Soil Water Conserv. 51:160–166.

Shea, P.J., L.N. Mielke, and W.D. Nettleton, 1992. Estimation of relative pesticide leaching in Nebraska soils. Res. Bull. 313-D. March 1992. Agric. Res. Div., IANR, Univ. of Nebraska, Lincoln.

Singh, K.K., T.S. Colvin, D.C. Erbach, and A.Q Mughal. 1990. Tilth index: an approach towards soil condition quantification. Pap. 90-1040. Am. Soc. of Agric. Eng., St. Joseph, MO.

Smith, J.L., and J.W. Doran. 1996. Measurement and use of pH and electrical conductivity for soil quality analyses. p. 169–185. *In* J.W. Doran and A.J. Jones (ed.) Methods for assessing soil quality. SSSA Spec. Publ. 49. SSSA, Madison, WI.

Visser, S., and D. Parkinson. 1992. Soil biological criteria as indicators of soil quality: Soil microorganisms. Am. J. Altern. Agric. 7:33–37.

3 Farmer-Based Assessment of Soil Quality: A Soil Health Scorecard[1]

Douglas E. Romig, M. Jason Garlynd, and Robin F. Harris

University of Wisconsin
Madison, Wisconsin

Soil quality is undergoing a dramatic process of redefinition. Its meaning has moved beyond soil productivity to encompass environmental quality, food safety, and animal and human health (Parr et al., 1992; Doran & Parkin, 1994; Karlen et al., 1997). Farmers also see the shortcomings of the old definition that limits soil quality to yield potential and nutrient levels. In listening meetings held in Wisconsin, many growers felt that the biological health of the soil was not receiving scientific attention (Kelling, 1989).

Soil health, a more integrative term preferred by some farmers to soil quality, is recognized and assessed by farmers using indicator properties of both soil and nonsoil target systems (Harris & Bezdicek, 1994). While a number of analytical parameters to measure soil quality have been proposed by scientists (Larson & Pierce, 1991; Karlen & Stott, 1994; Doran & Parkin, 1994), farmers' diagnosis of a soil's condition primarily uses qualitative or sensory means in addition to quantitative data (Romig et al., 1995). Granatstein and Bezdicek (1992) suggest that the tools or indices developed to measure soil quality, be they for research, farm, or regulatory purposes, need to integrate both sensory and quantitative aspects.

Descriptive and integrative approaches used by farmers to characterize soil health provide a mechanism for field assessment and monitoring of soil quality by scientists and farmers. Possible descriptive indicators to characterize and monitor soil quality are given by Arshad and Coen (1992) and Reganold et al. (1993), including surface crusting, evidence of erosion, ponding of water, vegetative cover, soil structure, friability, and consistence. Furthermore, the effectiveness of qualitative soil survey techniques has been demonstrated for describing changes in soil surface properties (Grossman & Pringle, 1987), and evaluating land suitability with descriptive indicators (Sharman, 1989).

---

[1] The Wisconsin Soil Health Program has been supported by the University of Wisconsin's Center for Integrated Agricultural Systems, and the Agricultural Technology and Family Farm Institute; the Wisconsin Department of Agriculture, Trade, and Consumer Protection's Sustainable Agriculture Program; the Wisconsin Fertilizer Research Council; the Wisconsin Liming Materials Research Council; and the Kellogg Foundation through the Wisconsin Integrated Cropping System Trial.

Copyright © 1996 Soil Science Society of America, 677 S. Segoe Rd., Madison, WI 53711, USA. *Methods for Assessing Soil Quality,* SSSA Special Publication 49.

Several authors have promoted the development of various analytical tools and indices to assess soil health and quality that would document changes that occur due to management (Haberern, 1992; Granatstein & Bezdicek, 1992; Karlen et al., 1994; Doran & Parkin, 1994). Assessment tools and strategies need to serve not only scientists and resource managers, but also should be measured by how well they meet the needs of farmers and enhance their analyses (Thrupp, 1989; Chambers, 1993; Acton & Padbury, 1993). To meet the challenge and develop a practical index for soil quality and health, the University of Wisconsin's Soil Health program has promoted a partnership between farmers and scientists. This chapter presents the outcome of structured farmer interviews, the *Wisconsin Soil Health Scorecard*, (hereafter referred to as the scorecard), an assessment tool for soil quality based on farmer knowledge of soil health. The scorecard is presented in full immediately following the explanatory text. Details concerning the methodological approaches are presented in Romig (1995) and interpretation of farmer knowledge of soil health is amplified in Romig et al. (1995).

INTERVIEW METHOD

A qualitative approach was taken to solicit and analyze farmers' knowledge of soil health through structured interviews. These interviews were given to 28 farmers in the summer of 1993, and explored the question: "How do you recognize a healthy soil?" Participants were associated with the Wisconsin Integrated Cropping Systems Trial (Cunningham et al., 1992) and operated conventional and low-input cash grain and dairy farms typical of southeastern Wisconsin. Agricultural soils of the region are formed in silt overlying glacial till or outwash.

Use of an interview guide ensured that each soil health interview covered the same material (Garlynd et al., 1994). Interviews were conducted on the farm and recorded on tape for further analysis. The soil health interviews were conducted in two stages. First, open-ended questions were used to explore how farmers recognize a healthy soil. This allowed for free association and brought to light characteristics of a healthy soil that were in the forefront of the farmer's mind. When it seemed that the participant had addressed all the indicator properties, questions became closed-ended. In the second phase of the interview, farmers were asked to consider if a soil's health was related to specific properties that had been collated in earlier research (Garlynd et al., 1994), but the farmer had not addressed in the first phase. Throughout the entire interview, probing questions were asked to gather descriptions of each soil health indicator, be they words, phrases or numerical values.

The interview responses were coded and these data were analyzed to determine the most important properties of soil health as perceived by farmers. Properties were considered greater in importance if they were: (i) used by the majority of the farmers, (ii) mentioned earlier and more frequently, and (iii) mentioned in the open-ended question period rather than being prompted by closed-ended questions. A procedure was developed based on these assumptions to rank the properties relative to one another and determine which indicator would be included in the scorecard (Romig, 1995). Additionally, all descriptions of healthy and unhealthy soil for each indicator property were cataloged.

A SOIL HEALTH SCORECARD

Conceptual Basis

The interview analysis created a database of farmer knowledge about soil health and quality, identifying, and characterizing important soil health properties. Transformation of these data into scaled items and their integration into an assessment tool for soil health introduces a measure of subjectivity. Indeed, any integration of data, be they analytical or descriptive, will implicitly confer a qualitative nature to an index (Granatstein & Bezdicek, 1992). To minimize the level of subjectivity, literature concerning human health and behavioral sciences was reviewed to define health and deal with issues of scaling.

Overviews of health definitions are provided by Larson (1991) and Bowling (1991). By far the most common and simplest definition of health is the medical model—the absence of disease and disability. There are shortcomings of the medical model, namely one can feel ill without having a disease and, similarly, one can be in good health even though a disease has been diagnosed. More holistic models extend to mental, social, and environmental aspects of health and often include positive dimensions: functional fitness, wellness, and quality-of-life (Larson, 1991; Bowling, 1991). Culyer (1983) cuts across many of the abstract differences in various health models to a pragmatic definition in terms of a set of relevant characteristics. Health indicators in this characteristics approach include measures of functional capacity (ability to perform normal tasks and activities), morbidity (illness), and vital statistics (attributes of a particular health state).

Through conversations with farmers, a set of soil health indicator properties was identified for soil health characterization. These included both the capacity of a soil to perform certain functions (infiltrate, decompose, cycle nutrients, and others) and attributes or vital statistics (soil color and structure, root morphology, earthworms, animal health, and others). Employing the characteristics definition of health (Culyer, 1983) allows soil health to be described and evaluated with respect to both functional ability and structures or attributes. In the scorecard, each soil health indicator is operationalized to conform to the following subjective rating scale:

1. Healthy: Performance of function is optimal and structure is normal.
2. Impaired: An abnormality in function and/or structure.
3. Unhealthy: Severe restriction or inability to perform function considered normal, severe deformity or loss of structure, disabled.

The scorecard's soil health scale rests on the assumption that the majority of indicator properties are subjectively measured by the senses (look, smell, or feel). Indicators of soil health vary in quality or character rather than in an specific measurable quantity. The scale used in the scorecard does not measure the exact magnitude of difference among healthy, impaired, and unhealthy categories, meaning the scale is at the ordinal-level of measurement.

Scaling methods for objects that have degrees of magnitude that are subjective or qualitative are addressed by psychologists Stevens (1975) and Torgerson (1958) who studied the characteristics of humans as measuring instruments. Their research demonstrated that humans can make accurate and internally consistent judgments of phenomena and partition stimuli into ordered, homogenous classes. In short, an individual has a relatively constant sensitivity to differences even when there is no objective, physical scale by which they would measure. Rating scales initially produced by people are at an ordinal level of measurement. That is, objects are ranked along a scale as to the degree they possess a given characteristic without a distinct and uniform distance between each class or category (Blalock, 1972). An ordinal scale can be mathematically transformed into interval scale values through various psychological scaling procedures, given enough data from trials that test the rating scale. Brown and Daniel (1990) review several of these scale transformation methods and demonstrate their application to the development of environmental rating scales. Psychological scaling techniques also have been widely used to develop rating scales for human health indices (Rosser, 1983; McDowell & Newell, 1987). Given the quantity of data necessary to construct interval scales from ordinal categories, and the limited empirical data that have been collected with the scorecard to date, the scale at which soil health indicator properties are measure by the scorecard remains at the ordinal level.

There are, however, simple techniques, such as partition or category scaling suitable to assign numerical values to equal appearing intervals linearly related to stimuli measured on a subjective scale of magnitude (Stevens, 1960; Torgerson, 1958). Farmers described soil health properties on an essentially dichotomous scale, providing qualitative descriptions of healthy or unhealthy extremes. Table 3–1 is a catalog of representative terms farmers used to describe the top 20 soil health indicators at healthy and unhealthy states, making these end-points relatively well defined. By employing a partitioning scaling method, descriptions of the impaired state of health for each indicator property were interpolated to give equal-appearing intervals between healthy and unhealthy. The scorecard previews the descriptive range of the indicator property and anchors the rating criteria to a numerical scale for unhealthy (0 to 1), impaired (1.5 to 2.5), and healthy (3 to 4) options. The numerical rating scale allows the scorecard user to grade each indicator with some degree of sensitivity. Yet it is important to remain cognizant that this rating scale is ordinal, thereby limiting the mathematical operations the data generated by the scorecard can support (Blalock, 1972).

Every attempt was made to keep indicator property descriptions used in the scorecard aligned with the language and intent of the farmers who were interviewed; however, some indicators (hardness and color) employ soil science techniques used to describe morphological properties. In other instances, statements regarding analytical features (yield, organic matter, and pH) were reviewed by extension soil scientists. Users also will note that a composite question relating to soil tests and primary nutrients (N, P, and K) is descriptive. Farmers did not state exact nutrient levels for healthy and unhealthy soils, rather soil tests served more to identify the amount of corrective action required to build or maintain soil fer-

Table 3-1. Representative descriptive terms for top 20 soil health properties (from Romig et al., 1995).

Rank	Soil health property	Healthy	Unhealthy
1	Organic matter	As high as possible, at soil's potential, manure, compost, 3%, 2%, 7–8%, putting more back.	Rough, lack of organic matter, less, low.
2	Crop appearance	Green, healthy, uniform, lush, dense stand, tall, larger, sturdy, stout, proper color, darker, good crop.	Yellow, stunted corn, small, poor color, poorer, lack of green, light green, streaks in field.
3	Erosion	Wouldn't erode, water and wind not taking soil, prevented, stays in place, less, slowed down, delayed.	Blows sooner, washes, topsoil's lost, erodes more, clouds of dust, ravines, runs bad, any, easier.
4	Earthworms	Fishing and red worms present, see after rain, a lot, angle worms, see holes and castings, see during plowing.	Not there, don't work, can't find, no holes, lack of, killed by insecticides or anhydrous, void.
5	Drainage	Water goes away, fast, better, no ponding, moves through, takes a lot of rain, drains properly, dries out.	Tight, waterlogged, drains too fast, ponding, no outlet for water, won't drain, slope, poor, saturated.
6	Tillage ease	One pass and ready, breaks up, mellow, easier, smooth, crumbles, flows, plow a gear faster, minimum.	Never works down, needs more disking, lumps, slabs, shiney, pulls hard, worked wet, overworked.
7	Soil structure	Won't roll out of hand, crumbly, loose, holds together, granular.	Hard, doesn't hold together, lumpy, falls apart, massive, cloddy, lumpy, clumpy, tight, compacted, powder.
8	pH	7.0, 6.7–6.8, 6.2–6.7, balanced, neutralize.	<6.0, high, nothing works, wrong, too low, high acidity.
9	Soil test	Up to recommendations, high, elevated, complete, where it belongs, every year or two, stay up with soil test.	Law or minimum at work.
10	Yield	150–180 bu corn, 60 bu beans, 30–40% higher, +10 bu acre^{-1}, better 5 yr average, significantly higher.	110 bu corn, 150 bu corn, 35 bu beans, 20–50% less, don't get much off, down, reduced, low.
11	Compaction	Doesn't pack down, not compacted, stays loose, not out there when wet.	Compacted, plow layer, packs down, hardpan, plowsoil, tight, can't get into it, packed.
12	Infiltration	Water doesn't stand, absorbs, water moves into soil, soaks, rapid, no ponding, fast, spongy.	Water runs off soil, sits on top, water stands, doesn't absorb, puddles, nonporous.
13	Soil color	Dark, black, dark brown, gray, holds dark color.	Orange, brown, light, white, red, blue-gray, subsoil color, bleached, sandy colored, light brown, pale, anemic, gray.
14	N	Put on less, manure, as required, compost, slurry, more available, organic N, organic matter.	Too much N, chemical N, commercial fertilizers burn ground, anhydrous, sludge.
15	Water retention	Holds moisture, get by with less, retains more, moisture travels, gives and takes water freely, conserving.	Too much water, doesn't hold water, dries out, too wet or dry, droughty, stays wet, runs out of moisture, poor.
16	P	As required.	
17	Nutrient deficiency	Has what it needs, no shortage of elements, no spots on leaves.	Yellow, purple, discoloration in leaves, lodging, crop falls off, stripping, brown streaks, firings on bottom, blight.
18	Decomposition	Breaks down, decays, rots in 4–5 mon, manure part of soil in 1 yr, disappears, not too fast, 2/3 gone in year.	See stalks from last year, doesn't break down, manure plows up next year.
19	K	As required.	
20	Roots	Larger, spread out, grow down, white, deep, numerous, good penetration, full, lots of feeders, branched out.	Don't penetrate, undeveloped, balled up, grow crossways discolored, diseased, at hard angles, shallow, short.

tility (Romig et al., 1995). For the scorecard, the soil test question reflects this farmer point of view.

Content and Structure

The scorecard is a field tool to assess and monitor soil quality and health. The scorecard is farmer-based, reflecting the priorities, language, and intent of the growers we interviewed. It is integrative in its evaluation, scoring soil attributes, and properties of the systems supported by the soil (plant, animal, and water systems) that contribute to a farmer's diagnosis of soil health. The scorecard uses primarily sensory-perceived or descriptive indicator properties rather than results from laboratory analyses.

Structurally, the scorecard is in the form of a booklet, modeled after descriptive questionnaires, schedules, and profiles used in human health and quality-of-life studies (Bowling, 1991; McDowell & Newell, 1987). Following a cover page, explanatory text reviews the procedures of the scorecard's use. Forty-three soil health indicator properties are described at the three levels of health and scored in the remaining pages. Associated with each indicator property is a superscript number that denotes its relative importance and rank with respect to the other properties in the scorecard. Users are encouraged to score each indicator anywhere along the 0 to 4 point scale in intervals preferably no smaller than 0.5 to reflect more accurately the observed condition of the soil being graded.

The scorecard integrates observations made throughout the growing season and is best completed near or just following harvest. Attributes that are expressed either seasonally or periodically (e.g., soil smell, seed germination, or infiltration) should be recorded when observed to increase the precision of the instrument.

On the final page is a guide for users to interpret the results of the scorecard. First distributions of soil health indicator properties within the three health categories – healthy (3 to 4), impaired (1.5 to 2.5), and unhealthy (0 to 1) – are tallied and percentages are calculated. Ideally, most, if not all, indicators should lie within the healthy category. Even if a high percentage of indicators were graded as healthy, it is important to note those properties that scored low and may be serious problems requiring attention. Those indicators that were scored as unhealthy probably need immediate corrective action. For indicators either in the unhealthy or impaired category, it is necessary to consider what caused the condition in the first place and how it might be attenuated. For indicators that were assessed as impaired, the users should judge from a historical perspective, whether that particular property's condition is deteriorating or improving. Users also may wish to focus more on those properties farmers considered most important as indicated by its relative rank in superscript.

Final scores could be calculated to assess the entire farming system or separate scores for each target system could be computed for comparative analysis between the scores for the soil system and the plant, animal, and water systems. This approach would provide a gross estimate of management effects on soil health over time. Such simplified procedures, however, may obscure certain properties that were graded low and may need attention. Furthermore, the computation of an overall final score is built on the assumption that all indicators are equally important to a soil health assessment. The scorecard in its present form does not recognize the relative importance properties have with respect to one another and weight the score of each indicator accordingly. The benefit of differential weighting schemes in indices similar to the scorecard has been a topic of discussion in the health and behavioral sciences for some time and with little resolve (Wang & Stanley, 1970; Wainer, 1976; Skinner & Lei, 1980; Streiner & Norman, 1989). The use of differentially weighting indicators in the scorecard will only become feasible as measurement models of complex soil processes related to soil quality become more fully developed.

Pretesting

Initial field tests in cropping system comparison trials have shown that the information generated by the scorecard holds promise. The scorecard identified several descriptive indicators in both soil and plant systems that demonstrated variable responses under different management regimes, even on soils considered very resilient (Garlynd et al., 1995). These indicators included earthworms, soil structure, decomposition, surface crusting, soil smell, aeration, crop appearance, yield (especially when the growing season was stressful), nutrient deficiencies, mature crop, plant stems, and plant resistance to disease and pests. Distributions of indicator properties within the health categories have exhibited a greater percentage of healthy indicators in low-input cropping systems as compared with continuous corn (*Zea mays* L.). For the top 12 indicator properties (Romig et al., 1995), soil health scores for continuous corn and lower-input cropping strategies were uniformly high, except for earthworms (*Lumbricus terrestris*) that scored lower in continuous corn treatments (Garlynd et al., 1995).

Pretests of the scorecard with farmers showed that farmers may be inherently biased when grading their own soils. Besides the affection they have for their farm, farmer bias may be due to the integrative nature of their knowledge. Farmer knowledge of soils is often intimately connected to management practices, both past and present. When judging any property they also may be judging the practices carried out on the field (Romig et al., 1995). Users need to be aware that without maintaining some level of objectivity, the effectiveness of the scorecard will be compromised. Furthermore, farmers are most familiar with the soils in their locality and may have limited experience with a number of different soil types. Full potential of the scorecard will be realized as users become familiar with the natural variability of soil descriptive indicators for the region where the scorecard was developed.

Initial reaction of farmers to the scorecard has been generally positive. One of the shortcomings of the scorecard, however, is that it is driven solely by the results of these 28 interviews, and may contain biases with respect to soils, farm type, and input strategy. Organic fruit and vegetable producers, for example, have commented on the scorecard's conventional audience and its focus on cow crops. Therefore, caution is advised when applying the scorecard to farming systems in locations unlike those from which it was developed. Modification of the scorecard for other cropping systems and other regions would require structured input from additional farmers.

Applications

The scorecard provides a farmer-based assessment of soil health that has inherent value as a field tool for farmers, extension agents, crop consultants, and scientists. The scorecard is flexible not only for these different user groups, but for different farming systems as well, because only the indicators relevant to the farming system need to be evaluated.

The scorecard helps farmers collect and assess information that is both pertinent to their operation and at their fingertips. Moreover, the scorecard provides the farmer with an integrative assessment of the overall effect of management on soil health, helps monitor soil health over time, and identifies the individual properties that are being impacted either negatively or positively by soil management. This, in turn, can help farmers conduct their own analyses of a soil's condition and make adjustments in management if necessary.

In a research setting comparing management strategies, the scorecard provides an initial appraisal of the effects of different cropping systems on individual properties as well as a more integrated understanding of a soil's health under different systems. The farmer knowledge embodied in the scorecard provides additional insight that may help ground scientific interpretations, predictions, and models of the effects of management on specific properties and functions of soil quality. The scorecard also can be used to identify descriptive indicators, and their corresponding analytical properties, that may be sensitive to changes in management. For example, preliminary use of the scorecard in trials monitoring the transition from continuous corn to more complex rotations indicates that the number of earthworms, their castings, and holes increase in more varied rotations (Garlynd et al., 1995). Another avenue researchers may wish to examine is the comparison of the scores of soil health indicators relevant to a given soil quality function (i.e., water relationships, crop performance, and nutrient cycling), perhaps in a framework similar to that presented in Karlen et al. (1994). By examining indicator properties that contribute to a given function, it may be possible to better articulate the relationship between certain management practices and soil functions. The scorecard has additional merit as a reference base for soil quality assessments and may assist with the interpretation of quantitative data by helping calibrate data from laboratory analyses to field conditions.

The scorecard has value in that it provides linkages between scientist and farmer assessment of soil health and quality. First, the integrative nature of the scorecard, assessing soil, plants, animals or humans, and water indicator properties, has conceptual value for scientists, particularly in the development and validation of indices addressing the functions of soil quality (Harris & Bezdicek, 1994; Romig et al., 1995). Moreover, the scorecard may help acclimate the scientific community to soil health and quality indices that qualitatively integrate analytical and/or descriptive data. Second, despite recognition of the need to move beyond disciplinary boundaries and pursue holistic, accessible indicators (Doran & Parkin, 1994), nonsoil and descriptive indicator properties for soil quality characterization have received little attention. In this regard, the scorecard provides a base to identify and develop both descriptive and analytical soil and nonsoil indicators for soil health and quality assessment. Furthermore, farmer use of descriptive soil health indicator properties supports the inclusion of corresponding analytical properties in quantitative soil quality indices. Third, the scorecard provides a source of descriptive data for integrated soil health and quality scorecards (Harris et al., 1996).

It is important to recognize that the Scorecard has limitations that were inherited from the interview data from which it was developed. One shortcoming is that the scorecard represents the unchallenged perceptions of a relatively small group of farmers. While confident that the scorecard addresses many of the crucial indicator properties of soil health and quality, continued dialogue between farmers and scientists is necessary to advance our collective understanding and will undoubtedly result in modifications of the structure and composition of the scorecard, particularly in the scope and nature of the plant, animal, and water indicators of soil quality.

A second limitation to the scorecard may be its narrow frame of reference, especially in light of the expanded definition of soil quality. Given that it reflects the nature of soil health that farmers perceive as important, the scorecard is inherently focused upon crop production and may fail to recognize critical aspects of soil quality with respect to environmental protection. This is evident in that only a small minority of soil health indicators identified by the farmers are directly associated with environmental quality, specifically erosion, surface water appearance and chemicals in groundwater. Similarly, a soil test with high nutrient levels for crop production would be graded high by the scorecard, while that same soil test might receive a much lower rating with regards to water quality.

Despite these concerns, the strength and utility of the *Wisconsin Soil Health Scorecard* should be emphasized. The scorecard provides a mechanism to assess and monitor individual descriptive soil quality indicators in the field, acts as a communication bridge between farmers and scientists, and gives valuable insights into meaningful analytical soil quality indicators.

REFERENCES

Acton, D.F., and G.A. Padbury. 1993. Introduction. p. 2.1–2.10. *In* D.F. Acton (ed.) A program to assess and monitor soil quality in Canada. Res. Branch, Agric. Canada. CLBRR No. 93-49. Agric. Canada, Ottawa, Ontario, Canada.

Arshad, M.A., and G.M. Coen. 1992. Characterization of soil quality: Physical and chemical criteria. Am. J. Altern. Agric. 7:25–30.

Blalock, H.M. 1972. Social statistics. McGraw-Hill, NewYork.

Bowling, A. 1991. Measuring health: A review of quality of life measurement scales. Open Univ. Press, Philadelphia, PA.

Brown, T.C., and T.C. Daniel. 1990. Scaling of ratings: Concepts and methods. U.S. Dep. of Agric. Forest Serv. Res. Pap. RM-293. Rocky Mountain For. and Range Exp. Stn., Fort Collins, CO.

Chambers, R. 1993. Methods of analysis by farmers. J. Farming Syst. Res. 4:87–101.

Culyer, A.J., 1983. Health indicators. St. Martin's Press, New York.

Cunningham, L., J. Doll, J. Hall, D. Mueller, J. Posner, R. Saxby, and A. Wood. 1992. The Wisconsin integrated cropping system trial first report. Univ. of Wisconsin, Arlington.

Doran, J.W., and T.B. Parkin. 1994. Defining and assessing soil quality. p. 3–21. *In* J.W. Doran et al. (ed.). Defining soil quality for a sustainable environment. SSSA Spec. Publ. 35. SSSA, Madison, WI.

Garlynd, M.J., A.V. Karakov, D.E. Romig, and R.F. Harris. 1994. Descriptive and analytical characterization of soil quality/health. p. 159–168. *In* J.W. Doran et al. (ed.) Defining soil quality for a sustainable environment. SSSA Spec. Publ. 35. SSSA, Madison, WI.

Garlynd, M.J., D.E. Romig, and R.F. Harris. 1995. Effect of a cropping systems shift from continuous corn on descriptive and analytical indicators of soil quality. p. 58 *In* Agronomy abstracts. ASA, Madison, WI

Granatstein, D., and D.F. Bezdicek. 1992. The need for a soil quality index: Local and regional perspectives. Am. J. Altern. Agric. 7:12–16.

Grossman, R.B., and F.B. Pringle. 1987. Describing surface soil properties – their seasonal changes and interpretations for management. p. 57–75. *In* Soil survey techniques. SSSA Spec. Publ. 35. SSSA, Madison, WI.

Haberern, J. 1992. A soil health index. J. Soil Water Conserv. 47:6.

Harris, R.F., and D.F. Bezdicek. 1994. Descriptive aspects of soil quality/health. p. 23–35. *In* J.W. Doran et al. (ed.) Defining soil quality for a sustainable environment. SSSA Spec. Publ. 35. SSSA, Madison, WI.

Harris, R.F., D.L. Karlen, and D.J. Mulla. 1996. A conceptual framework for assessment and management of soil quality and health. p. 61–82. *In* J.W. Doran and A.J Jones (ed.) Methods for assessment of soil quality and health. SSSA Spec. Publ. 49. SSSA, Madison, WI.

Karlen, D., and D. Stott. 1994. A framework for evaluating physical and chemical indicators of soil quality. p. 53–72. *In* J.W. Doran et al. (ed.) Defining soil quality for a sustainable environment. SSSA Spec. Publ. 35. SSSA, Madison, WI.

Karlen, D.L., M.J. Mausbach, J.W. Doran, R.G. Cline, R.F. Harris, and G.E. Schuman. 1997. Soil quality: Concept, rationale, and research needs. Soil Sci. Soc. Am. J. 60:4–10.

Karlen, D.L., N.C. Wollenhaupt, D.C. Erbach, E.C. Berry, J.B. Swan, N.S. Eash, and J.L. Jordahl. 1994. Crop residue effects on soil quality following 10 years of no-till corn. Soil Tillage Res. 31:149–167.

Kelling, K.A. (ed.). 1989. Proc. of the University of Wisconsin sustainable agriculture "Listening" meetings. Wisconsin Agric. Exp. Stn. and Wisconsin Coop. Ext. Serv., Madison.

Larson, J.S., 1991. The measurement of health: Concepts and indicators. Greenwood Press, New York.

Larson, W.E., and F.J. Pierce. 1991. Conservation and enhancement of soil quality. In Evaluation for sustainable land management in the developing world. p. 175–203. *In* Proc. of the Int. Workshop on Evaluation for Sustainable Land Management in the Developing World, Chiang Rai. 15–21 Sept. 1991. Int. Board of Soil Res. and Manage., Bangkok, Thailand.

McDowell, I., and C. Newell. 1987. Measuring health: A guide to rating scales and questionnaires. Oxford Univ. Press, New York.

Parr, J.F., R.I. Papendick, S.B. Hornick, and R.E. Meyer. 1992. Soil quality: Attributes and relationship to alternative and sustainable agriculture. Am. J. Altern. Agric. 7:5–11.

Reganold, J.P., A.S. Palmer, J.C. Lockhart, and A.N. Macgregor. 1993. Soil quality and financial performance of biodynamic and conventional farms in New Zealand. Science (Washington, DC) 260:344–349.

Romig, D.E. 1995. Farmer knowledge of soil health and its role in soil quality assessment. M.S. thesis. Univ. of Wisconsin, Madison.

Romig, D.E., M.J. Garlynd, R.F. Harris, and K. McSweeney. 1995. How farmers assess soil health and quality. J. Soil Water Conserv. 50:225–232.

Rosser, R. 1983. Issues of measurement in the design of health indicators: A review. p. 34–43. *In* A.J. Culyer (ed.) Health indicators. St. Martin's Press, New York.

Sharman, A.K. 1989. Qualitative and quantitative land evaluation for rainfed maize in subhumid tropical and subtropical climates. Adv. Soil Sci. 9:147–156.

Skinner, H.A., and H. Lei. 1980. Differential weights in life change research: Useful or irrelevant. Psychosomatic Medicine 42:367–370.

Stevens, S.S. 1960. Ratio scales, partition scales, and confusion scales. p. 49–66. *In* H. Gulliksen and S. Merrick (ed.) Psychological scaling: Theory and applications. John Wiley & Sons, New York.

Stevens, S.S. 1975. Psychophysics. John Wiley & Sons, New York.

Streiner, D.L., and G.R. Norman, 1989. Health measurement scales: A practical guide to their development and use. Oxford Univ. Press, New York.

Thrupp, L.A. 1989. Legitimizing local knowledge: From displacement to empowerment for third world people. Agric. Human Values 6(3):13–24.

Torgerson, W.S. 1958. Theory and methods of scaling. John Wiley & Sons, New York.

Wainer, H. 1976. Estimating coefficients in linear models: It don't make no nevermind. Psychol. Bull. 83:213–217.

Wang, M.W., and J.C. Stanley. 1970. Differential weighting: A review of methods and empirical studies. Rev. Educ. Res. 40:663–705.

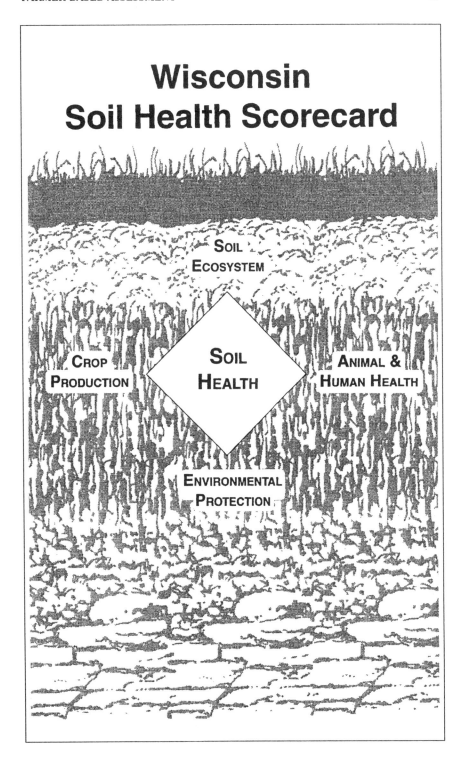

FARMER-BASED ASSESSMENT

SCORECARD INSTRUCTIONS

The Wisconsin Soil Health Scorecard assesses a soil's health as a function of soil, plant, animal and water properties identified by farmers. The scorecard is a field tool to monitor and improve soil health based on field experience and a working knowledge of a soil.

The scorecard is best completed near or just following harvest. Periodic and seasonally expressed properties (soil smell, seed germination, infiltration, etc.) should be recorded during the growing season to increase its effectiveness. When scoring your soil's health, please:

1. Read each question completely. Focus only on the property being graded.

2. Choose the answer that best describes the property and enter score between 0 and 4 in the box provided. The scale corresponds to healthy (3-4 pts.), impaired (1.5-2.5), and unhealthy (0-1).

3. Answer as many questions as possible to ensure an accurate evaluation of your soil's health.

4. Enter NA (not answered) if a question does not apply to your farm, and go to the next question.

The scorecard was developed by the University of Wisconsin's Soil Health Program[1] from structured interviews with 28 farmers in conjunction with the Wisconsin Integrated Cropping System Trial[2]. Superscript numbers indicate the relative importance and rank of the property. Farmers who were interviewed operated conventional and low-input cash grain and dairy farms typical of southeast Wisconsin. Typical soils are formed in silt over glacial till or outwash. Applying this scorecard to other locations should be done with caution. Modification of the scorecard for other cropping systems and other regions requires structured input from additional farmers.

---

[1] Supported by the UW Center for Integrated Agricultural Systems, and Agricultural Technology and Family Farm Institute; the WI Department of Agriculture, Trade, and Consumer Protection's Sustainable Agriculture Program; the WI Fertilizer Research Council; the WI Liming Materials Research Council; and the Kellogg Foundation through the Wisconsin Integrated Cropping System Trial.

[2] D. E. Romig, M. J. Garlynd, and R. F. Harris. 1994. Farmer-based soil health scorecard. p. 288. Agronomy abstracts. ASA, Madison, WI.

Please go to next page ▶▶▶

SOIL - Questions refer primarily to the plow layer

Descriptive Properties   SCORE

1. EARTHWORMS[3]
   0 Little sign of worm activity
   2 Few worm holes or castings
   4 Worm holes and castings numerous

2. EROSION[4]
   0 Severe erosion, considerable topsoil moved, gullies formed
   2 Moderate erosion, signs of sheet and rill erosion, some topsoil blows
   4 Little erosion evident, topsoil resists erosion by water or wind

3. TILLAGE EASE[5]
   0 Plow scours hard, soil never works down
   2 Soil grabs plow, difficult to work, needs extra passes
   4 Plow field in higher gear, soil flows & falls apart, mellow

4. SOIL STRUCTURE[7]
   0 Soil is cloddy with big chunks, or dusty and powdery
   2 Soil is lumpy or does not hold together
   4 Soil is crumbly, granular

5. COLOR (MOIST)[13]
   0 Soil color is tan, light yellow, orange, or light gray
   2 Soil color is brown, gray, or reddish
   4 Soil color is black, dark brown, or dark gray

6. COMPACTION[11]
   0 Soil is tight & compacted, cannot get into it, thick hardpan
   2 Soil packs down, thin hardpan or plow layer
   4 Soil stays loose, does not pack, no hardpan

7. INFILTRATION[12]
   0 Water does not soak in, sits on top or runs off
   2 Water soaks in slowly, some runoff or puddling after a heavy rain
   4 Water soaks right in, soil is spongy, no ponding

Please go to next page ▶▶▶

SOIL - Questions refer primarily to the plow layer

Descriptive Properties   SCORE

8. DRAINAGE[6]
- 0 Poor drainage, soil is often waterlogged or oversaturated
- 2 Soil drains slowly, slow to dry out
- 4 Soil drains at good rate for crops, water moves through

9. WATER RETENTION[14]
- 0 Soil drys out too fast, droughty
- 2 Soil is drought prone in dry weather
- 4 Soil holds moisture well, gives and takes water easily

10. DECOMPOSITION[16]
- 0 Residues and manures do not break down in soil
- 2 Slow rotting of residues and manures
- 4 Rapid rotting of residue and manures

11. SOIL FERTILITY[20]
- 0 Poor fertility, nutrients do not move, potential is very low
- 2 Fertility not balanced, needs help
- 4 Fertility is balanced, nutrients available, potential is high

12. FEEL[21]
- 0 Soil is mucky, greasy, or sticky
- 2 Soil is smooth or grainy, compresses when squeezed
- 4 Soil is loose, fluffy, opens up after being squeezed

13. SURFACE CRUST[24]
- 0 Soil surface is hard, cracked when dry, compacted
- 2 Surface is smooth with few holes, thin crust
- 4 Surface does not crust, porous, digs easily with hand

14. SURFACE COVER[23]
- 0 Soil surface is clean, bare, residue removed or buried following harvest
- 2 Surface has little residue, mostly buried
- 4 Surface is trashy, lots of mulch left on top or cover crop used

Please go to next page ▶▶▶

SOIL - Questions refer primarily to the plow layer

Descriptive Properties   **SCORE**

15. HARDNESS[28]
- 0 Soil is hard, dense or solid, will not break between two fingers
- 2 Soil is firm, breaks up between fingers under moderate pressure
- 4 Soil is soft, crumbles easily under light pressure

16. SMELL[25]
- 0 Soil has a sour, putrid or chemical smell
- 2 Soil has no odor or a mineral smell
- 4 Soil has an earthy, sweet, fresh smell

17. SOIL TEXTURE[31]
- 0 Texture is a problem, extremely sandy, clayey or rocky
- 2 Texture is too heavy or too light, but presents no problem
- 4 Texture is loamy

18. AERATION[35]
- 0 Soil is tight, closed, almost no pores
- 2 Soil is dense, has a few pores
- 4 Soil is open, porous, breaths

19. BIOLOGICAL ACTIVITY[36]
- 0 Soil shows little biological activity, no signs of soil microbes
- 2 Moderate biological activity, some wormlike threads, moss, algae
- 4 Biological activity high, white wormlike threads, moss, algae plentiful

20. TOPSOIL DEPTH[38]
- 0 Subsoil is exposed or near surface
- 2 Topsoil is shallow
- 4 Topsoil is deep

Please go to next page ▶▶▶

FARMER-BASED ASSESSMENT

SOIL - Questions refer primarily to the plow layer

Analytical properties   SCORE

Values are for typical soils of southeast Wisconsin

21. ORGANIC MATTER[1]
- 0 Organic matter less than 2% or greater then 8%
- 2 Organic matter 2 to 4% or 6 to 8%
- 4 Organic matter between 4 and 6%

22. pH[8]
- 0 Soil pH less than 6.4 or greater than 7.2
- 2 Soil pH 6.4 to 6.7 or 7.0 to 7.2
- 4 Soil pH between 6.7 and 7.0

23. SOIL TEST - N, P & K[9]
- 0 Two or more nutrient levels very low, law of minimum at work
- 2 Soil test values are below recommended levels, need extra inputs
- 4 All nutrient levels at recommended levels

24. MICRONUTRIENTS[30]
- 0 Severe shortages of trace minerals (magnesium, zinc, sulfur, boron, etc.)
- 2 Micronutrients at a minimal level or not balanced
- 4 Levels of micronutrients high and balanced

Please go to next page ▶▶▶

PLANTS - Questions concern typical years with adequate rainfall and temperatures

Descriptive Properties   SCORE

25. CROP APPEARANCE[2]
  0 Overall crop is poor, stunted, discolored, in an uneven stand
  2 Overall crop is light green, small, in a thin stand
  4 Overall crop is dark green, large, tall, in a dense stand

26. NUTRIENT DEFICIENCY[15]
  0 Crop shows signs of severe deficiencies (blighted, streaky, spotty, discolored, leaves dry up)
  2 Crop falls off or discolors as season progresses
  4 Crop has what it needs, shows little signs of deficiencies

27. SEED GERMINATION[34]
  0 Seed germination is poor, hard for crop to come out of ground
  2 Germination is uneven, seed must be planted deeper
  4 Seed comes up right away, good emergence

28. GROWTH RATE[19]
  0 Crop slow to get started, never seems to mature
  2 Uneven growth, late to mature
  4 Rapid, even growth, matures on time

29. ROOTS[17]
  0 Plant roots appear unhealthy (brown, diseased, spotted), poorly developed, balled up
  2 Plant roots are shallow, at hard angles, development limited, few fine roots
  4 Plant roots are deep, fully developed with lots of fine root hairs

30. STEMS[40]
  0 Stems are short, spindly, lodging often a problem
  2 Stems are thin, leaning to one side
  4 Stems are thick, tall, standing, straight

Please go to next page ▶▶▶

FARMER-BASED ASSESSMENT   57

PLANTS - Questions concern typical years with adequate rainfall and temperatures

Descriptive properties   SCORE

31. LEAVES[33]
   0 Leaves are yellow, discolored, few in number
   2 Leaves are small, narrow, light green
   4 Leaves are full, lush, dark green

32. RESISTS DROUGHT[27]
   0 Plants dry out quickly, never completely recover
   2 Plants suffer in dry weather, slow to recover
   4 Plants withstand dry weather, fast to recover

33. RESISTS PESTS AND DISEASE[29]
   0 Plants damaged severely by diseases & insects
   2 Plants stressed by diseases & insects
   4 Plants tolerate pests & disease well

34. MATURE CROP[18]
   0 Seedhead or pod misshapened, grain is not ripe, shrivelled, poor color
   2 Seedhead small, unfilled, grain slow to ripen
   4 Seedhead large, grain full, ripe, with good color

Analytical Properties
Values are typical for soils of southeast Wisconsin

35. YIELD[10]
   0 Corn: less than 85 bushel/acre, Alfalfa: less than 2 ton/acre
   2 Corn: 85 to 130 bushel/acre, Alfalfa: 2 to 6 ton/acre
   4 Corn: greater than 130 bushel/acre, Alfalfa: greater than 6 ton/acre

36. FEED VALUE[41]
   0 Feed has poor nutritional value (energy, protein, minerals), supplements must be used
   2 Feed is unbalanced in energy, protein, or minerals, may require supplements
   4 Feed is balanced, high in nutritional value, supplements used infrequently

Please go to next page ▶▶▶

PLANTS - Questions concern typical years with adequate rainfall and temperatures

Analytical Properties SCORE

37. TEST WEIGHT[32]
- 0 Grain test weight is low, takes a deduction
- 2 Grain test weight is average
- 4 Grain test weight is high

38. COST OF PRODUCTION AND PROFIT[26]
- 0 Production and input costs high yet profits are low
- 2 Profits are variable, yields maintained with high input costs
- 4 Profits are dependable, high, yields maintained with low input costs

ANIMALS - Questions should not relate to improper housing, poor water or inclement weather.

Descriptive Properties SCORE

39. HUMAN HEALTH[37]
- 0 Human health is poor, recurrent health problems, recovery is difficult and long
- 2 Occasional health problems, slow recovery time
- 4 Human health is excellent, resists diseases, long life, quick recovery time

40. ANIMAL HEALTH[42]
- 0 Continuous animal health problems, poor performance and production
- 2 Occasional animal health problems, performance average
- 4 Animal health excellent, exceptional performance and production

41. WILDLIFE[43]
- 0 Signs of wildlife rare, animals do not appear healthy
- 2 Infrequent signs of wildlife; songbirds, deer, turkey, etc. uncommon
- 4 Wildlife is abundant, gulls behind plow, songbirds, deer, turkey, etc. are common

Please go to next page ▶▶▶

WATER

Analytical Properties   SCORE

42. CHEMICALS IN GROUNDWATER[22]
   0 Chemicals found in groundwater above allowable levels
   2 Chemicals found in groundwater below allowable levels
   4 No chemicals present in groundwater

Descriptive Properties
43. SURFACE WATER[39] (open water flowing from fields - lakes, marshes, streams, etc.)
   0 Surface water is very muddy or slimy
   2 Surface water is brownish with dirt and silt
   4 Surface water is clear and clean

Please go to next page ►►►

INTERPRETING THE SOIL HEALTH SCORECARD'S RESULTS

Review the scorecard and tally the number of indicator properties that reside within the three categories of health listed below. Divide the number in each health category by the total number of questions answered (a maximum of 43) and multiply by 100% for the percentage within each category.

HEALTH CATEGORY	NUMBER	%
HEALTHY (SCORE OF 3 - 4)		
IMPAIRED (SCORE OF 1.5 - 2.5)		
UNHEALTHY (SCORE OF 0 - 1)		
TOTAL		100%

Scorecard users should examine the distribution of indicator properties within the three categories of health. Ideally, one would prefer to see all of the properties score in the *healthy* category. Even if 90% or more of the indicators you scored are *healthy*, your soil may still have serious problems with the remaining properties. For indicators either in the *impaired* and *unhealthy* categories, careful consideration is necessary to identify what caused the property to be in a less-than-optimum condition. *Impaired* indicator properties should be closely monitored over time to determine whether they are deteriorating or improving. *Unhealthy* properties need immediate attention and corrective action. You may also wish to give higher priority to those properties farmers considered more important as indicated by their relative rank in superscript.

4 A Conceptual Framework for Assessment and Management of Soil Quality and Health

Robin F. Harris
University of Wisconsin
Madison, Wisconsin

Douglas L. Karlen
USDA-ARS and National Soil Tilth Laboratory
Ames, Iowa

David J. Mulla
University of Minnesota
St. Paul, Minnesota

Diverse field and laboratory approaches varying in cost and labor intensiveness are currently available for characterizing soil quality and health. In this chapter, our objective is to present a conceptual framework within which to locate these different approaches and provide a basis for their practical application to site specific assessment and management of land for integrated land use and environmental protection. First, we review (i) the functional role of soil in terrestrial ecosystems, (ii) the direct and indirect connections of soil to the environmental and biological systems supported by soil, and (iii) the need to qualify site specific soil quality and soil health indicator properties as a function of land use, landscape and climate characteristics. This leads to an interchangeable, functional definition of soil quality and soil health (collectively called soil quality and health) that recognizes the fitness of soil to carry out biological production and environmental protection functions within specified land use, landscape, and climate boundaries. Experimental implications of this functional definition are discussed, and basic approaches for characterizing soil quality and health are reviewed. Finally, a protocol for assessing and managing soil quality and health is presented.

Copyright © 1996 Soil Science Society of America, 677 S. Segoe Rd., Madison, WI 53711, USA.
Methods for Assessing Soil Quality, SSSA Special Publication 49.

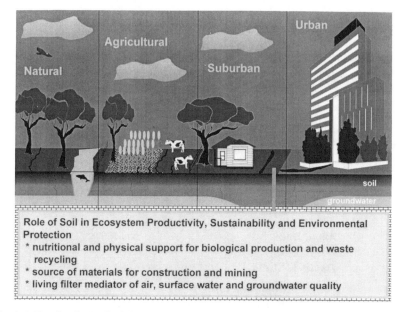

Fig. 4–1. Functional role of soil in ecosystems.

ECOSYSTEM ROLE OF SOIL AND FUNCTIONAL DEFINITION OF SOIL QUALITY AND SOIL HEALTH

Ecosystem Role of Soil

The functional role of soil in terrestrial ecosystem productivity, sustainability and environmental protection is illustrated in Fig. 4–1. In this figure, natural, agricultural, suburban, and urban ecosystems are depicted from left to right in accordance with the degree of development (least to most). All ecosystems share air, surface water and groundwater environmental components, and also a common mantle of living soil. Even without explicit definition of the terms soil quality and soil health, it can be seen from Fig. 4–1 that for all ecosystems, the quality and health of soil dictates the fitness of soil to (i) provide nutritional and physical support for biological production and waste recycling, (ii) act as source of materials for construction and mining, and (iii) act as a living filter mediating the quality of interfacing air, surface water and ground water. The nature of the biological systems supported by soil and the extent of industrial use of soil materials varies markedly between ecosystems, but managers of all ecosystems must carry out the mandate of protecting the quality of the air and water resources interfacing with soil.

The vertical lines separating the ecosystems in Fig. 4–1 are a reminder of traditional views. They could be interpreted to imply that competitive and profitable agriculture are the major focus of managers of agricultural ecosystems, and that environmental stewardship is only the concern of those persons interested in

CONCEPTUAL FRAMEWORK

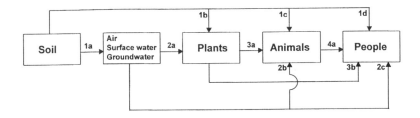

Fig. 4–2. Direct (1a–1d) and indirect (2a–4a) quality and health connections of soil to air, water, plants, animals, and people.

natural ecosystems. Although in many ways these views may still exist, they are obviously no longer tenable. A broad-based, environmental literacy is needed to understand integrated land use (conservation and development) and environmental protection.

Understanding the concept of soil quality and soil health requires a whole-systems approach that integrates soils, plants, and animals (including humans) to minimize leakage of natural and synthetic chemicals to surface water, groundwater, and air resources. Understanding and improving soil quality also requires cross-cutting expertise from persons who understand different ecosystems to optimize different conservation and development options. From a general communication standpoint, Fig. 4–1 illustrates in pictorial form understandable by all age groups and interest areas, the important message that the soil resource provides a common link and plays a key role in sustainable land use and environmental protection across all terrestrial ecosystems.

Soil Quality and Soil Health Connections to Systems Supported by Soil

The multiple quality and health connections of soil to the environmental and biological systems supported by soil (Cihacek et al., 1996, this publication) are summarized in Fig. 4–2. This figure shows that direct (1a–1d) and indirect food chain (2a–4a) connections exist between soil properties and (i) environmental (groundwater, surface water, and air) quality, (ii) health and productivity of plants and quality of plant products, (iii) health and productivity of animals consuming plants and the quality of animal products, and (iv) the human component of the food chain. The existence of these connections has given rise to simplistic sayings such as "healthy soil—clean air and water—healthy plants—healthy animals—healthy people" (Fig. 4–2) that trigger different and often deep-seated responses from different people. One of the major challenges and educational opportunities in soil quality and soil health work is to rigorously describe the direct and indirect connections, and put them in the appropriate perspective.

Emphasis to date has focussed on direct connections, particularly those between soil and plant productivity, surface water quality, or groundwater quality. Indirect food chain connections have been assumed, but have yet to be systematically and rigorously examined by the broad-based interdisciplinary teams needed to accomplish such work.

Interaction between Soil Quality and Health Indicators, and Land Use, Landscape, and Climate Characteristics

As recognized in Fig. 4–3, functional assessment of soil quality and soil health indicator data requires integration of soil quality and soil health indicator properties with land use (natural as compared with agricultural, suburban or urban ecosystems and kinds of plants and animals supported by the soil), three-dimensional landscape characteristics such as (slope steepness and length, topsoil and rooting depth, depth to groundwater, and distance to streams and lakes), and climate characteristics (such as precipitation, growing degree days, and frost). Such integration of soil quality and soil health indicator properties with land use, landscape, and climate characteristics provides a site-specific[1] data base for assessment of the fitness of the soil to carry out its quality and health support functions. This data base provides the information needed to develop site-specific management plans (such as residue cover and tillage, tile drainage, and irrigation, crop rotation, fertilizer, and pesticide practices) for improving or maintaining soil quality and soil health and thereby accomplishing site-specific assessment and management for integrated land use and environmental protection.

Functional Definition of Soil Quality and Health

Explicit and implicit consideration of the ecosystem role of soil (Fig. 4–1; Blum & Santileses, 1994; Doran et al., 1996), the direct and indirect connections between soil and the systems supported by soil (Fig. 4–2), and the interaction between soil quality and soil health indicators, land use, landscape, and climate (Fig. 4–3), has led to a variety of separate and combined definitions of soil quality and soil health (Acton & Gregorich, 1995; Doran & Parkin, 1996, this publication; Karlen et al., 1996). We suggest that the definitions of soil quality and soil health become synonymous for a role of soil confined to supporting biological production and environmental protection, and not considering the soil as a source of materials for construction and mining (Fig. 4–1). Accordingly, taking into consideration Figs. 4–1 to 4–3, we propose the following definition: *Soil quality and*

---

[1] The term site-specific is commonly used for differential fertilizer management of a field for maximized crop production efficiency based on mapping sites within the field recognizing different soil fertility properties. We use an expanded application of site-specific that includes other types of land use and environmental protection (e.g., Mulla et al., 1996), and gives rise to an integrated site-specific assessment and management approach with roots in the concept of soil quality and health. This approach dictates that a land area designated for a specified land use should (i) be assessed and mapped to identify management options for maximized land use, (ii) be assessed and mapped to identify management options for maximized environmental protection, and (iii) the resulting site-specific maps and potentially conflicting management options should be coordinated for development of a management plan to optimize integrated land use and environmental protection. By definition, soil quality and health indices represent core site-specific assessment properties for integrated land use and environmental protection management.

CONCEPTUAL FRAMEWORK

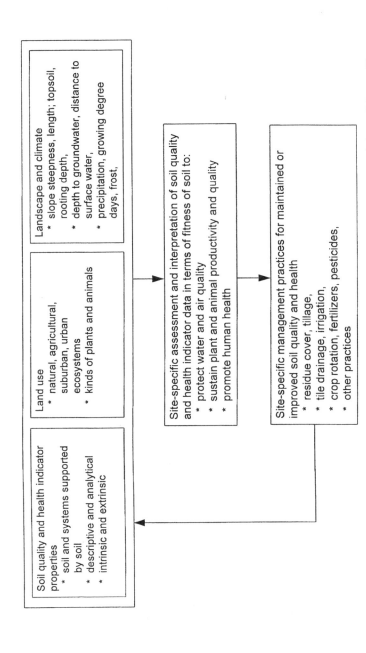

Fig. 4–3. Interrelationships between soil quality and health indicator properties, land use, landscape and climate characteristics, and site-specific assessment and management of soil quality and health.

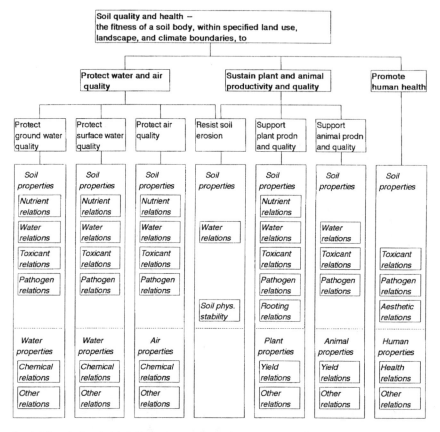

The definition of soil quality is in bold print; the expanded functions are in normal print, and the categories of indicator properties supporting each function are in italics.

Fig. 4-4. Functions of soil quality and health, and categories of indicator properties supporting each function.

health reflect the fitness of a soil body, within land use, landscape and climate boundaries, to protect water and air quality, sustain plant and animal productivity and quality, and promote human health.* In this definition, and hereafter in the text, we use the term *soil quality and health* rather than the more cumbersome term *soil quality/health* (Harris & Bezdicek, 1994).

Practical Implications of the Functional Definition

Categories of Soil Quality and Health Indicator Properties

The functional definition of soil quality and health identifies the numerous functions that indicator properties must address. Figure 4-4 identifies common categories of indicators (such as *nutrient relations*) essential for supporting each function. The category of *aesthetic relations* is placed in the promote human health function to recognize the deep feelings of some regarding the role of soil quality and health. An example of this is the sense of well being and harmony with nature that is engendered in some by dark, rich earthy smelling humus, and friable, crumb structured soil tilth. Figure 4–4 is intentionally constructed so that each category is located in a fixed vertical position to facilitate comparisons of shared or unique contributions of specific soil-based indicators to different soil functions. It also shows that some categories support more than one function (e.g., nutrient relations), while others support one (e.g., rooting relations).

In addition to soil-based indicator properties, indicator properties of target systems also are recognized in Fig. 4–4. This provides ground truth linkages between soil quality indicator properties and their postulated functional outcomes.

Soil Quality and Health Indicators

Indicators within each category, and the mechanistic linkages between them and soil quality functions, vary with the focus, priorities, degree of completeness and experimental procedures used for assessment; however, as illustrated in Table 4–1, the following generalizations can be made: (i) some indicators will occur in different categories supporting different functions (e.g., soil nitrate will occur in the nutrient relations category supporting sustain plant production, protect surface water quality, protect groundwater quality functions); (ii) some will occur in different categories supporting the same function (e.g., soil pH will occur in the nutrient relations and toxicant relations categories supporting the sustain plant production function); and (iii) each indicator supporting a given function must be related quantitatively to the function it supports. Indicator properties may be expressed in quantitative, analytical terms, and/or in qualitative, descriptive terms. Soil-based indicators may be intrinsic, such as clay or organic matter content, or extrinsic, such as toxic contaminants.

All soil quality and health indicators must address one or more of the functions of soil quality and health. Organization of data sets in accordance with the categories and functions identified in Fig. 4–4 (e.g., Table 4–1) provides a check on the completeness of the information needed to evaluate critical functions and may identify the need for more specific or targeted soil quality and health assessment. Finally, it should be stressed that the functional interpretation of indicators located under more than one function (as illustrated in Table 4–1) may, and in many cases probably will, be different (e.g., high soil nitrate is good for crop production but poor for groundwater protection).

APPROACHES FOR INTERPRETING SOIL QUALITY AND HEALTH INDICATOR DATA

The properties and processes measured to assess soil quality and health are commonly expressed in standard (soil quality and health neutral) mass and volume units. After defining the functions and indicator variables that will be used for making the soil quality assessments, a decision is required with regard to how the indicator data will be interpreted in functional, value-based terms. During this decision-making process, properties or processes that cannot be interpreted in

Table 4–1. Example of a minimum analytical data set of indicator properties organized within categories directly supporting each function of soil quality and health.†‡

Soil quality and health indicator properties		Functions of soil quality and health						
		Protect water and air quality			Resist soil erosion	Sustain plant and animal productivity and quality		Promote human health
		Ground water	Surface water	Air		Support plant and animal production and quality		
Property	Category					Plant	Animal	
Physical properties								
Texture (sand, silt, clay)	Nutrient relations	x	x	x		x		
	Water relations	x	x	x	x	x		
	Toxicant relations	x	x	x		x		
	Pathogen relations	x	x			x		
	Rooting relations							
	Soil phys stability		x	x	x			
Bulk density	Water relations	x	x		x	x		
	Rooting relations					x		
Infiltration	Nutrient relations					x		
	Water relations	x	x		x	x		
	Rooting relations					x		
Water holding capacity§	Water relations	x	x	x	x	x		
	Rooting relations					x		
Chemical properties								
Total organic C	Nutrient relations	x	x	x		x		
	Water relations	x	x		x	x		
	Toxicant relations					x		
	Soil phys stability		x		x		x	
	Aesthetic relations							x
Total organic N	Nutrient relations	x	x			x		
pH	Nutrient relations	x	x	x		x		x
	Toxicant relations							
Electrical conductivity	Nutrient relations	x	x	x		x	x	x
	Toxicant relations	x	x			x		
Extractable NH$_4^+$	Nutrient relations	x	x			x		
NO$_3^-$	Nutrient relations		x			x		
	Toxicant relations	x						
Extractable P	Nutrient relations	x	x			x		
Exchangeable K	Nutrient relations	x	x			x		
Biological properties								
Microbial biomass C	Nutrient relations	x	x			x		
Microbial biomass N	Nutrient relations	x	x			x		
Potentially mineralizable N	Nutrient relations	x	x			x		
Soil respiration¶	Nutrient relations	x	x	x		x		

† Indicator properties of Doran and Parkin (1996). Also included in this data set are landscape characteristics of depth of soil and topsoil, and rooting depth.
‡ The properties and categories address the direct connections between soil and systems supported by soil (Fig. 4–3, 1a-d; Fig. 4–4, soil properties). The assignment of indicator properties to specific categories supporting specific functions was done by the authors based on information provided in Doran and Parkin (1996). The "x" symbol identifies that the property plays a function–dependent role with respect to the fitness of the soil to support the identified function(s) of soil quality and health.
§ Calculated from bulk density, texture, and organic C.
¶ Interpretation qualified by water content and temperature.

value-based terms, or those found to be quantitatively very minor components, are prime candidates for deletion from the measurement list. This is especially true if the analytical measurements are labor and/or cost intensive.

Analytical measurements do not inherently include a value-based dimension that can be used for assessing soil quality and health. Identifying how individual soil quality and health indicator properties should be translated into qualitative soil quality and health ratings or scores and used to assess components of and overall soil quality with respect to the critical functions is a major challenge for current and future soil quality and health research.

A variety of approaches can be used to interpret measurement data into value-based scores that can then be used to quantify soil quality and health. As part of the interpretation or scoring of individual measurements, the importance of a particular function, process, or indicator with regard to the assessment being made must be established. This requires a value judgement and documenting the reason for the decision is a critical part of the assessment process.

It should be recognized that the following examples of specific approaches that have been used for interpretation of soil quality and health were not planned within the context of the functional definition of soil quality and health used in this chapter. The approaches are confined largely to the historical sustaining plant productivity function of soil quality and health. In addition, although the target crop is identified, the approaches do not explicitly recognize the need to integrate site and climate characteristics into their functional assessment of soil quality and health. The basic methodology, however, is applicable in principle to ultimate development of procedures accomplishing integrative assessment of the fitness of soil to protect and water and air quality, sustain plant and animal productivity and quality, and promote human health.

Descriptive Approaches for Soil Quality and Health Assessment

A descriptive screening approach for use by farmers or researchers to assess soil quality and health is described in Romig et al. (1996, this publication). This approach involves use of the Wisconsin Soil Health Scorecard, which was developed from structured interviews with farmers centering on the question "how do you recognize a healthy soil" (Romig et al., 1995). The Wisconsin Soil Health Scorecard provides a high-to-low scoring mechanism for each soil health indicator property with respect to the fitness of a soil to support the crop productivity function. An example of a typical experimental data set, organized according to the categories in Fig. 4–4, is presented in Table 4–2. The indicator properties were recorded within three categories: healthy, impaired, and unhealthy. Ideally, one would prefer to see all of the properties score in the healthy category. Even if 90% or more of the indicators scored are healthy, the soil may still have serious problems with the remaining properties. For indicators in either the impaired or unhealthy categories, careful consideration is necessary to identify what caused the property to be in a less-than-optimum condition. Impaired indicator properties should be closely monitored over time to determine whether they are deteriorating or improving. Unhealthy properties need immediate attention and corrective action (Romig et al., 1996, this publication).

Table 4–2. Example of a descriptive data set of soil health indicator properties, with integrated value–based scoring of fitness to support the crop production function of soil health, using a healthy impaired unhealthy soil health scoring approach.†

Property		Soil health characteristics‡	Score range	R1 Continuous corn	R2 Narrow row beans	R5 Oats and alfalfa	R3 Wide row beans	R6 Continuous pasture
	Status	Description						
		Soil-Based Indicators						
Nutrient relations								
Organic matter[1,21]	H	Organic matter 4–6%	3.0–4.0	4.0	3.6	3.6	3.7	4.0
	I	Organic matter 2–4% or 6–8%	1.5–2.5					
	U	Organic matter <2% or >8%	0.0–1.0					
Soil test–N,P,K[9,23]	H	All nutrient levels at recommended levels	3.0–4.0	4.0	4.0	4.0	4.0	4.0
	I	Soil tests below recommended levels, need extra inputs	1.5–2.5					
	U	Two or more nutrients very low, law of minimum	0.0–1.0					
Decomposition[16,20]	H	Rapid rotting of residues and manures	3.0–4.0					4.0
	I	Slow rotting of residues and manures	1.5–2.5	2.0	2.0	3.0	2.3	
	U	Residues and manures do not break down in soil	0.0–1.0					
Average (7 properties)	H		3.0–4.0	3.6	3.5	3.6	3.6	3.9
Water relations								
Drainage[6,8]	H	Soil drains at good rate for crops, water moves through well	3.0–4.0	3.3	3.3	3.3	4.0	3.3
	I	Soil drains slowly, slow to dry out	1.5–2.5					
	U	Poor drainage, soil often waterlogged or over saturated	0.0–1.0					
Infiltration[12,7]	H	Water soaks right in, soil is spongy, no ponding	3.0–4.0	3.3	3.3	3.3	4.0	3.3
	I	Water soaks in slowly, some runoff/puddling after heavy rain	1.5–2.5					
	U	Water doesn't soak in, sits on top or runs off	0.0–1.0					
Water retention[14,9]	H	Soil holds moisture well, gives and takes water easily	3.0–4.0	4.0	4.0	4.0	4.0	4.0
	I	Soil is drought prone in dry weather	1.5–2.5					
	U	Soil dries out too fast, droughty	0.0–1.0					
Average (4 properties)	H		3.0–4.0	3.7	3.7	3.7	4.0	3.7

CONCEPTUAL FRAMEWORK

Indicator		Description	Range					
Rooting relations								
Earthworms[3,1]	H	Worm holes and castings numerous	3.0–4.0		3.3	4.0	3.3	4.0
	I	Few worms or castings	1.5–2.5	2.3				
	U	Little sign of worm activity	0.0–1.0					
Erosion[4,2]	H	Little erosion evident, topsoil resists water, wind erosion	3.0–4.0	4.0	4.0	4.0	4.0	4.0
	I	Moderate erosion, signs of sheet, rill erosion, topoil blows	1.5–2.5					
	U	Severe erosion, considerable topsoil loss, gullies form	0.0–1.0					
Tillage ease[5,3]	H	Plow field in higher gear, soil flows and falls apart, mellow	3.0–4.0	4.0	4.0	4.0	4.0	
	I	Soil grabs plow, difficult to work, needs extra passes	1.5–2.5					
	U	Plow scours hard, soil never works down	0.0–1.0					
Average (6 properties)			3.0–4.0	3.3	3.4	3.9	3.8	3.1
Total Soil Relations								
Average (5 categories)	H			3.4	3.5	3.8	3.7	3.6
Crop-Based Indicators								
Crop appearance[2,25]	H	Overall crop is dark green, large, tall, in a dense stand	3.0–4.0	3.0	4.0	3.5	3.7	3.7
	I	Overall crop is light green, small, in a thin stand	1.5–2.5					
	U	Overall crop is poor, stunted, discolored, uneven stand	0.0–1.0					
Yield[10,35]	H	Corn greater than 130 bu acre^{-1}	3.0–4.0	4.0	4.0	3.5	4.0	4.0
	I	Corn 85 to 130 bu acre^{-1}	1.5–2.5					
	U	Corn less than 85	0.0–1.0					
Nutrient deficient[17,26]	H	Crop has what it needs, show little deficiency signs	3.0–4.0	3.3	3.5	1.7	2.5	4.0
	I	Crops falls off or discolors as season progresses	1.5–2.5					
	U	Crop severely: blighted, streaky, spotty, discolored	0.0–1.0					
Average (10 properties)			3.0–4.0	3.6	3.7	3.7	3.7	4.0

† Soil health scores for 1994 WICST rotations at Arlington (Garlynd et al., 1995). Averages are for all properties within a given category. The first superscript number qualifying each property corresponds to the farmer-based priority ranking, the second corresponds to the numbering system of the Wisconsin Soil Health Scorecard (Romig et al., 1995, 1996, this publication), which focusses on crop production. Abreviations: H, Healthy; I, Impaired; U, Unhealthy.

‡ Scores are the average of three plots for each systems.

The main strengths of the descriptive approach include ease of data characterization and collection, and compatibility with Whole Farm Planning approaches. The main weakness is the inability to evaluate quantitative impacts on water and air quality of observations on soil quality within the context of site and climatic characteristics.

Analytical Approaches for Soil Quality and Health Assessment

Soil quality and health indicator data have been used directly for assessment of soil quality. For example, Reganold (1988) compared A horizon thickness on adjacent conventional vs. organic farms to evaluate effects of long-term differences in soil management on soil erosion. A simple scoring approach, patterned after that used in the descriptive Soil Health Scorecard (Romig et al.,1996, this publication), and based on analytical data interpreted on a high-marginal-poor scale with respect to the functions of sustaining crop growth and protecting water quality, may have applicability as a relatively simple method for assessing the impact of toxicants, N, P, or other nutrients on soil quality and health.

Comparison of a few key indicators is usually insufficient for a comprehensive evaluation of all the functions associated with the soil quality and health concept. In addition, some soil quality and health indicators such as microbial biomass, organic C, aggregate stability, or respiration may have to be evaluated or scored using integrative approaches. Comparison of more than a few indicators becomes complicated without a framework for determining the relative importance of each indicator, and without evaluating these indicators within the context of site and climatic characteristics. For these reasons, assessment of soil quality and health is most efficiently achieved using a modeling framework that is based upon collecting and synthesizing an array of soil quality and health indicators and site or climatic characteristics.

Analytical models for evaluation of soil quality data may be function-based or process-based. In the function-based model approach, it is likely that weighting factors and functional equations will vary from region to region, and that considerable difference of opinion could arise over specification of values for weighting factors.

Functional Models: Productivity Index

In the Productivity Index (PI) approach of Larson and Pierce (1994), pedotransfer functions are used to estimate the sufficiency of bulk density, pH, and available water holding capacity from a minimum data set of soil quality indicator properties. The PI is a weighted summation over soil horizons of sufficiencies for each property. The advantage of this approach is that it is semiquantitative. The disadvantage is that it focusses only on the productivity function of soil quality, and neglects the functions pertaining to environmental protection and promotion of human health.

Functional Models: Soil Quality Index

In the Soil Quality Index (SQI) approach of Doran and Parkin (1994), soil quality is estimated as the product of weighted subfactors for each of the functions of soil quality. The sub-factors are estimated from regression equations obtained by simulating the relationship between soil quality indicator properties and each of the functions of soil quality. This approach has the advantage of being able to quantitatively estimate the relative importance to society for each of the three soil quality functions in view of local policies, societal concerns, and environmental or economic constraints. Its main disadvantage is the lack of simulation-based regression equations relating soil quality indicator properties to soil quality functions

Functional Models: Scoring Functions

The use of mathematical scoring functions is another approach for quantifying changes in soil quality and health indicators. This was demonstrated for various tillage or crop residue management practices by Karlen et al. (1994a,b).

To compute an overall soil quality or health index for a particular location, all critical soil functions (i.e., sustaining plant and animal productivity, protecting water and air quality, and promoting human health) must be considered. The process, however, is iterative so we have chosen to illustrate the concept by starting with just one component of an overall soil quality or health assessment. Following the organizational diagram in Fig. 4–4, the user selects the functions that will be considered and the processes (e.g., protect groundwater, surface water, and/or air quality; resist soil erosion; support plant and animal production and quality) that will be evaluated. For each process, the user organizes the available data, preferably at least those factors included in the minimum data set (Doran & Parkin, 1996, this publication) into sets that describe water relations, nutrient relations, toxicant relations, or other appropriate processes. A very simple example for the process of supporting plant productivity and quality beneath the function of sustaining plant and animal production and quality is shown in Table 4–3.

While grouping the parameters, the user must document the expected or desired response for each parameter based on knowledge or data for the kind of soil being evaluated. Generally, a more is better, more is worse, or optimum response curve is chosen. Lower and upper threshold values and baseline (50%) values for each parameter are established. These are generally based on literature values or expert opinion (Karlen et al., 1994a,b) developed through years of experience in a given location. The most critical requirement is to clearly document the reasons that some particular value was chosen as a threshold or baseline value.

The entire scoring function approach is iterative and structured, so scoring functions can be programmed using computer software such as SAS[2], LOTUS 123, QUATRO PRO, or EXCEL. To illustrate the procedure, data from a Conservation Reserve Program (CRP) site in southwestern Iowa will be used to compute a simple index for the process of supporting plant production and quality. An

---

[2] Mention of a trademark, proprietary product, or vendor does not constitute a guarantee or warranty of the product by the USDA, University of Wisconsin, or University of Minnesota, and does not imply its approval to the exclusion of the other products or vendors that also may be suitable.

Table 4–3. A sample grouping of indicators used to calculate the plant production and environmental quality components of soil quality using a scoring function approach.

Data set	Parameter	Expected range	Plant production	Environmental quality	Integrated
Nutrient availability	Bray P	7.5 to 150 mg kg^{-1}	Optimum (SSF 5)†	Optimum (SSF 5)†	Optimum (SSF 5)†
	Exchangeable K	45 to 525 mg kg^{-1}	Optimum (SSF 5)	Optimum (SSF 5)	Optimum (SSF 5)
	pH	3.5 to 9.5	Optimum (SSF 5)†	Optimum (SSF 5)†	Optimum (SSF 5)†
	Organic C	5 to 65 g kg^{-1}	More is better (SSF 3)	More is better (SSF 3)	More is better (SSF 3)
	NO$_3$–N	3 to 50 mg kg^{-1}	Optimum (SSF 5)	Optimum (SSF 5)	Optimum (SSF 5)
Water availability	Surface residue	1000 to 18 000 mg ha^{-1}	Optimum (SSF 5)	Optimum (SSF 5)	Optimum (SSF 5)
	Porosity	20 to 80%	Optimum (SSF 5)	Optimum (SSF 5)	Optimum (SSF 5)
	Organic C	5 to 65 g kg^{-1}	More is better (SSF 3)	More is better (SSF 3)	More is better (SSF 3)
	Aggregate stability	15 to 70%	More is better (SSF 3)	More is better (SSF 3)	More is better (SSF 3)
Rooting environment	pH	3.5 to 9.5	Optimum (SSF 5)	Optimum (SSF 5)	Optimum (SSF 5)
	Bulk density	1.2 to 2.1 g cm^{-3}	Less is better (SSF 9)	Less is better (SSF 9)	Less is better (SSF 9)
	Rooting depth	60 to 250 cm	More is better (SSF 3)	More is better (SSF 3)	More is better (SSF 3)
	Organic C	5 to 65 g kg^{-1}	More is better (SSF 3)	More is better (SSF 3)	More is better (SSF 3)

† Shape of the standard scoring function (SSF) as described by Wymore (1993).

overall soil quality or health index would be computed by using the same procedures for all of the processes and functions outlined in Fig. 4–4.

The study from which these data are taken is farmer-controlled with supporting data being collected by various researchers. The focus for this evaluation is entirely on returning this land to crop production, so the assessment is intended to help predict potential effects of using moldboard plowing or no-till practices to return the land to row crop production when the CRP contracts expire.

Ideally, all parameters suggested for the minimum data set would have been collected, but only a limited amount of mean data are available from the surface 30 cm under CRP, plowed, and no-till treatments (Table 4–4). To compute an index based on nutrient, water, and plant rooting relationships for the various management practices, the parameters affecting each are grouped as shown in Table 4–3. It should be noted that some parameters (i.e., organic C) are included in more than one of the critical relationships. The only requirement as the parameters are grouped is that the rationale for doing so should be documented, just as the user does for the type of scoring function, expected ranges, baseline, and threshold values. A relative weight or importance for each parameter associated with a specific relationship is then chosen and entered into a spreadsheet (Table 4–5). Measured values for each management practice (Table 4–4) are entered into the spreadsheet value column. This value is scored based on the mathematical equations that describe the various standard scoring functions (Wymore, 1993). A score is returned and then multiplied by the weighting factor (Table 4–5). After each parameter associated with a specific relationship is scored, the weighted values are summed. These values are transferred to the column entitled Previously Computed Score and multiplied again by the weight assigned to each data set. This process returns plant production index values of 0.87, 0.50, and 0.44 for CRP, no-till, and plowed treatments, respectively. This limited set of data suggests that returning this soil resource to row crop production resulted in a lower plant productivity index than leaving the area in permanent grass cover, probably because this was a fragile soil that had originally qualified for the CRP. If return-

Table 4–4. Soil quality indicator data, averaged for the upper 30 cm of the soil profile, for evaluation of alternative post-Conservtion Reserve Program (CRP) land management practices in southwest Iowa.

Indicator	CRP	Plowed	No-till
Observed rooting, cm	150	100	125
Surface residue, kg ha^{-1}	7620	2000†	4570†
Aggregate stability, %	40	19‡	26‡
Bulk density, g cm^{-3}	1.07	1.39	1.31
Porosity, %	0.60	0.48	0.51
Extractable P, mg kg^{-1}	38	6	13
Exchangeable K, mg kg^{-1}	267	137	167
pH	6.3	6.4	6.4
NO^3-N, mg kg^{-1}	1.0	2.2	1.4
Organic N, mg kg^{-1}	1.9	1.4	1.5
Organic C, mg kg^{-1}	23.9	16.6	16.0

† Plowed areas had been planted to soybean in 1994, while no-till area had been planted to corn.
‡ Estimated from data collected from tilled and no-till fields in Iowa as part of a survey condected to assess CRP effects on various soil quality indicators.

Table 4-5. Spreadsheet layout for computing the plant productivity component of a soil quality index using scoring function approach.

SSF†	Parameter	Weight	Lower threshold	Upper threshold	Lower baseline	Lower slope	Upper baseline	Upper slope	Optimum value	Value	Previously computed score	Score	Score*weight
5	Bray P	0.15	7.5	150	16	0.235	100	−0.040	30	38		0.999	0.150
5	Exch. K	0.15	45	525	85	0.050	450	−0.027	175	267		1	0.150
5	pH	0.20	3.5	9.5	5.3	1.67	7.70	−1.67	6.50	6.30		0.999	0.200
3	Organic C	0.40	5	65	20	0.133		0.000		23.9		0.889	0.356
5	NO$_3$	0.10	3	50	12	0.222	38	−0.167	25	1.00		0	0.000
5	Residue	0.25	1 000	18 000	3 000	0.001	14 000	−0.0005	10 000	7 620		1	0.250
5	Porosity	0.20	20	80	40	0.200	60	−0.200	50	60		0	0.000
3	Organic C	0.20	5	65	20	0.133		0.000		23.9		0.889	0.178
3	Aggregate stability	0.35	15	70	30	0.133		0.000		40		0.995	0.348
5	pH	0.35	3.5	9.5	5.3	1.67	7.70	−1.67	6.50	6.30		0.999	0.350
9	Bulk density	0.35	1.20	2.10	1.45	−8.00		0.000		1.07		1	0.350
3	Rooting depth	0.20	60	250	125	0.031		0.000		150		0.956	0.191
3	Organic C	0.10	5	65	20	0.133		0.000		23.9		0.889	0.089
Nutrient relations		0.33									0.855587	0.856	0.282
Water relations		0.33									0.776196	0.776	0.256
Rooting relations		0.34									0.980120	0.980	0.333
Plant productivity index for CRP treatment												0.872	

† Standard scoring function (SSF) as described by Wymore (1993).

ing this land to row crop production was the desired land use, however, the production index suggests that using no-tillage practices would be better than moldboard plowing.

A complete soil quality or health index could be developed by adding additional factors to these critical functions or by adding additional functions to the spreadsheet. This may or may not change the relative rankings for the three management treatments depending on how they affect the other critical functions. What is most important, however, is that the rationale for choosing any particular function or indicator variable should be agreed upon by all stakeholders and well documented.

An advantage of using scoring functions rather than the SQI approach discussed previously is that simulation modeling is not needed to estimate the functional relationships between soil properties and soil quality or health. This does not imply that output from simulation models could not be used, but rather if the amount of information available is insufficient to fully calibrate simulation models, the scoring functions provide a more simplistic approach for combining various physical, chemical, and biological indicators into an assessment of soil quality or health.

Functional Models: Fuzzy Logic

Another method for interpreting multiple types of data for a soil quality evaluation is the use of fuzzy logic set theory. The procedures were developed primarily for analyzing complex biological and social science systems, but have been primarily used to control industrial processes or operation of consumer products. It also was used with minimum success to estimate crop yields for precision farming (Ambuel et al., 1994).

Fuzzy logic involves establishing sets and then determining relative membership of various data points within the established categories. It has been demonstrated as a tool for characterizing uncertainty in soils information so that a risk-based method of soil interpretations can be implied (Mays et al., 1995). In their study, map unit data from a soil survey are used to demonstrate the appropriateness of soils for septic tank filter fields and tillage. As for all methods being used to quantify soil quality and health, further development is required to ensure incorporation of factors that will directly reflect the quality of the data in the final rating and to refine the understanding of how soil properties interact to affect the final rating.

Process Models

Mechanistic process modeling has been proposed to make direct estimates of soil quality from indicator data without the use of empirical weighting factors or functional equations. An example of the process approach to evaluate C and N cycling in soil is the NCSOIL model of Molina et al. (1983, 1994).

Process models also could be applied for estimating the impacts of soil quality on water quality. Examples of such models include the NLEAP model for N (Shaffer et al., 1991), the P-index model for P (Lemunyon & Gilbert, 1993), the Attenuation Factor (AF) model for pesticides (Mulla et al., 1996), the RUSLE model for water erosion (Busacca et al., 1993), and the EPIC model for crop productivity (Williams & Renard, 1985). The strength of such models is their ability to quantify the effects of soil quality indicator properties on crop productivity, erosion, or water quality in the context of landscape and hydrologic characteristics. No empirical weighting factors are needed to specify the relative impact of soil quality indicator properties on water quality. The main weakness of these process models is that they require extensive amounts of quantitative input data. In addition, none of them evaluates all of the functions for soil quality in a comprehensive fashion. The latter issue has been addressed by the recent development of an integrated model (Hawkins et al., 1995) to evaluate the impacts of management practice on productivity, environmental quality, and economic profitability using subroutine elements from the NLEAP, P-index, AF, RUSLE, and EPIC models.

STRATEGIC PLANS FOR ASSESSMENT AND MANAGEMENT OF SOIL QUALITY AND HEALTH

A plan for assessing soil quality and health and determining whether or not soil quality and health is responding to management strategy involves five basic steps (Fig. 4–5). These are: (i) identification of socio-economic constraints and goals for soil quality and health management, (ii) determination of soil quality and health indicators and site and climatic characteristics, (iii) assessment of soil quality and health using models, (iv) temporal or spatial comparisons of model output, and (v) modification of management strategies.

Socio-Economic Constraints and Goals for Soil Quality and Health Management

The socio-economic goals for soil management are determined from a variety of constraints, including societal policies, laws, and financial incentives, economic costs, and the attitudes, education, and training of the land manager. These factors determine, in large part, the relative value placed upon the soil resource for its ability to sustain productivity, attenuate pollution, or promote animal and human health. In addition, each land manager works within a unique set of social, economic, behavioral, and site or climatic characteristics that constrain the choice and adoption of various management strategies. These characteristics largely determine whether descriptive, analytical, or process-based approaches are most suitable for evaluating soil quality.

Soil Quality and Health Indicators and Site and Climatic Characteristics

Soil quality and health is dependent upon many physical, chemical, and biological properties that comprise a data set of soil quality and health indicators. These indicators may be analytical (Tables 4–1, 4–3 to 4–5) or descriptive (Table 4–2). Soil quality and health indicators represent a wide range of categories pertaining to various soil functions, and although focussing on soil properties also

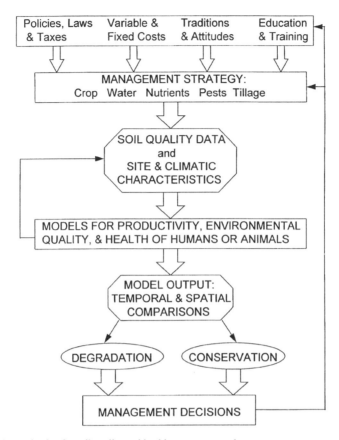

Fig. 4–5. Strategic plan for soil quality and health assessment and management.

may include properties of the environmental and biological systems supported by soil (Fig. 4–4). Auxiliary information about the site and climate is essential for evaluating the ability of a soil to carry out its support functions (Fig. 4–2). Auxiliary information includes ecosystem characteristics such as the nature of the ecosystem and the kinds of agronomic and other plants and animals it supports, landscape features such as slope length and steepness, hydrologic features such as depth to groundwater or proximity to surface water, climatic features such as precipitation and temperature, and management features such as tillage, residue cover, tile drainage, irrigation, and agrichemical applications (Fig. 4–2). For example, functional soil quality and health could be quite different for two soils having identical soil quality indicator values but located in sites with different precipitation regimes, depths to groundwater, and management practices.

Models for Assessing Soil Quality and Health

Assessment of soil quality and health with models involves (i) modeling the effect of soil quality indicator data and site and climatic characteristics on particular functions of soil quality, and (ii) value-based classification of model output for various functions of soil quality into categories that can be interpreted across a range of high to low soil quality and health. The structure, and strengths and weaknesses of representative descriptive and analytical function and process-based models were reviewed earlier. It should be recognized that there is an interaction between the model selected for assessment of soil quality and health and the amount of data collected for input to the model. In general, quantitative and mechanistic models require more extensive input data than descriptive models.

Temporal and Spatial Comparisons of Model Output

Output from models used to interpret soil quality and health indicator data is not meaningful unless a baseline for comparison is available. The baseline may be the same site at a previous time, or a reference site at the same time. These two contrasting approaches have been termed the dynamic (changes over time) or the comparative (reference site) approaches to soil quality and health evaluation (Larson & Pierce, 1994). By evaluating temporal or spatial changes in soil quality and health, it is possible to determine whether the net effect of management strategies is degradation or conservation of soil quality and health.

Management Strategy

If temporal and spatial changes in soil quality and health identify that the current management strategy leads to conservation of soil quality and health, it is not necessary to change to an alternative management system. If, however, degradation of soil quality is indicated, it is advisable to undertake a planning process for implementation of alternative management practices that can restore soil quality and health. In this process, it is first necessary to determine which function or functions of soil quality and health are impaired, and then to determine which types of management changes could restore the proper functioning of the soil. For instance, if the ability of the soil to store and transmit water is impaired because of soil compaction it may be advisable to make changes in the timing and type of mechanical operations. If the ability of the soil to adsorb and degrade pesticides has been diminished by loss of organic matter rich topsoil due to erosion, adoption of conservation tillage practices may be advisable. If crop productivity has decreased due to soil borne diseases in a continuous monoculture rotation, it may be advisable to adopt a more diverse cropping rotation. Likewise, the presence of excessively high soil test phosphorus levels, salinization, or acidification are reasons to consider changes in management.

Once a set of management options has been identified to improve soil quality and health, a process is initiated to determine which of the practices is most suitable for implementation. This involves consideration of the existing system of policies, laws, and taxes, as well as the land manager's economic status, traditions and attitudes, and education and training. In addition, the effect of the improved strategy on all other practices in the existing management system must be considered. The most suitable management practice must be economically viable, compatible with the existing suite of other management practices, and acceptable for implementation by the land manager.

ACKNOWLEDGMENTS

Appreciation is extended to John Doran, William Larson, Maurice Mausbach, and Louis Schipper for comments and suggestions during the development of the chapter, and to Jerry Tyler for providing the graphics base for construction of Fig. 4–1.

REFERENCES

Acton, D.F., and L.J. Gregorich. 1995. Understanding soil health. p. 5–10. *In* D.F. Acton and L.J. Gregorich (ed.) The health of our soils: Toward sustainable agriculture in Canada. Centre for Land and Biol. Resour. Res., Res. Branch, Agric. and Agri-Food, Ottawa, ON.

Ambuel, J.R., T.S. Colvin, and D.L. Karlen. 1994. A fuzzy logic yield simulator for prescription farming. Trans. ASAE 37:1999–2009.

Blum, W.E.H., and A.A. Santelises. 1994. A concept of sustainability and resilience based on soil functions. p. 535–542. *In* E.J. Greenland and I. Szabolcs (ed.) Soil resilience and sustainable land use. CAB Int., Wallingford, Oxon, England.

Busacca, A.J., C.A. Cook, and D.J. Mulla. 1993. Comparing landscape-scale estimation of erosion in the Palouse using Cs-137 and RUSLE. J. Soil Water Conserv. 48:361–367.

Cihacek, L.J., W.L. Anderson, and P.W. Barak. 1996. Linkages between soil quality and plant, animal and human health. p. 9–23. *In* J.W. Doran and A.J. Jones (ed.) Methods for assessing soil quality. SSSA Spec. Publ. 49. SSSA, Madison, WI.

Doran, J.W., and T.B. Parkin. 1994. Defining and Assessing soil quality. p. 1–21. *In* J.W. Doran et al. (ed.) Defining soil quality for a sustainable environment. SSSA Spec. Publ. 35. SSSA, Madison, WI.

Doran, J.W., and T.B. Parkin. 1996. Quantitative indicators of soil quality: A minimum data set. p. 25–37. *In* J.W. Doran and A.J. Jones (ed.) Methods for assessing soil quality. SSSA Spec. Publ. 49. SSSA, Madison, WI.

Doran, J.W., M. Sarrantonio, and M.A. Liebig. 1996. Soil health and sustainability. Adv. Agron. 56:1–54.

Garlynd, M.J., D.E. Romig, and R.F. Harris. 1995. Effect of a cropping system shift from continuous corn on descriptive and analytical indicators of soil quality. p. 58. *In* Agronomy abstracts. ASA, Madison, WI.

Harris, R.F., and D.F. Bezdicek. 1994. Descriptive aspects of soil quality. p. 23–35. *In* J.W. Doran et al. (ed.) Defining soil quality for a sustainable environment. SSSA Spec. Publ. 35. SSSA, Madison, WI.

Hawkins, R., R. Craven, K. Klair, R. Loppnow, D. Nordquist, W. Richardson, and J. Whitney. 1995. PLANETOR users manual. Ctr. for Farm Financial Manage., Univ. Minnesota, St. Paul, MN.

Karlen, D.L., M.J. Mausbach, J.W. Doran, R.G. Cline, R.F. Harris, and G.E. Schuman. 1997. Soil quality: Concept, rationale, and research needs. Soil Sci. Soc. Am. J. 61:Jan–Feb.

Karlen, D.L., N.C. Wollenhaupt, D.C. Erbach, E.C. Berry, J.B. Swan, N.S. Eash, and J.L. Jordahl. 1994a. Crop residue effects on soil quality following 10-years of no-till corn. Soil Tillage Res. 31:149–167.

Karlen, D.L., N.C. Wollenhaupt, D.C. Erbach, E.C. Berry, J.B. Swan, N.S. Eash, and J.L. Jordahl. 1994b. Long-term tillage effects on soil quality. Soil Tillage Res. 32:313–327.

Larson, W.E., and F.J. Pierce. 1994. The dynamics of soil quality as a measure of sustainable management. p. 37–52. *In* J.W. Doran et al. (ed.) Defining soil quality for a sustainable environment. SSSA Spec. Publ. 35. SSSA, Madison, WI.

Lemunyon, J.L., and R.G. Gilbert. 1993. Concept and need for a phosphorus assessment tool. J. Prod. Agric. 6:483–486.

Mays, M.D., C.S. Holzhey, I. Bogardi, and A. Bardossy. 1995. Managing risk and variability with fuzzy soil interpretations. p. 187–198 *In* P.C. Robert et al. (ed.) Site-specific management for agricultural systems. Misc. Publ. ASA, SSSA, and CSSA, Madison, WI.

Molina, J.A.E., C.E. Clapp, M.J. Shaffer, F.W. Chichester, and W.E. Larson. 1983. NCSOIL, a model of nitrogen and carbon transformations in soil: Description, calibration and behavior. Soil Sci. Soc. Am. J. 47:85–91.

Molina, J.A.E., B. Nicolardot, S. Houot, R. Chaussod, and H.H. Cheng. 1994. Biologically active soil organics: A case of double identity. p. 169–178. *In* J.W. Doran et al. (ed.) Defining soil quality for a sustainable environment. SSSA Spec. Publ. 35. SSSA, Madison, WI.

Mulla, D.J., C.A. Perillo, and C.G. Cogger. 1996. A site-specific farm-scale GIS approach for reducing groundwater contamination by pesticides. J. Environ. Qual. 25:419–425.

Reganold, J.P. 1988. Comparison of soil properties as influenced by organic and conventional farming systems. Am. J. Alter. Agric. 3:144–155.

Romig, D.E., M.J. Garlynd, and R.F. Harris. 1996. Farmer-based assessment of soil quality: A soil health scorecard. p. 39–60. *In* J.W. Doran and A.J. Jones (ed.) Methods for assessing soil quality. SSSA Spec. Publ. 49. SSSA, Madison, WI.

Romig, D.E., M.J. Garlynd, R.F. Harris, and K. McSweeney. 1995. How farmers assess soil health and quality. J. Soil Water Conserv. 50:225–232.

Shaffer, M.J., A.D. Halvorson, and F.J. Pierce. 1991. Nitrate leaching and economic analysis package (NLEAP): Model description and application. p. 285–322. *In* R.F. Follett et al. (ed.) Managing nitrogen for groundwater quality and farm productivity. SSSA, Madison, WI.

Williams, J.R., and K.G. Renard. 1985. Assessments of soil erosion and crop productivity with process models (EPIC). p. 68–105. *In* R.F. Follett and B.A. Stewart (ed.) Soil erosion and crop productivity. ASA, CSSA, and SSSA, Madison, WI.

Wymore, A.W. 1993. Model-based systems engineering. An introduction to the mathematical theory of discrete systems and to the tricotyledon theory of system design. CRC Press, Boca Raton, FL.

5 On-Farm Assessment of Soil Quality and Health

Marianne Sarrantonio

Slippery Rock University
Slippery Rock, Pennsylvania

John W. Doran and Mark A. Liebig

USDA-ARS & University of Nebraska
Lincoln, Nebraska

Jonathan J. Halvorson

Washington State University
Pullman, Washington

For the concept of soil health or quality in agriculture to become more than an esoteric scientific notion, assessments of soil health must be made under the conditions of active, working farms. At a broad scale, implemented on carefully selected arrays of real farms, soil health assessment will aid researchers in evaluating the impact of existing or new farming practices on the cultivated soil and in identifying which soil characteristics are the most important determinants of soil health as related to soil productivity and environmental quality. At the individual farm level, methodologies for soil health assessment should enable farmers to not only evaluate the overall impact of a new farm management system on their soil, but also to identify potential soil problems in their fields, or to assess the extent of recognized soil constraints, ultimately with the goal of achieving the most profitable and environmentally sound long-term management system for fields and farms as a whole.

To include farmers as active participants in the quantitative assessment of soil health on their own farms, tools, and methodologies used by researchers must be adapted for application by farmers themselves. The tests should be simple to perform, use inexpensive equipment, and give rapid results. Additionally, the tests should measure soil characteristics that are meaningful to the farmers' understanding of soil and soil processes, and give results that are reliable, accurate within an acceptable range, and interpretable with minimal training. Tests that can be related to farmers' qualitative, or sensory, assessments of soil health, as described in Romig et al. (1996, this publication), will probably be of the greatest value. Because most farmers are unlikely to have sufficient time to devote to

---

Copyright © 1996 Soil Science Society of America, 677 S. Segoe Rd., Madison, WI 53711, USA. *Methods for Assessing Soil Quality,* SSSA Special Publication 49.

regular testing of their fields to assess soil quality, the types of tests described in this chapter may more likely be carried out by consultants and conservationists. Nevertheless, a concerted effort to make the tests farmer-friendly will probably result in a higher degree of farmer interest and participation in the assessment.

In the quest to define and characterize soil health, the challenges presented by testing real farm fields become quickly apparent: it is on the farm that the tools of agricultural science are perhaps both most relevant and most challenged. Fields on working farms add another level of variability to the inherent topographical and soil variability found on experiment stations: a history of unavoidable differential farm management, such as dead furrows or nonuniform manure or fertilizer spreading, as well as possible crop rotation divergences across one field at some point in the past. Farms rarely have the detailed soil mapping or climatic information that is common at experiment stations, and field histories are more likely to be anecdotal than systematically recorded. Given that on-farm testing for soil health assessment is likely to be based on a limited number of samples, care in developing a logical sampling scheme is very critical.

To obtain meaningful information from on-farm sampling for soil health indicators, it is necessary to clearly define questions that can be answered on a rather gross scale. For instance, it may be possible, over the course of 3 or 4 yr, to determine whether compost is having an overall positive effect on soil physical parameters, such as water infiltration rates, but nearly impossible to establish annual nitrate availability patterns following compost additions. Of primary interest to many farmers will be questions of how management changes will affect soil health, what the corresponding effects on plant growth will be, and how long it will take for such changes to take place. Farmers generally have a good sense of soil physical changes related to different management options, and perhaps a less confident sense of chemical or biological changes, as these changes are often less apparent. While the temptation will undoubtedly exist to concentrate on measuring soil parameters that will most predictably change with new management, assessors should take the time to run the full array of biological, chemical, and physical tests in order to emphasize the holistic nature of soil health. Unanticipated patterns that emerge in seemingly unrelated soil parameters can serve as important discovery mechanisms for researchers and farmers alike.

ON-FARM SOIL HEALTH TESTS

Tests that are appropriate for on-farm soil health assessment should be a subset of the screening parameters listed in Doran and Parkin (1996, this publication). Dr. John Doran (USDA-ARS, Univ. of Nebraska, Lincoln) has put together a prototype soil quality test kit (Cramer, 1994) that is currently being tested and adapted at 50 to 60 locations in the USA and several countries around the world, and may serve as a model for on-farm soil health assessment. The kit is used to measure water infiltration, bulk density, soil respiration at field moisture and at field capacity, soil water content, water holding capacity, water-filled pore space, soil temperature, soil pH, electrical conductivity, and soil nitrate. Preliminary results obtained with the test kit compare well to standard laboratory procedures that are more time consuming and costly (Liebig et al., 1996). Assessments conducted with the test kit are described in detail later in this chapter.

Site Characterization

The following minimum set of site characteristics should be included in any on-farm assessment of soil health:
- soil series
- soil texture
- signs of erosion
- description of past and present land and crop management, including notations on differences in crop growth across the field
- slope and topographical features
- geographic location of the field and sampling areas (longitude and latitude)
- climatic information such as precipitation and high and low temperature averages for each month (data from a county or watershed level will often be sufficient)
- location of ponds, creeks, wetlands or other environmentally fragile sites adjacent to the field

Sampling

When few samples can be taken, a completely randomized sampling scheme is not appropriate. The choice of where to sample most efficiently then depends largely on what questions are being asked. For example, if the overall effect of a new field management system is of primary interest, the field can be sampled in areas that most represent the average of the whole field. If management effects on problem areas, such as seasonally wet spots, are of concern, sampling can be concentrated in the problem areas with perhaps a few samples taken in the non-problem areas for comparison. The farmer's knowledge of the field variability will be a valuable resource in the informal grouping or concentration of sampling to test the desired aspects of the field. The total sample number needed to make a meaningful assessment will depend both on field variability and on the number of test variables (Cambardella et al., 1994). A minimum of three samples should be collected on any soil type and management combination, but a greater number will increase the reliability of the data and allow for statistical analysis. For tests that are performed on bulked soil samples, one option is to composite (or uniformly mix) eight or more cores taken from a wide sampling area, then remove as little as two samples for analysis from the composited soil sample. In fields with pronounced gradients of soil texture, soil moisture, slope or other factors, the choice of average or representative areas may be irrelevant. In such cases, ranges or fixed levels across the gradient may serve well as secondary test variables. In addition to testing the effects of cover cropped vs. non-cover cropped areas, then, the assessor may be grouping the sampling sites into top, mid, and bottom of a sloped area, as done for data collected at a North Dakota site, as presented later in the chapter.

At times it may make sense to test the effects of management on localized areas of the field. For instance, a switch from whole-field plowing to ridge tillage or permanent bed management will concentrate field traffic in fixed wheel lanes, leading to long-term wheel lane characteristics that are significantly different from the area between lanes (Doran & Linn, 1994). Thus it may be logical to sample wheel lanes separately from inter-wheel areas. In the ridge-till example, the comparison is not entirely straightforward, as you must decide what is the meaningful comparison—the ridge top, the ridge side, the nonwheel inter-ridge area, or all three. A transect across the four areas would give an interesting, albeit time-consuming, snap shot of soil changes that plant roots will experience as they explore the soil volume available to them.

Spatially recurrent patterns of variability such as fertilizer bands also will locally influence soil health indicators. Different approaches can be used to deal with such patterns. One approach would be to sample in a pattern that includes an atypical area, the furthest point from the atypical area, and possibly one point in between. This is a very thorough scheme that would yield interesting information, but may be too extensive for this type of evaluation. Another approach would be to sample from areas that represent an average of the soil volume. In the case of fertilizer bands this would represent a point one-fourth of the way between one fertilizer band and the next. Samples could be composited within each area to reduce the overall number of samples. This is another variation of the stratified composite sampling approach as discussed by Dick et al. (1996, this publication).

In fields where, for instance, residue cover varies, again the question you are asking will help determine the sampling scheme. If you want to know what extent residue cover affects soil characteristics, consider residue cover as a treatment. Sample from several high-residue areas and several low or no-residue areas. If you are interested in the affects of some other factor and variability in residue cover is a nuisance, sample from a variety of residue levels (high sampling number), or sample from areas that have an average amount of residue for that field (lower sampling number).

Measurements with the soil quality test kit are conducted on the surface 3 in. (7.6 cm) of soil. As the use of the kit is meant to be a screening procedure, evaluating the soil surface is appropriate to get a general idea of soil condition; however, to obtain information that is needed for management decisions it is necessary to sample to at least 1 ft (30.5 cm) with a soil core sampler. This represents the depth of soil generally affected by conventional tillage and fertilizer operations and is the zone of maximum rooting intensity for most crop plants. If care is taken in the collection of the core sample, it can be partitioned into different depth increments for separate analyses.

Scheduling soil health sampling for optimum timing is as important as establishing a meaningful spatial sampling pattern, particularly if the number of samples at a given location will be limited. Some of the screening tests proposed for on-farm soil health assessment lend themselves more to infrequent sampling than others. Since it is recommended to perform all the tests together, the assessor will need to determine the optimum sampling scheme for the suite of tests overall, based on the questions being asked and such factors as field operations and the weather, rather than concentrating on the ideal timing for individual tests. In general, the objective for overall soil health assessment will be to take an annual snapshot of the field soil to detect tangible long-term changes or to identify areas of immediate concern. In cases where specific research answers are sought and key soil parameters are being monitored closely, sampling schemes must be tailored to the questions being asked. Below is a brief discussion of the time frame for change of each soil health parameter tested with the kit and recommendations for timing and frequency of sampling. A summary of this information is presented in Table 5–1.

Soil pH and Electrical Conductivity

In most soils, these will change slowly, with the exception of localized changes in and near a fertilizer band or those related to applications of manure or liming materials. Yearly sampling is sufficient; time of year is not particularly important.

Bulk Density

Bulk density will obviously change most immediately following tillage-induced soil disturbance. Because bulk density is a moving target, it is best to sample at least twice during the year—once following tillage and once at the end of the cropping cycle, or prior to the next primary tillage operation; however, bulk density measurement is needed each time soil samples are taken for physical, chemical, and biological analyses to permit adjustment of results to a volumetric basis. This will allow for valid comparison of management effects (Doran & Parkin, 1996, this publication).

Table 5–1. Recommendations for on-farm sampling for soil health.

Test	Relative variability (spatial and temporal)	Recommended sampling frequency	Recommended timing
Soil pH	Low	Once per year	Anytime
Electrical conductivity	Low to moderate	Once per year; area where EC is high may require more frequent sampling	With soil pH measurement, but may change with inorganic N availability in areas with high EC
Bulk density	Moderate	Depends on tillage - at least twice/year in tilled systems	Soon after tillage, and at least once more in the season as soil settles or is compacted
Infiltration rate	High	With bulk density	With bulk density
Water-holding capacity	Low to moderate	At least once per year	With bulk density
Soil respiration rate	High	Often, depending on questions asked	After tillage; after residue addition; during warm months
Soil nitrate	Moderate to high	Depends on questions asked	Early summer for crop availability; fall for potential loss by leaching

Infiltration Rate

Infiltration will be most affected in the short-term by tillage and wheel traffic; however, it also will change in the medium-term by development of plant roots, earthworm burrows, and soil aggregation, and in the long-term by overall increases in stable organic matter. Because spatial variability of this measurement is often greater than whole-field changes, year-to-year analyses of infiltration rates will rarely show dramatic changes in similarly-managed soils (Peters et al., 1996). It can be, however, very useful for within-year comparisons of differently-managed soils, or of spatial patterns, such as wheel lane vs. nonwheel lane effects. Infiltration rates can be logically performed at the same time as bulk density measurements.

Water Holding Capacity

This will change slowly with changes in organic matter, but may change rapidly with tillage events that are vigorous enough to pulverize the soil. In general, this can be measured once a year, at any time the soil is not frozen, but also can be conveniently performed at the same time as bulk density measurements.

Soil Respiration

Microbial respiration peaks immediately following tillage or residue addition, and during warm, moist periods. If all three coincide, sampling at this time will amplify differences in soil management. Keep in mind that the differences are unlikely to be sustained throughout the year. Standardizing soil respiration levels for 25°C and 60% water-filled pore space as described in the methodologies section will help to overcome much of the weather-related and daily fluctuations in soil respiration related to changes in water content and soil temperature. Plant root respiration will increase with root growth and may overwhelm microbial respiration as plants begin to form significant ground cover.

Soil Nitrate

Soil nitrate levels will generally change significantly during the course of the year, and may change significantly during the course of a few days. Nitrate level will tend to be low (<5 ppm or mg/kg) in the early spring in humid regions, as excess nitrate will generally have been leached below the sampling zone. As soils warm in April and May, nitrate levels will often begin to increase as organic matter mineralization begins and nitrifiers become active. While presidedress nitrate tests may be quite useful to determine crop N needs in conventionally managed soils (Magdoff, 1991), addition of high-N plant residues, manures, or composts will often result in dynamic changes in nitrate levels for a month or more as the microbial population increases, decreases, and shifts in response to substrate availability (Sarrantonio & Scott, 1988). In such cases single point assays of soil nitrate in early summer may be unreliable. Soil nitrate levels measured during active crop growth must be interpreted with associated information on plant N uptake.

In order for a single annual assessment of soil health to be meaningful, it is most important to be consistent in timing from year to year. An early spring assessment in fields primarily cropped to summer annuals has the advantage of allowing time for management decision changes, and may occur at a time that will allow farmers to devote attention to the process. In early spring, however, soils are biologically less active and respiration and nitrate tests are unlikely to yield useful information. Tests run in June may best detect management differences in physical changes due to tillage, as well as in soil microbial respiration and nitrate, but may be problematic for farmer participation, and results also may be very weather dependent. Mid to late summer testing may adequately test average yearly effects of soil physical properties, but respiration and nitrate levels may be difficult to interpret due to plant growth. Assessments run in the fall can be useful in pinpointing inefficient N management and represent acceptable timing for most of the other tests. With more complex rotations involving winter annuals, spring grains, and perennials, soil health parameters can still be tested at a fixed time of year, but soil changes will be best measured against the recurring points in the full rotation [i.e., soil parameters in corn (*Zea mays* L.) after soybean [*Glycine max* (L.) Merr.] should be compared with the same parameters the last time corn followed soybeans in the rotation].

Making Comparisons

There is no absolute standard of soil health against which to judge an individual soil's status. Nevertheless, it is still important to have some basis for comparison to determine whether relative improvement or degradation has taken place. There are several options for comparison, each of which may be appropriate under different circumstances.

Comparing Different Management Schemes on the Same Farm

This option is the most useful when trying to determine whether a new management system is superior to an old one in terms of its effect on soil health. It requires little or no baseline data but does require the maintenance of side-by-side management of old and new schemes for many years.

Comparing the Same Field Over Time

This is a workable option when a new management scheme will be imposed on a field that was managed differently for many years. To be most useful, this option requires substantial baseline data before the new management scheme is initiated, preferably from more than 1 yr prior to the start of the new system. Collecting data from more than one field undergoing the new management scheme will increase the reliability of the information gathered. Spatial variability within a field can be minimized over time by sampling from fixed sites. Although it is difficult to maintain markers in a cultivated field, sites can be relocated yearly using geographic landmarks such as trees, or points along a fence.

Comparisons to Neighboring Farms

This approach is most useful to researchers, who may be able to perform sufficient paired-farm comparisons to draw reliable conclusions. The paired farms must have similar soil and site characteristics. Neighboring farms with similar economic operations but different management operations will be the most useful comparisons (e.g., conventional vs. organic cash grain operations). For individual farmers concerned with their own farms, a single paired comparison of this sort may be too confounded by management and inherent differences in site characteristics to be useful.

Comparing with Natural Ecosystem, or Undisturbed Areas

Although farm field conditions are unlikely to ever return entirely to their original undisturbed state, this comparison may provide some insights to natural limits, such as the highest organic matter level likely to be achieved under similar soil type and climatic combinations. The area for comparison should be nearby, of similar parent material and topography, and be under the type of ecosystem that the cultivated field was originally under, as best as can be determined. A native prairie or forest may serve as an excellent comparison in some cases, found in local parks or refuge areas, but an undisturbed fencerow also may be appropriate.

Sampling Scheme

The suite of measurements in the soil quality test kit developed by Doran uses a large cylinder (5 in. length by 6 in. diam.; 12.7 cm by 14.9 cm i.d.) cut from aluminum irrigation pipe as a focal point in each sampling area. The pipe is used to measure infiltration rate, respiration, field water holding capacity, water-filled pore space, and in some cases, soil bulk density. Where possible, tests results are expressed in units that are familiar to farmers.

To help standardize and facilitate interpretation of results, soil respiration should be measured before and after soil wetting at each site where infiltration is run. Preirrigation soil water content and bulk density measurements are obtained in areas adjacent to the irrigation rings using a small aluminum sampling tube (5 in. length by 3 in. diam.; 12.7 cm by 7.3 cm i.d.). If the soil is too hard or stony to use the tube, a hand trowel can be used to excavate a hole 3 in. (7.6 cm) deep, taking care to not compact the soil surrounding the hole. The volume of soil removed is then determined by lining the hole with plastic wrap and measuring the volume of water it will contain using the gas sampling syringe to add and measure the water. The soil in the tube, or that removed from the hole, is placed in a bag and mixed. The soil in the bag is weighed and a subsample is removed for determination of soil water content by drying in a microwave. Soil bulk density can then be calculated by determining the dry weight of soil occupying the sample volume. A second subsample is removed for analysis of electrical conductivity, soil pH, and soil NO_3–N as described in detail later. If desired, the remainder of the sample can be sent to a laboratory for analysis of additional soil quality indicators, such as soil texture, total organic C and N, microbial biomass, and potentially mineralizable N.

Since only the surface 3 in. (7.6 cm) of soil for a given management system is evaluated in a soil quality screening test, it is wise to take a composite sample of six to eight cylindrical soil cores to 1-ft depth (30.5-cm) with a soil core sampler. These cores are randomly sampled nearby the infiltration ring, within specific areas of each management situation and soil type that is being evaluated. These cores can be sectioned into 0- to 3-, 3- to 6-, and 6- to 12-in. (0- to 7.6-, 7.6- to 15.2-, and 15.2- to 30.5-cm) depth increments and saved in sealed plastic bags for later analysis of water content, soil texture, soil pH, electrical conductivity, soil NO_3–N, total organic C and N, microbial biomass, and potentially mineralizable N. If carefully taken and sectioned, these cores also can be used to estimate soil bulk density using the calculated volume of soil contained in each core, as determined by the inside diameter of the auger coring tip, and the oven-dry weight of soil sampled.

Recommended Schedule for Efficient Sampling

Phase 1

Install infiltration rings to a 3 in. (7.6 cm) depth in the field and measure the initial soil respiration before wetting the soil. For two rings, it will take about 40 min to complete this phase.

Phase 2

Next, take the infiltration measurement for the first 1 in. (2.54 cm) of water. This requires about 10 to 15 min for two rings, but time for infiltration may vary from less than 1 min to more than 1 h. Irrigation with a second inch of water can be done as soon as the first 1 in. has infiltrated. Need for this second irrigation is determined by the soil water content at time of first irrigation. If soil was already wet and became saturated after the first irrigation, the second irrigation is usually unnecessary unless a ponded infiltration rate is needed. Use your best judgment.

Phase 3

Prior to the time of the respiration measurement or after irrigation, soil samples for other soil quality measures are taken using the small aluminum tube or a trowel to sample soil adjacent to the infiltration rings. During this time, soil cores from a larger area can be taken to a depth of 1 ft (30.5 cm) and composited by selected depth increments, if desired. A deep core (40 to 60 in.; 1.0 to 1.5 m) also can be taken with a soil auger during this time to identify depth of topsoil, depth of rooting, soil textural changes, evidence of compacted layers, and presence or absence of carbonates.

Phase 4

A post-irrigation respiration measurement is made at least 6 h, but preferably 16 to 24 h, after soil wetting. During this time, soil in the infiltration ring is protected from rain or evaporation losses by capping the ring loosely with a respiration chamber lid. After the respiration measurement, soil within the infiltration ring is sampled for water content, bulk density, and field water-holding capacity (which is equal to the soil volumetric water content 16 to 24 h after the soil is saturated with water).

Methodologies for Soil Quality and Health Tests

Selection of Sampling Area

As reviewed above, selection of a sampling area depends on what questions are being asked. It is important to choose the most representative area within the treatment of interest. If sampling is conducted within row crops, it is advisable to set up comparisons between row and interrow locations. Soil conditions between locations are likely to vary considerably, and are important to quantify. If not sampling within row crops, an 1/8 in. (0.3 cm) diameter brazing rod can be used to probe the surface soil to identify compacted areas that should be assessed separately.

Soil Respiration

A detailed diagram illustrating use of equipment described below is given by Parkin et al. (1996, this publication).

1. Install infiltration ring (beveled edge down) to a 3 in. (7.6 cm) soil depth using a hardwood block and hand sledge. If the ring is installed correctly, there should be 2 in. (5.1 cm) from the soil surface to the top of the ring around its entire circumference.
2. Cap the infiltration ring with a lid possessing septums for gas sampling and soil temperature determination. Record time of capping.
3. At 25 min after capping, insert a soil thermometer through the center septum on the lid to a depth of approximately 1.5 in. (4 cm). Prepare a 0.1% CO_2 Dräger gas detection tube for sampling by breaking off each tube tip using the hole drilled in the handle of the 140 mL syringe (Drägerwerk, AG, Lübeck, Germany)[1]. Each end of the detection tube should then be connected to a piece of latex tubing. With the arrow on the side of the detector tube pointing towards the syringe, connect the latex tubing to the syringe. A needle (18 gauge) should then be connected to the other piece of latex tubing.
4. At 29 min after capping, insert a needle (18 gauge) into one of the septums on the lid. This will keep pressure within the chamber headspace constant as the sample is taken.
5. At 30 min after capping, insert the needle connected to the syringe assembly into the remaining septum on the lid. Draw out 100 mL of headspace sample during a 15-s time period through the gas detection tube using the gas sampling syringe for suction.

---

[1] Mention of commercial products and organizations in this chapter is solely to provide specific information. It does not constitute endorsement over other products and organizations not mentioned.

6. Read CO_2 as % by volume off the $N =1$ scale (100 mL) of the Dräger tube as indicated by the furthest advance of a violet color change down the tube. If the advancing color line is not parallel with the gradation lines, guesstimate its average position. Record soil temperature and % CO_2 at time of sampling.
7. A postirrigation measurement of soil respiration should be repeated at least 6 h, but preferably 16 to 24 h after infiltration measurements.

Water Infiltration

1. Gently firm the soil around inside wall of infiltration ring if the surface was disturbed during installation or by the removal of the soil thermometer.
2. Line the inside of the ring with plastic wrap so that the soil is completely covered.
3. Using a calibrated pint bottle, add 444 mL of distilled water to the plastic-lined surface of the ring. This volume of water results in a 1 in. (2.54 cm) water depth in the ring.
4. Measure and record the distance from top of the water surface to the top edge of ring if you use this ring for measuring soil bulk density or soil respiration.
5. Gently pull the plastic wrap out of the ring and record the time it takes for the 1 in. (2.54 cm) of water to infiltrate the soil (end time is when the soil surface is just glistening).
6. If the soil was not saturated after adding the first 1 in. (2.54 cm) of water you should add a second 1 in. (2.54 cm) of water and record the time for infiltration.
7. Soil bulk density and water holding capacity can be measured by taking the wet weight of soil and water content of soil in the ring 16 to 24 h after the last wetting.

Soil Water Content and Soil Bulk Density

Pre-Irrigation

1. Select a sampling area adjacent to the infiltration ring.
2. Install the small aluminum sampling tube to a 3 in. (7.6 cm) soil depth. Measure the distance from the soil surface to lip of the tube [it should be 2 in. (5.1 cm) if properly installed].
3. Carefully lift the tube from the soil with a shovel or trowel. Remove any excess soil clinging to the outside of the tube with a gloved hand, and then cover the tube top with your hand and invert to enable trimming excess soil from bottom with a knife.
4. Remove the soil in the 3 in. (7.6-cm) segment of the tube into a sealed plastic bag and mix thoroughly.
5. Weigh the soil in the bag using a scale with a resolution of 0.1 g.
6. Remove a 20- to 30-g subsample from the bag for determination of soil water content. Place the subsample in a glass container and dry in a

microwave oven for two 4-min cycles at medium power (400 W). To facilitate efficient drying, open microwave door and vent humid atmosphere in oven between cycles.
7. Determine soil bulk density from the dry weight of soil occupying the sample volume (reviewed in Calculations Section).

Post-Irrigation

1. This sample is taken within the infiltration ring using the small aluminum sampling tube. Start by pushing the tube to a 3 in. (7.6 cm) soil depth in the ring. As before, measure the distance from the soil surface to lip of the tube as a check.
2. Lift the large ring out of the soil with a shovel or by digging a trough around the ring with a trowel and undercutting ½ to 1 in. (1.27 to 2.54 cm) below the ring. Pivot the ring on its side, cup one hand over the top of the tube to prevent soil loss, and push the tube up and out of the ring of soil. Invert and trim the excess soil flush with the tube bottom with a knife.
3. Remove the soil in the 3 in. (7.6 cm) segment of the tube and place in a sealed plastic bag.
4. Determine soil water content and bulk density as described above. If sampling was conducted 16 to 24 h after irrigation, soil volumetric water content is equal to the field water-holding capacity. (Note: Postirrigation water content and bulk density also can be determined using the large infiltration ring if desired).

Electrical Conductivity, Soil pH, and Soil Nitrate

1. The preirrigation sample for soil water content and soil bulk density also is used for determining electrical conductivity, soil pH, and soil nitrate. Start by taking a subsample of mixed soil using an (1/8 cup) 29.5 cm^3 measuring scoop. Collect the subsample so the scoop is full of soil and level across the top (gently tamp soil if necessary).
2. Weigh the soil in the scoop and then transfer to a 4 oz (120 mL) plastic vial.
3. Add a level scoop of distilled water to the vial. This will create a 1:1 soil–water mixture by volume.
4. Cap the vial and shake vigorously 25 times.
5. Let the soil–water mixture stand for 5 min.
6. Open the vial and measure electrical conductivity in topmost solution with a standardized EC pocket meter. Divide reading on display by 10 to get reading in dS/m.
7. *After* reading electrical conductivity, take soil pH reading in the topmost solution using a standardized pH pen. Rinse pocket meters with distilled water after use.
8. Filter soil–water mixture through a doubled Whatman no. 1 or 5 paper (11 and 2.5 µm particle retention, respectively) folded to nest in a plastic funnel (Whatman International Ltd., Maidstone, England).

9. Catch the filtered extract in a test tube or catsup cup.
10. Using an eyedropper, transfer 1 or 2 drops of the extract to an AquaChek Nitrate/Nitrite water quality test strip (Environmental Test Systems, Elkhart, IN). Compare nitrite test pad with color chart on AquaChek bottle after 30 s. After 60 s, compare nitrate test pad with color chart and record NO_3–N value for the extract as parts per million (milligrams per liter). Estimate value if color falls between two color patches. This value is the NO_3–N concentration in the extract that approximates the concentration in soil (ppm or micrograms per gram) since a 1:1 soil to water dilution (by volume) was used. A more precise method of calculating soil nitrate concentration is described in the Calculations Section. If the NO_3–N concentration exceeds the highest scale (50 mg/L), the sample can be diluted with distilled water using the eye dropper for measurement (i.e., one volume of 100 mg/L extract + one volume distilled water (mix thoroughly in catsup cup) will read 50 mg/L on test strip).

Calculations for Soil Quality and Health Tests

Soil Water Content

Soil Water Content (g/g) =

$$\frac{\text{(weight of moist soil} - \text{weight of oven dry soil)}}{\text{weight of oven dry soil}}$$

Soil Bulk Density

$$\text{Soil Bulk Density (g/cm}^3\text{)} = \frac{\text{oven dry weight of soil}}{\text{volume of soil}}$$

where,

volume of soil (cm^3) = 324 cm^3 for a 3-in. soil depth with small tube or 1333 cm^3 for 3-in. soil depth with infiltration ring.

$$\text{oven dry weight of soil (g)} = \frac{\text{weight of moist soil from field}}{(1 + \text{decimal soil water content})}$$

(Note: 1 mL = 1 cm^3)

Soil Water-Filled Pore Space (WFPS)

$$\text{WFPS (\%)} = \frac{\text{(volumetric water content} \times 100)}{\text{soil porosity}}$$

where,

volumetric water content (cm^3/cm^3) = (decimal soil water content × soil bulk density)

soil porosity (cm^3/cm^3) = (1 − (soil bulk density ÷ 2.65))

(Note: 1 g of water has a volume of 1 cm^3)

Soil Nitrate

Estimated

Soil NO_3–N (lb NO_3–N/acre) = [ppm (mg/L) extract NO_3–N]

× (cm depth of soil sampled ÷ 10) × soil bulk density × 0.89

Exact. A more accurate estimate for soil nitrate content can be obtained using the exact weight of dry soil contained in the extraction scoop as calculated from the soil water content. In this case,

Soil NO_3–N (lb NO_3–N/acre) = [exact ppm (mg/L) soil NO_3–N]

× (cm depth of soil sampled ÷ 10) × soil bulk density × 0.89
where,
  exact ppm (µg/g) soil NO_3–N =

$$\frac{[\text{ppm (mg/L) extract } NO_3\text{–N}] \times (\text{volume of water in extract and soil})}{\text{dry weight of soil extracted}}$$

dry weight of soil extracted (g) = (weight of soil in scoop) ÷ (1 + decimal soil water content)
volume of water (cm^3) = 29.5 + (dry weight of soil × decimal soil water content)
(Note: 1 in. = 2.54 cm)

Soil Respiration

Soil Respiration (lb CO_2–C/acre/d) = PF × TF × (%CO_2 – 0.035) × 116.4

where,
  Pressure factor (PF) = (raw barometric pressure, inches of mercury) ÷ (29.9)
(Note: This adjustment factor is only necessary at altitudes where elevation >3000 ft.)
Temperature factor (TF) = (soil temperature in °C + 273) ÷ 273

Soil Temperature and Water Status Adjustments

Soil respiration is sensitive to soil temperature and water status. Between 15 and 35°C, biological activity generally doubles with each 10°C rise in temperature. Therefore respiration rates measured at different temperatures should be adjusted to a common temperature, for example 25°C, before being compared. To do this, use the following relationships:

For measured soil temperatures (TM) of 15 to 25°C:
  Respiration at 25°C = (respiration at TM) × (1 + [(25 – TM) ÷ 10])
For measured soil temperatures (TM) of 25 to 35°C:
  Respiration at 25°C = (respiration at TM) ÷ (1 + [(TM – 25) ÷ 10])

Likewise, biological activity in soil will increase with soil water content until a point is reached where higher water content interferes with soil aeration and the activity of O_2 requiring organisms (both plants and microorganisms). For an illustration of this relationship see Fig. 14–2 (Parkin et al., 1996, this publication). Past research indicates that the activity of O_2 requiring microorganisms (majority of soil organisms) increases in a consistent manner with increasing soil water content above 30% WFPS and reaches a maximum at 60% WFPS (Linn & Doran, 1984; Doran et al., 1990). Further increases in water content above 60% WFPS result in a decline in measurable respiratory activity to 40% of maximum at 80% WFPS. To compare readings at different water contents, calculate the non-limited respiration rate at 60% WFPS using the following relationships:

For measured WFPS between 30 and 60%:
Respiration at 60% WFPS = measured respiration rate × (60 ÷ measured %WFPS)

For measured WFPS between 60 and 80%:
Respiration at 60% WFPS = measured respiration rate ÷ [(80 − %WFPS) × 0.03] + 0.4 or
Respiration at 60% WFPS = measured respiration rate ÷ [2.8 − (0.03 × %WFPS)]

CASE STUDIES

Two research sites were chosen on farmer-operated fields in Stutsman County, North Dakota for evaluation with the soil quality test kit on 21 to 23 June 1994. Sites were selected so that sampling could be conducted within one soil map unit between farms under organic and conventional management, as well as on land that was under native grass and enrolled in the Conservation Reserve Program (CRP). Details on site location, plot establishment, and sampling protocol are reviewed elsewhere (Liebig et al., 1996).

Soil Bulk Density, Water Holding Capacity, and Water Infiltration

Bulk density of the surface 3 in. (0 to 7.6 cm) was substantially lower in native grass as compared with organic and conventional management at Site 1 (Table 5–2). Bulk density did not vary among management systems at Site 2 (1.10 to 1.21 g/cm^3). Water holding capacity was influenced by management system at both sites. Water holding capacity for the footslope position in native grass was 0.09 cm^3/cm^3 greater (29%) than that observed in organic and conventional management at Site 1. At Site 2, water holding capacity averaged 0.12 cm^3/cm^3 greater (37%) in organic management as compared to conventional management and CRP land. On average, water infiltration rates for the first inch (2.54 cm) of water were under 3 min for all management systems except for organic management at Site 2. Similarly, infiltration rates for the second inch of water averaged <30 min for all treatments except for organic management at Site 2, where an additional inch of water took on average >111 min to infiltrate. Slow infiltration

Table 5–2. Effect of management system on selected soil quality indicators for top 3 in. (7.6 cm) as measured by the soil quality test kit.

Treatment	Landscape position/row location	Pre-irrigation											Post-irrigation			
		Bulk density	Respiration			EC	Soil pH	Soil NO$_3$–N		Infiltration time		WHC	Bulk density	Respiration		
			Soil temperature	WFPS	CO$_2$–C			Measured	Estimated†	1st	2nd			Soil temperature	WFPS	CO$_2$–C
		g/cm³	°C	%	lb/ac/d	dS/m		lb/ac		min/in.		cm³/cm³	g/cm³	°C	%	lb/ac/d
								Site 1								
Organic	Summit	1.09	31	19	12(15)‡	0.2	7.7	3(4)§	5	1.8	19.5	0.29	1.16	22	52	37(56)
Organic	Shoulder	1.07	33	17	10(11)	0.2	7.6	4(5)	5	1.0	6.3	0.25	1.08	22	41	23(44)
Organic	Footslope	0.95	32	18	40(47)	0.2	6.9	5(6)	5	1.8	17.5	0.32	1.06	22	52	23(35)
Organic Mean		1.04¶	32	18	21(24)	0.2	7.4	4(5)	5	1.6	14	0.29	1.10	22	48	28(45)
CV%#		7	3	6	81(81)	0	6	25(20)	0	31	50	12	5	0	13	29(23)
Conv.	Summit	1.11	30	28	30(40)	0.2	7.7	14(15)	5	2.3	30.0	0.29	1.43	23	63	58(76)
Conv.	Shoulder	1.24	32	23	11(13)	0.6	7.6	49(50)	53	4.3	32.0	0.30	1.27	23	58	29(36)
Conv.	Footslope	1.18	30	17	25(33)	0.6	6.0	43(50)	50	1.5	22.5	0.30	1.08	23	51	45(64)
Mean		1.18	31	23	22(29)	0.5	7.1	35(38)	36	2.7	28	0.30	1.26	23	57	44(59)
CV%		6	4	24	45(49)	49	13	53(53)	75	52	18	2	14	0	11	33(35)
Native	Summit	0.54	18	11	24(82)	0.0	6.8	2(2)	0	1.7	2.8	0.28	0.84	17	41	37(97)
Native	Shoulder	0.67	17	14	46(166)	0.1	7.6	1(2)	0	1.8	4.3	0.21	0.74	17	30	39(140)
Native	Footslope	0.61	16	51	51(114)	0.5	5.9	17(11)	20	3.8	34.0	0.38	0.61	17	49	40(88)
Mean		0.61	17	25	40(121)	0.2	6.8	7(5)	7	2.4	14	0.29	0.73	17	40	39(108)
CV%		11	6	88	36(35)	132	13	134(104)	173	49	129	29	16	0	24	4(26)

		Site 1							Site 2							
Organic	Row	1.21	25	34	60(106)‡	0.2	6.5	16(15)§	6	3.5	19.0	0.36	1.17	32	65	110(76)
Organic	Notrack	1.08	25	25	31(62)	0.3	6.2	20(20)	15	2.0	127.0	0.44	1.09	30	74	75(86)
Organic	Track	1.32	25	52	31(36)	0.3	6.1	26(20)	19	167.0	>167.0	0.44	1.22	30	81	61(102)
Weighted mean		1.14¶	25	30	37(67)	0.3	6.2	20(19)	14	24.2	>111	0.42	1.12	30	73	80(86)
CV%#		11	0	45	46(53)	21	3	25(15)	49	391	69	11	6	4	11	32(15)
Conv.	Row	1.18	24	26	37(81)	0.9	6.5	88(90)	84	1.0	5.0	0.32	1.04	26	54	32(32)
Conv.	Notrack	1.17	25	28	37(74)	0.9	5.3	85(90)	83	1.8	8.0	0.30	1.15	28	53	41(36)
Conv.	Track	1.27	25	33	31(56)	0.8	5.4	96(90)	78	7.0	36.0	0.38	1.20	27	70	45(54)
Weighted mean		1.18	24	27	37(76)	0.9	5.9	87(90)	83	1.8	8	0.31	1.10	27	55	37(35)
CV%		5	2	13	9(17)	6	11	7(0)	4	185	205	13	7	4	17	18(33)
CRP	Row	1.25	22	28	37(96)	0.1	6.4	1(1)	0	1.5	29.0	0.27	1.13	30	48	82(68)
CRP	Notrack	1.23	22	34	37(85)	0.1	6.9	2(2)	0	0.7	2.0	0.29	1.19	28	53	77(67)
CRP	Track	1.15	23	24	50(120)	0.1	6.3	1(1)	0	2.0	24.0	0.33	1.22	29	60	70(50)
Mean		1.21	22	29	41(100)	0.1	6.5	1(1)	0	1.4	18	0.30	1.18	29	54	76(62)
CV%		4	3	18	18(18)	0	5	43(43)	0	48	78	10	4	3	11	8(16)

† Estimated soil nitrate (ppm) = [EC − (background)] × 140 = (EC − 0.15) × 140

‡ Value in parenthesis is soil respiration calculated at 25°C and 60% WFPS.

§ Value in parentheses is extract NO_3-N (ppm).

¶ Landscape positions at Site 1 weighted evenly when calculating means. Means for organic system at Site 2 weighted by proportion of row location in a 30 ft tractor pass with 30 in. rows (row width = 6 in., interrow width = 24 in., two wheel tracked interrows per pass). Means for conventional system at Site 2 weighted by proportion or row location in a 30 ft tractor pass with 8 in. rows (both row and interrow width = 4 in, six wheel tracked interrows per pass).

%Coefficient of variation = (standard deviation ÷ mean) × 100

for this management system was apparently related to soil disturbance and compaction caused by late planting of the field crop in wet conditions. Despite the low infiltration rates for this treatment, they may not represent a serious potential for erosion and runoff in this climatic zone, as the expected frequency of a 1 in. (2.54 cm) rainfall of 30 min duration is 2 to 5 yr and for a 2 in. (5.08 cm) rainfall of 2-h duration is 5 to 10 yr (Hershfield, 1961). However, the extremely slow infiltration rate in the wheel track position with organic management is cause for concern as runoff could occur in this location and result in channeling and soil loss by erosion.

Soil pH, Electrical Conductivity, and Soil Nitrate

Soil pH in the surface 3 in. (7.6 cm) did not differ among management systems at both sites (Table 5–2); however, soil pH under conventional management at Site 2 was quite low (≤5.4) in wheel track and nonwheel track locations, which may have been caused by acidification associated with nitrification of ammoniacal fertilizer N. As discussed by Smith and Doran (1996, this publication), such acidification may be an indicator of inefficient N use and nitrate loss from surface soils by leaching (Bouman, 1995). Electrical conductivity differed among management systems at both sites. Lowest electrical conductivity values (0.1 to 0.3 dS/m) were observed under organic management, native grass, and CRP land, while highest values (0.5 to 0.9 dS/m) were found under conventional management. Soil nitrate was greatly influenced by management system. At Site 1, soil nitrate in the surface 3 in. (7.6 cm) under conventional management was approximately 30 lb/acre (34 kg/ha) greater than that observed under organic management and native grass. At Site 2, the disparity in soil nitrate levels among management systems was even greater, with values under conventional management more than 60 lb/acre (67 kg/ha) greater than that found under organic management and CRP land. This observation, when coupled with the short infiltration times for this treatment, reflects a high potential for nitrate leaching to lower soil horizons under normal rainfall conditions. Soil nitrate levels as estimated from soil electrical conductivity closely approximated measured soil nitrate levels ($R^2 = 0.97$). In all management systems, average estimated soil nitrate values were within 6 lb/acre (7 kg /ha) of measured values, showing the utility of using electrical conductivity as a front-line indicator of soil nitrate concentration (reviewed in Smith & Doran, 1996, this publication).

Soil Respiration

Under water-limiting conditions, soil respiration under native grass was approximately two-fold greater than that observed under organic and conventional management at Site 1 (Table 5–2). After wetting by irrigation, differences in soil respiration among management systems were not present due largely to low soil temperature and WFPS under native grass. Soil respiration calculated at 25°C and 60% WFPS was two- to three-fold greater under native grass as compared with organic and conventional management before and after irrigation. At Site 2, soil respiration before irrigation did not differ among management systems. After irrigation, however, soil respiration under organic management and

CRP land was more than two-fold greater than that observed under conventional management, reflecting greater microbial responsiveness to water addition in the former two management systems. Calculated values for soil respiration at 25°C and 60% WFPS did not follow a clear trend before and after irrigation among management systems, as would be expected if water status and temperature were the major factors determining soil respiration rates. Soil respiration was greatest under CRP land and lowest under organic management before irrigation, and greatest under organic management and lowest under conventional management after irrigation. The lack of responsiveness of respiration to a more optimal soil water status in the conventional management system may indicate a lower level of soil quality due to excessive tillage and chemical inputs.

Supplementary Indicators

Supplementary indicators were evaluated in the laboratory from the surface 12 in (30.5 cm) of each treatment (Table 5–3). Bulk density under conventional management at Site 1 was 0.18 g/cm^3 and 0.34 g/cm^3 greater than that found in organic management and native grass, respectively. At Site 2, bulk density in organic management averaged 0.10 g/cm^3 greater than that under conventional management and CRP land. Electrical conductivity values tended to be quite low (≤0.3 dS/m) in all treatments at both sites. Soil pH values were generally greater than those found with the test kit due to the likely inclusion of carbonate from the 3- to 12-in. (7.6- to 30.5-cm) depth. Trends in soil pH among treatments, however, were similar between laboratory and test kit evaluations. At both sites, soil nitrate levels were considerably greater in the conventional management system as compared with other forms of management. This was especially true at Site 2, where the conventional system had 107 lb/acre (119 kg/ha) and 170 lb/acre (190 kg/ha) more soil nitrate in the surface 12 in. (30.5 cm) than organic management and native grass, respectively. Mineralizable N (anaerobic incubation) was greatest under grass-based management systems and lowest under conventional management at both sites. There was a greater proportion of mineralizable N as compared with soil nitrate in organic management. This is important, for it may reflect greater N use efficiency as compared with the conventional system, where much of the N was in the form of nitrate. Soil nitrate in excess of crop requirements could easily be lost to the environment due to leaching or denitrification. Total and organic C, total N, microbial biomass, and the microbial quotient at Site 1 were greatest in native grass and lowest in conventional management. At Site 2, total and organic C, total N, and microbial biomass levels were similar in organic and conventional management systems and lowest in CRP land. The microbial quotient did not differ among management systems at Site 2.

SUMMARY

Differences in soil quality indicators among management systems indicated two distinct trends. Potential nutrient loss was greatest under conventional management among all management systems evaluated. This was reflected by (i)

Table 5–3. Effect of management system on selected soil quality indicators for top 12 inches (30.5 cm) from composited samples.

Treatment	Landscape position	Bulk density	Soil water content	EC	Soil pH	Soil NO$_3$-N	Soil NH$_4$-N	Mineralizable N (anaerobic)	Total C	Total† organic C	Total N	Microbial biomass C	Microbial‡ biomass N	Microbial§ quotient
		g/cm^3	g/g	dS/m					lb/ac					%
								Site 1						
Organic	Summit	1.36	0.11	0.1	7.5	17	1	125	69 359	53 131	4 788	1 285	151	2.4
Organic	Shoulder	1.26	0.11	0.1	7.7	19	2	102	71 376	61 539	5 111	1 870	220	3.0
Organic	Footslope	1.25	0.13	0.1	6.9	33	2	142	88 047	87 787	6 861	2 470	290	2.8
Mean		1.29	0.12	0.1	7.4	23	2	123	76 261	67 486	5 587	1 875	220	2.8
CV%¶		5	10	0	6	38	35	16	13	27	20	32	32	11
Conventional	Summit	1.45	0.15	0.2	7.9	66	1	92	95 753	42 639	4 034	854	100	2.0
Conventional	Shoulder	1.41	0.14	0.1	7.8	97	0	111	70 463	44 745	4 204	735	87	1.6
Conventional	Footslope	1.56	0.15	0.1	6.8	67	1	72	62 164	62 093	5 021	1 306	154	2.1
Mean		1.47	0.15	0.1	7.5	77	1	92	76 127	49 826	4 420	965	114	1.9
CV%		5	4	43	8	23	0	21	23	21	12	31	31	13
Native	Summit	1.10	0.10	0.0	7.3	10	0	187	101 709	92 560	7 040	1 404#	165	5.2
Native	Shoulder	1.13	0.14	0.0	7.6	14	3	155	82 837	81 592	6 644	1 901	224	2.3
Native	Footslope	1.17	0.34	0.5	6.7	23	2	155	111 093	110 933	8 667	4 707	554	4.2
Mean		1.13	0.19	0.2	7.2	16	2	166	98 546	95 028	7 450	2 671	314	3.9
CV%		3	67	173	6	42	92	11	15	16	14	67	67	37

ON-FARM ASSESSMENT

Site 2

	Replication												
Organic	1	1.48	0.19	0.2	6.9	68	0	101	82 141	79 676	6 884	1 328	156	1.7
Organic	2	1.50	0.21	0.2	6.9	70	0	67	85 232	83 005	7 441	1 544	182	1.9
Organic	3	1.50	0.22	0.2	7.0	60	1	90	89 371	88 191	7 649	1 193	140	1.4
Mean		1.49	0.21	0.2	6.9	66	0	86	85 581	83 624	7 325	1 355	159	1.6
CV%¶		1	7	0	1	8	0	20	4	5	5	13	13	16
Conventional	1	1.41	0.14	0.2	7.2	189	1	59	77 394	69 660	6 060	1 546	182	2.2
Conventional	2	1.36	0.15	0.3	6.3	147	6	48	75 832	75 646	6 787	1 103	130	1.5
Conventional	3	1.43	0.15	0.3	6.3	182	1	54	76 973	76 592	7 007	1 618	190	2.1
Mean		1.40	0.15	0.3	6.6	173	3	54	76 733	73 966	6 618	1 422	167	1.9
CV%		3	4	22	8	13	108	10	1	5	7	20	20	21
CRP	1	1.38	0.10	0.0	6.5	4	0	104	56 977	56 968	4 705	1 259	148	2.2
CRP	2	1.39	0.12	0.1	7.6	4	0	99	57 951	53 890	4 734	322††	38	2.1
CRP	3	1.40	0.11	0.1	7.1	2	0	104	68 189	67 847	5 985	544‡‡	64	1.2
Mean		1.39	0.11	0.1	7.1	3	0	102	61 039	59 568	5 141	708	83	1.8
CV%		1	9	87	8	35	0	3	10	12	14	69	69	30

† Total Organic Carbon = Total Carbon − Carbonates
‡ Microbial biomass nitrogen calculated from ratio (8.5 Biomass Carbon: 1 Biomass Nitrogen).
§ Microbial Quotient = (Microbial Biomass C/Total Organic C) × 100
¶ %Coefficient of variation = (standard deviation ÷ mean) × 100
\# Microbial biomass C and N and microbial quotient refer to 0 to 7.5 cm depth only.
†† Microbial biomass C and N and microbial quotient refer to 7.5 to 15 cm depth only.
‡‡ Microbial biomass C and N quotient refer to 0 to 15 cm depth only.

soil loss from upland landscape positions as indicated by a greater amount of soil carbonate in the surface 12 in. (30.5 cm); (Site 1, exclusively), and (ii) soil nitrate levels in excess of crop needs that could be lost by leaching or denitrification (Site 2, primarily). This *leakiness* indicated a greater likelihood of non-point source pollution from the conventional system. Associated with these nutrient losses was a reduction in nutrient cycling efficiency in the conventional system. Conventional management possessed the highest levels of soil nitrate and lowest levels of anaerobic (mineralizable) N at both sites. This was in contrast to organic management, where soil nitrate levels were relatively low, but mineralizable N levels (anaerobic incubation) were probably high enough to meet crop needs.

From a methodological standpoint, results from this study indicate the utility of in-field tests to discriminate among management-induced differences in soil quality indicators. Results are strengthened by the fact that trends among management systems were apparent even though landscape positions and row locations were used for replicates. This permitted evaluation of the relation of management systems to one another, and provided information on landscape and interrow variation within each management system. This study also demonstrated the utility of supplementary indicators from the surface 12 in (30.5 cm) of treatments to provide useful information regarding soil condition.

REFERENCES

Bouman, O.T., D. Curtin, C.A. Campbell, V.O. Biederbeck, and H. Ukrainetz. 1995. Soil acidification from long-term use of anhydrous ammonia and urea. Soil Sci. Soc. Am. J. 59:1488–1494.

Cambardella, C.A., T.B. Moorman, J.M. Novak, T.B. Parkin, D.L. Karlen, R.F. Turco, and A.E. Konopka. 1994. Field-scale variability of soil properties in central Iowa soils. Soil Sci. Soc. Am. J. 58:1501–1510.

Dick, R.P., D.R. Thomas, and J.J. Halvorson. 1996. Standardized methods, sampling, and sample pretreatment. p. 107–121. *In* J.W. Doran and A. J. Jones (ed.) Methods for assessing soil quality. SSSA Spec. Publ. 49. SSSA, Madison, WI.

Doran, J.W., and D.M. Linn. 1994. Microbial ecology of conservation management systems. p. 1–27. *In* J.L. Hatfield and B.A. Stewart (ed.). Soil biology effects on soil quality. Advances in Soil Science. Lewis Publ., Boca Raton, FL.

Doran, J.W., L.N. Mielke, and J.F. Power. 1990. Microbial activity as regulated by soil water-filled pore space. p. 94–99 *In* Trans. Int. Congr. Soil Sci. 14th, Kyoto, Japan. 12–18 Aug. 1990. ISSS, Wageningen, the Netherlands.

Doran, J.W., and T.B. Parkin. 1996. Quantitative indicators of soil quality: A minimum data set. p. 25–37. *In* J.W. Doran and A. J. Jones (ed.) Methods for assessing soil quality. SSSA Spec. Publ. 49. SSSA, Madison, WI.

Hershfield, D.M. 1961. Rainfall frequency atlas of the United States. Weather Bureau Tech. Pap. no. 40. U.S. Gov. Print. Office, Washington, DC.

Liebig, M.A., J.W. Doran, and J.C. Gardner. 1996. Evaluation of a field test kit for measuring selected soil quality indicators. Agron. J. 88:683–686.

Linn, D.M., and J.W. Doran. 1984. Effect of water-filled pore space on carbon dioxide and nitrous oxide production in tilled and nontilled soils. Soil Sci. Soc. Am. J. 48:1267–1272.

Magdoff, F.R. 1991. Understanding the Magdoff pre-sidedress nitrate test for corn. J. Prod. Agric. 4:297–305.

Parkin, T.B., J.W. Doran, and E. Franco-Vizcaino. 1996. Field and laboratory tests of soil respiration. p. 231–245. *In* J.W. Doran and A. J. Jones (ed.) Methods for assessing soil quality. SSSA Spec. Publ. 49. SSSA, Madison, WI.

Peters, S.E., M.M. Wander, L.S. Saporito, G.H. Harris, and D.B. Friedman. 1996. Management impacts on soil organic matter and related soil properties in a long term farming systems trial, 1981–1991. *In* E.A. Paul et al. (ed.) Soil organic matter in temperate agroecosystems: Long term experiments in North America. Lewis Publ., Boca Raton, FL.

Romig, D.E., M.J. Garylynd, and R.F. Harris. 1996. Farmer-based assessment of soil quality: A soil health scorecard. p. 39–60. *In* J.W. Doran and A. J. Jones (ed.) Methods for assessing soil quality. SSSA Spec. Publ. 49. SSSA, Madison, WI.

Sarrantonio, M., and T.W. Scott. 1988. Tillage effects on availability of nitrogen to corn following a winter green manure crop. Soil Sci. Soc. Am. J. 52:1661–1668.

Smith, J.L. and J.W. Doran. 1996. Measurement and use of pH and electrical conductivity for soil quality analysis. p. 169–185. *In* J.W. Doran and A. J. Jones (ed.) Methods for assessing soil quality. SSSA Spec. Publ. 49. SSSA, Madison, WI.

6 Standardized Methods, Sampling, and Sample Pretreatment

Richard P. Dick and David R. Thomas
Oregon State University
Corvallis, Oregon

Jonathan J. Halvorson
USDA-ARS and Washington State University
Pullman, Washington

Soil is a medium with properties that vary widely, both spatially and temporally. Because soil ecosystems are complex, understanding soil quality requires integration of many kinds of data over a broad range of spatial and temporal scales. Soil quality sampling programs must, therefore, be carefully planned. With proper design and implementation, sampling and pretreatment protocols will provide accurate, reliable estimates of soil quality indicators as well as characterization of spatial and temporal variability. The objectives of a soil quality sampling program include collection of data from sampling units that are of a size, shape, and orientation that result in an efficient tradeoff between the gain of useful information and cost of sampling. Further, data must be collected so that the resultant soil quality index is useful for real world applications. These objectives influence both the choice of analyses and the sampling plan.

This chapter will provide some general guidelines on soil sampling and statistical considerations. For more in depth discussions of these topics, the reader is referred to Kempthorne and Allmaras (1986), Petersen and Calvin (1986), Bates (1993), Wollum (1994), Halvorson et al. (1997), and to Crépin and Johnson (1993) for details of sampling environmentally contaminated soils.

STANDARDIZED METHODS FOR MEASURING SOIL PROPERTIES

Prior to collecting soil samples it is important to determine which analyses will be done and if possible use methods that have been standardized and widely reported in the literature. Traditional procedures for routine soil testing (e.g., nutrient availability, cation-exchange capacity, base saturation, pH) are generally standardized and calibrated, and most private or public soil testing laboratories would use the same analytical protocol for these procedures; however, many other methods that are not routinely run by soil testing laboratories, particularly biological measurements, are not likely to have widely-recognized standardized procedures.

Development of standardized soil quality data bases across a wide range of soil types and soil management systems is needed to determine soil quality thresholds or at least provide information on the variability and range of values that can be expected. This could provide a basis for future soil quality interpretations. A major goal of this book is to provide a set of procedures so that all investigators follow the same analytical protocols for key soil properties.

VARIATION OF SOIL PROPERTIES

Vertical and Horizontal Spatial Variability of Soil Properties

Soil is characterized as a diverse set of chemical, physical, and biological properties that have both vertical and horizontal dimensions. Soil classification units have been delineated based on morphology that resulted from long-term soil genesis and development processes. These units or soil types (soil series) have some degree of homogeneity both in horizontal and vertical dimensions. On the landscape, there usually is a gradual transition from one soil series to another and soil surveys may not have the resolution to identify the small changes in soil types at a sampling site (Fig. 6–1).

Vertical morphology is delineated by soil horizons that are layers of soil approximately parallel to the soil surface with characteristics produced by soil-forming processes. Properties that characterize soil horizons are color, texture (distribution of clay, silt, and sand size fractions), physical structure, cutans (concentration of soil constituents, e.g., clay coatings, Fe stains), concretions, voids, pH, boundary characteristics, and horizon continuity (Buol et al., 1973). Although there are many factors involved in classifying soils, the basis for the USDA taxonomic system (Soil Survey Staff, 1975) is the presence or absence of certain diagnostic soil horizons (Fig. 6–1). As a general rule, soil organic matter and biological activity decrease with depth.

Horizontal boundaries between classification units generally are not distinct and transitions are gradual between soil series. Soils tend to be shallow on the crest of knolls and deeper on the upland flats and lower slopes; however, local variations can be abrupt due to natural factors such as vegetation or topography or to agricultural or forestry activities, that may include chemical alteration (e.g., fertilizers), tillage, land leveling, and clearing of vegetation.

These factors should be kept in mind when sampling soils. For a given area, the soils should be subdivided into both horizontally and vertically homologous sampling units to account for natural and man-made variability. For example, it is preferable to sample by horizon than strictly by a certain depth, because ignoring this may result in some samples containing soil from another horizon that increases sampling variability.

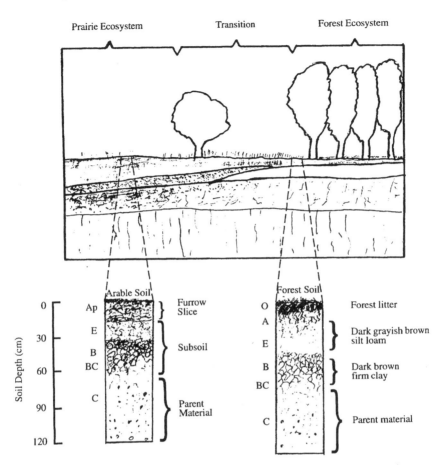

Fig. 6–1. Generalized characterization of transition from prairie soil to forest soil and associated soil profiles.

The intensity of sampling depends on the degree of variation within the sampling area and this will vary with the soil property being measured. Unfortunately, for many soil properties there is little information on landscape level variability. To some extent, properties related to plant nutrition (e.g., pH, inorganic N, extractable P, cations) have been characterized for their variability in agricultural soils (see Westerman et al., 1990).

An important resource that can assist sampling design is the regional–local soil survey. From this, you can determine the major soil series in the sampling area and associated descriptions of horizons, landscape position of the soil series, and land use categories; however, this should be used as a tool and can not substitute for field inspections in designing soil sampling protocols. Because of the scale of soil surveys, the resolution of soil maps are often unable to show smaller areas where changes in soil series may occur.

Temporal Variability of Soil Properties

For a few soil properties such as texture, temporal, or seasonal variability is not important because these properties will remain constant in the short run (on the order of decades or much longer periods). Many other soil factors vary on a seasonal basis, some more so than others.

Environmental factors, particularly moisture and temperature, affect chemical and biological reactions in soils. Soil biological activity is affected directly by the environment or indirectly via plants, which causes seasonal fluctuations of water, nutrients, C inputs, and other factors in the rhizosphere or bulk soil. Thus any soil property that is related to biological activity, will probably be highly variable.

One obvious way to overcome this is to sample several times during the year, which may be appropriate for certain situations; however, this greatly increases the labor and analytical costs of measuring a soil property. Another strategy, when the goal is to determine long-term trajectory of soil quality, is to measure soil quality at a time of the year when the climate is most stable and when there have been no recent soil disturbances, e.g., at the end of the growing season. If there are multiple-year samplings, measurements should be made at an equivalent time each year. The exact time may shift to ensure similar moisture or temperature conditions (e.g., growing degree days) each year.

Sources of Error

Three types of error are associated with soil sampling:sampling, selection, and analytical errors (Das, 1950). These errors are cumulative so it is important to take them into consideration in developing a sampling plan.

Sampling error occurs because the entire soil mass can not be measured and collection of field soil samples represents an extremely small subset of the entire soil sampling site. Increasing the number of soil samples increases the precision and decreases the impact of sampling error in estimating the true mean of a soil property.

Selection error arises from taking samples that do not represent the soil sampling site. For example, avoiding areas where it is difficult to obtain a sample such as rocky or highly vegetated areas would result in selection error, since only the soil in the easily obtained area would be sampled.

Measurement error results from sample handling, storage, preparation, and/or from the measurement technique. To reduce this error, it is important to follow the correct pre-analysis handling protocol and to use the best analytical technique possible.

Normally the measurement error is small and selection error can be greatly reduced by following a sampling plan that allows for every sampling point to have equal chance of being selected. Thus, the major source of variation usually is sampling error. Although sampling error can not be reduced, increasing the number of soil samples drawn from the sampling site can improve precision.

Inevitably the sampling plan is dictated by a compromise between precision and cost.

SAMPLING PLAN

Sampling Design

The first decision is whether the site should be selectively sampled (judgement sampling), randomly sampled (simple random sampling), or divided into subsites for sampling (stratified random sample). Here any history or knowledge of the sampling site is important. A good first step is to consult a soil survey to identify major soil types within the sampling site. Historical factors to consider for agricultural soils include kind, amount, and method of fertilization (e.g., banding vs. broadcast), prior tillage, and land leveling or erosion that may have exposed the subsoil (for a detailed discussion of sampling plans for soil testing–fertility see James & Wells, 1990). In forest systems, factors such as plant species mix, previous harvest methods (e.g., clear cuts), or fire history may affect the sampling plan. Although field inspection prior to sampling can not identify many facets of soil heterogeneity, it can be useful in identifying major changes in soil type or other obvious factors that could modify the sampling plan. This could be done by systematically inspecting soil cores to a specified depth on a grid of the sampling area and checking for any obvious changes in soil color or texture.

Judgement Sampling

Judgement sampling attempts to select a site that is typical and avoids areas that are thought to be nonrepresentative of the larger area. This is a highly biased approach and is entirely dependent on the skill of the investigator. This may be appropriate for certain situations where there is clear evidence that a representative area can be identified and there is no interest in characterizing the whole site. In general, this plan is not recommended because there is no way of assessing the accuracy of the results because they are entirely dependent on the sampler's judgement.

Simple Random Sampling

Simple random sampling requires that each soil sample must have equal opportunity to be selected. This is best done before going out to the field to avoid personal sampling bias. One approach is to make a grid over the sampling area at specified distances between the vertical and horizontal parallel lines and number each line (see Fig. 6–2 as an example of a rectangular area). Then pairs of random numbers are drawn from a random number table (see Fisher & Yates, 1963; or almost any statistics book). Alternatively random numbers can be obtained from many computer statistical or spreadsheet software programs using the random number function. If, in the course of selection, a pair has already been selected, it can be discarded and another pair can be selected. These coordinates establish the points on the grid to be sampled.

The drawback of random sampling is that it may provide limited information on the spatial distribution of the soil property being measured.

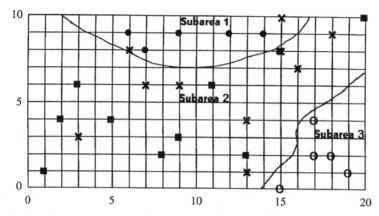

Fig. 6–2. Map of random sampling plan with coordinate pairs generated by a computer spread-sheet random number function for entire sampling of site (X) or stratified random sampling of subareas (●, ■, ○).

Stratified Random Sampling

Stratifies random sampling is obtained by collecting a random sample within each subarea (Fig. 6–2) as outlined for simple random sampling. Subareas are delineated because there maybe unique characteristics that can be readily identified. This could be due to soil series, topography, soil pollution, or obvious changes in soil management practices and vegetation. The advantage of this approach is that it allows the investigator to characterize each subarea and improve the precision of estimating the entire sampling area. The disadvantage is greater labor for sampling and analytical costs.

Systematic Sampling

Systematic sampling or grid sampling ensures that the entire area is sampled from predetermined points (Fig. 6–3). These points are selected at regular

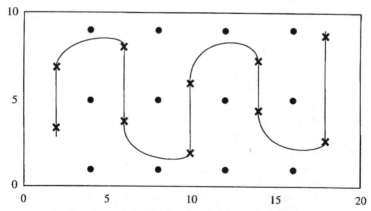

Fig. 6–3. Map of systematic grid sampling (●) or serpentine random sampling (X).

intervals that makes it easier than random sampling. Webster and Oliver (1990) report that empirical evidence shows that systematic sampling is often more precise than simple random sampling and somewhat more precise than stratified random sampling.

Systematic sampling requires consideration of periodic trends (e.g., banding of fertilizer in rows, equipment traffic patterns). This effect can be reduced by choosing a grid pattern that is unrelated in spacing and orientation to the periodic trend and increasing the number of samples.

Systematic sampling is advantageous for geostatistical methods like kriging and for identifying the location of high and low values, which is not possible with random sampling or composite sampling (outlined below). This is an important consideration for soil quality assessment. For example, this approach is effective at identifying exposed subsoil due to land leveling or erosion or to characterize contaminated soils.

Composite Sampling

Composite sampling consists of taking a number of field samples adequate to represent the area under consideration and thoroughly mixing these samples to form one composite or bulk sample. This bulk sample or a well-mixed representative subsample, is then analyzed for a given soil property. This procedure can be used only on soils that are unaffected by physical disturbance (most chemical and biological measurements).

This approach assumes that a valid estimate of the mean can be obtained for a soil property from a single analysis of the composite sample. Consequently, it is most appropriate when the sampling area is known to be relatively uniform.

Composite sampling does not enable determination of an estimate of the variance of the mean and consequently it is not possible to estimate precision; however, it is possible to estimate the variance between similar composite samples.

Composite sampling can significantly reduce the number of analytical analyses. To be valid: all samples must be drawn from the same soil type; each soil sample must contribute the same amount of soil to the composite sample; and the only objective is to provide an unbiased estimate of the mean. It is important to first determine the number of individual sampling units that are required to make up a composite sample. This can be done by Eq. [1] as described below.

Stratified Composite Sampling

Stratified composite sampling combines composite sampling with sampling by subarea. This can include sampling discrete subareas (Fig. 6–2) or periodic trends such as inside wheel track vs. outside wheel track in row crop fields (Fig. 6–4) or compositing samples from the same landscape position. This enables a reduction in the number of analyses within a stratification but still provides separate estimates of stratified regions that are known to be distinctly different. (This approach also is discussed in Sarrantonio et al., 1996, this publication.)

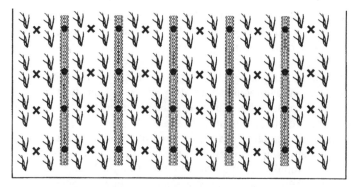

Fig. 6–4. Map of composite stratified random sampling plan where all samples within wheel track (●) or outside wheel track (X) are separately composited.

Depth of Sampling

Before collection of soil samples, it is important to determine what soil depth(s) will be sampled. In general, the surface soil would be the most sensitive region for monitoring changes in soil quality due to soil management; but, it may be important to identify changes with depth. Nutrient availability for plant growth and soil biological properties such as microbial biomass and activity, soil faunal activity, and enzyme activity would be most important in the surface soils (approximately 0 to 30 cm; 12 in.). Conversely the subsoil layers (30 to 100 cm; 12 to 40 in.) may control water storage and contain: hard pans; accumulation zones of nutrients and other salts; and pH changes that may be important in assessing soil quality.

In soils under native conditions, sampling the surface layer should be based on the depth of horizons (e.g., O or A horizons). In agricultural soils, the surface layer, which typically has maximum soil mixing, may be the most important soil depth to sample. Moldboard plowing homogenizes soil to a depth of about 20 cm (8 in.). Research on the distribution of chemical properties suggests minimum tillage and no-till systems have less-defined boundaries of soil mixing but organic matter, nutrients (James & Wells, 1990), pesticides, and biological properties (Dick & Daniel, 1987) tend to accumulate in the top 7.5 cm (3 in.). Thus for these systems, sampling the surface layer should be split to two depths, 0 to 7.5 cm (0 to 3 in.) and 7.5 to 15 cm (3 to 6 in.).

A rapidly growing conservation tillage practice that presents a challenge for sampling soils is ridge-till. This is particularly true for nutrient analysis because surface-applied fertilizers are swept onto the middle of the ridge during the ridging process. Moncrief et al. (1984) recommended that for soil fertility testing, the 0- to 15-cm (0- to 6-in.) depth be sampled after planting, but before ridging occurs. If this is not possible, sample half way up the ridge.

Permanent sod for hay, pasture, or grass seed production also accumulates nutrients near the surface. James and Wells (1990) recommended that for soil testing purposes the 0- to 10-cm (0- to 4-in.) depth should be sampled in soils under permanent sod.

Depending on the goal, there may be several methods of sampling by depth to consider. *Composite depth sampling* involves collection of a whole soil core, which is thoroughly mixed prior to removal of subsample for laboratory analysis. This provides an inexpensive means to obtain average values of soil properties of a soil profile, but does not provide any information about variability with depth. *Discrete depth sampling* uses predetermined depths that can be set by natural horizons or at constant increments. This is the preferred depth sampling method for most situations because it will provide information about changes in the soil property with depth but it has greater labor and analytical costs.

Volumetric or Bulk Density Sampling

At the time of sampling, bulk density of the soil should be estimated. Measuring bulk density can be important not only as a physical measurement of soils but also to enable converting other measurements to a volumetric basis. A soil measurement based on a volumetric basis can provide a more realistic estimate of field conditions because soils vary in bulk density. Furthermore, it will provide a more accurate extrapolation of results to a per hectare (acre) basis to a given depth. Measurement of soil bulk density is mandatory for accurate comparison of C and nutrient storage between soils of contrasting management (Ellert & Bettany, 1995).

Arshad et al. (1996, this publication) presents methods for determining soil bulk density.

Defining the Soil Sample and Its Size

Obviously in the case of soil there is a near infinite amount of sampling units at most sampling sites and it is impossible to withdraw the entire soil mass. Consequently, the investigator must first decide what constitutes a sampling unit at a given sampling point. The larger the sample the more representative it should be. In reality, what constitutes a sample and its size is dictated by pragmatic considerations such as soil property under investigation, handling, and storage capacities. For some measurements that require intact cores (e.g., bulk density), the sample size is controlled by the method and equipment available. For other procedures that do not require intact cores, the amount of soil is based on the suite of measurements to be determined. For some measurements such as a nutrient extraction, <50 g (2 oz.) may be needed whereas some macrofaunal procedures may require several kilograms (lb) of soil sample. It is always better to collect more soil than the minimum requirement to allow for sieving losses, extra analyses, etc.

Even if composite sampling is not the general approach for sampling across an area, it still is a good way to sample each sampling point. For many soil samples this can be done by taking about 10 cores [e.g., 2.5 cm (1 in.) diam.] for compositing within a 1 m^2 area. This provides a better representative sample of the sampling point and yields enough soil for many soil analyses. To even better represent the area, a larger composite sample is taken, placed in a bucket, thoroughly mixed, and then a subsample is drawn and brought back to the lab.

Statistical Concepts and Number of Samples

The sample mean ($\bar{x}$) or arithmetic average of a variable is an important statistic and is simply calculated as the sum of the observations divided by the number of observations. It also is necessary to have an idea of the variability or variance (s^2), which is the degree of dispersion around the mean. Another way to express variance is on a relative basis (percentage of the mean), the coefficient of variation (CV), which is the standard deviation (s) (square root of variance) divided by the mean and multiplied by 100%. These calculations can be found in any standard statistics book and many calculators or computer spreadsheets have these functions. If you use a calculator or computer program check whether variance is calculated as n or $n - 1$. It is important to use $n - 1$ as this provides an unbiased estimate of the variance.

Sampling and analytical work is time consuming and expensive. Therefore it is important to take only as many samples as needed. The number of samples is determined by the desired precision and the expected or actual degree of variation within the sampling area. Using the formula below, the number of samples needed to produce an estimate within a certain value of the population mean (margin of error) can be determined (assumes the distribution of a measurement is normal-bell shaped distribution around mean). The variance can be obtained from previous experience, from published sources, or be estimated by $s^2 = (R/4)^2$ (Freese, 1981) where R is the estimated range of measurements to be encountered in sampling. The sample size (N) needed is calculated as follows:

$$N = \frac{t^2 s^2}{D^2} \qquad [1]$$

where: s^2, is the variance, D is the specified margin of error of the estimated mean that the investigator will accept (same units as standard deviation), and t is a number obtained from a t table which can be found in any statistics book. To determine which t-value to use, select the degrees of freedom ($n - 1$ where n is the number values used to calculate s^2) and the level of confidence. Often the 95% confidence level (0.05 in a t table) is used.

Generally, the biggest challenge in soil sampling is to reduce the number of soil samples to an acceptable level of sampling and analytical costs. The number of soil samples can be reduced by: developing a sampling approach that reduces variability (e.g., composite sampling); accepting a lower level of precision or by lowering the confidence limit.

A more in-depth approach for describing the spatial relationship of soil properties is *Geostatistics*. This topic is covered in this chapter only to introduce the concept and interested readers are referred to detailed publications that fully discuss the applications to soils (Oliver, 1987; Warrick et al., 1986; Webster & Oliver, 1990). Geostatistics are a suite of methods that analyze and characterize spatially correlated data. The goal of this approach is to enable estimates of the properties being considered between sampling points. Geostatistics for soils is based on the general observation that sites more nearby are more closely related than those sites further away. This is called spatial dependence. The first step is to develop a model, called a semi-variogram, which describes the spatial relationship of close and faraway sample points at each site. The second step of geostatistics is kriging, which uses the semi-variogram to estimate each value in the study area. Thus kriging models the estimates within the study area that were not sampled and can produce detailed maps with interpolated isopleths (i.e., lines showing estimated distribution or gradient of a soil property over the landscape).

Practical Considerations of Soil Sampling

It is always important to have a sampling plan that reflects specific objectives for conducting the soil quality measurements and provide a specified precision at the lowest cost or the greatest precision at a fixed cost. Because soil properties generally have high variability, every plan usually succumbs to pragmatic compromises between precision and the funds available for sampling and analysis.

Experience suggests that variance of soil properties can be as high within a relatively short distance as at larger distances. For example, one-quarter to half of the total variance of many soil properties in areas of 10 to 10 000 ha (25 to 25 000 acres) can occur within a few square meters (Webster & Oliver, 1990). This would indicate that unless there is strong evidence for nonuniformity at the sampling site (e.g., obvious changes in soil type, topography, landuse) a simple random sampling is appropriate. In practice, small uniform areas (<0.5 ha or <1.2 acres) can be sampled with as few as 5 to 10 samples and that larger areas gain little in precision when numbers are greater than 25 (Webster & Oliver, 1990).

One statistically valid but straight forward approach for random sampling is to develop a serpentine sampling plan (Fig. 6–3). In this method, the distance of the serpentine path is first determined by counting the number of paces from beginning to end (Fig. 6–3) and dividing the distance by the number of samples to be taken to determine the distance between sampling points. To incorporate randomness the investigator randomly chooses the starting point of the serpentine sampling scheme. To do this, a random number from 1 to 10 is chosen using a random number generator table or computer program. Alternatively, 10 numbered pieces of paper can be put in a hat and one number drawn at random. The investigator then takes that many paces into the field and takes the first sample at that point with the remaining samples being taken at the calculated intervals as outlined above.

If the site clearly has nonuniformity and subareas can be identified, then the stratified approach is best. But within each subarea, the serpentine random sampling approach can be used.

When costs are a major consideration, composite sampling can be used in conjunction with stratification (*stratified composite sampling*) or uniform sampling to reduce the number of samples. This can provide a good average of each soil property. The disadvantage is that within the composite sample the variance of the mean is unknown.

In conclusion, development of a sampling plan is a critical component of measuring soil quality. Having as much historical information (e.g., management history, soil survey information) as possible about the site, combined with preliminary field inspection is important in guiding the sampling procedure. It is important to recognize the limitations of each sampling plan to improve interpretation of the results.

SOIL SAMPLING EQUIPMENT

Selection of appropriate soil sampling equipment depends on several questions. Are intact cores required (e.g., bulk density would require intact coring procedures)? How large a sample is required from each sampling site (e.g., some macro faunal procedures such as earthworms require large sample sizes that would not allow the use of small soil probes)? What is the moisture content of the soil to be sampled?

The type of sampling equipment may vary as a function of the soil property to be determined. A wide array of soil sampling equipment is available that can be as simple as a shovel or as sophisticated as vehicle-mounted hydraulic-driven power probes or augers. For most soil sampling, the standard hand-held soil probe that has a diameter of about 2.5 cm (1 in.) and of varying length is appropriate.

A major challenge can be to sample dry soils, particularly heavy clay soils. Augers are superior to probes, particularly powered augers. Alternatively, if probes are required, use of heavy-duty equipment that is made for hammering or slam driving should be used. Other equipment not specifically designed for this will quickly become damaged and unusable.

Labeling of samples is an important factor to carefully consider. Commercial tags with a wire twist or string can be used on plastic bags. Also, soil science supply houses can provide standard paper soil testing bags that are lined with plastic and the labeling can be put directly on the bag. These bags will typically hold about 500 g (1 lb). If a large number of samples is to be taken, it is best to label and organize the bags in a logical sequence according to the sampling plan before going to the field. Unless the soil is dry, plastic bags or plastic lined bags are preferable because paper bags with moist soil will lose their integrity during handling and storage.

Plastic or paper bags or tags should be labeled with either permanent markers or with a pencil. Ink labeling will run or smear if it gets wet. To avoid confusion later, each sample should have a unique sample number and enough information to enable later determination of sampling dates, location, and other pertinent information. Always bring extra bags!

For most soil sampling, it is often useful for each worker to have a bucket to collect composite soil samples. This should be a clean, plastic bucket because it is light, easy to clean, and is unlikely to be a source of contaminants. For example, metal containers can be cause for contamination of micronutrients (particularly galvanized pails and coffee cans for Zn analysis).

A field book should be available to record specifics of date of sampling, directions, or coordinates for sampling sites, and any other general observations on the sampling site or sampling process that may be important for later interpretation of the data.

SAMPLE HANDLING AND PRETREATMENT

To determine the best method for sample handling and pretreatment, it is important to know what analyses will be run and what special precautions may be required for each of these. Should the sample be air dried or sieved? How soon does the analysis need to be run? Are there any specific storage requirements in terms of containers or temperature?

When composite samples are taken, it is important that the sample be completely mixed. Samples too wet or too dry may be difficult to mix readily. Ultimately the sample should be crushed to pass a 2 to 5 mm sieve (1/8 to 1/4 in.) to remove roots and stones. It is important once a sample has been taken for sieving that all the sample be passed through the sieve. For example, if aggregates that are difficult to force through the sieve are always discarded with each sieving, the sample will be biased because not all the soil sample will be represented after sieving. Dry soils that do not readily pass through the sieve should first be crushed. In the case of wet soils that can be air-dried, the samples should be dried to a soil moisture when sieving is easiest. It is important that heavy clay soils not be dried completely as this will result in hard clods that will be difficult to sieve. After sieving, the soil can be completely dried. Normally spreading a soil sample thinly (2 to 5 mm; 1/8 to 1/4 in.) on paper and air-drying soil samples for 48 h will be adequate to stabilize soil properties that can be run on dried samples. Alternatively, wet samples that are to be dried can be placed in an oven at 35 to 50°C (~100 to 120°F).

Many commercial soil testing laboratories use mechanical grinders and with proper procedures, this pretreatment has been found to be appropriate for soil fertility measurements. Stainless steel grinders and screens are preferred to prevent contamination for most metals. It is important that there be adequate consideration of cleaning between samples to avoid cross contamination. Also, large scale grinders do not lend themselves to small samples as there can be significant losses of soil during the grinding process. Small samples can be passed through small scale grinders or be crushed with mortar and pestle and passed through the proper size sieve. Some analyses, like total elemental analysis, require a finer size than 2 mm. For example, total N determinations should be based on samples that have passed a 0.149-mm screen.

Generally for many chemical properties (e.g., soil testing procedures such as pH, cation-exchange capacity, nutrient status including nitrate ammonium, and total C) and physical properties, air drying is appropriate if drying begins within a day of sampling; however, most biological measurements must be run on field moist samples and require specific storage conditions.

When sampling for biological properties it is best if the samples are maintained at ambient or cooler temperatures (4°C) during transport to the laboratory. As a general rule biological analyses should be made within 24 h of sampling; however, Wollum (1994) suggested that some soil biological properties remain stable for at least 7 d (microbial biomass, available N, and certain enzyme activities). Storing samples at 4°C helps preserve biological properties in soil but freezing should be avoided. Prolonged storage at 4°C can cause changes in community structure and activities, and may induce anaerobis (Wollum, 1994). Unfortunately, there is relatively little information on storage effects on soil biological properties, so at this time the best approach is to develop a sampling plan and laboratory arrangements that allow biological measurements to be run as soon as possible after sampling.

During sieving or other handling of samples for determination of soil biological properties it is important that the sample does not become dry or be subject to high temperatures and that the proper sieve size be used for a specific biological measurement.

Dried soil samples can be stored for many years and for some analyses like total elemental analysis, extractable nutrients, and other soil fertility measurements, valid results can be obtained decades later (Bates, 1993). Storage conditions for air-dried samples are best where extreme temperatures are avoided and humidity remains low; however, for any chemical or physical measurement, storage time should not be taken for granted, it may depend on the particular element, extractant, or other aspects of a procedure.

QUALITY ASSURANCE AND QUALITY CONTROL

Quality assurance (QA) and quality control (QC) is a means to assess and reduce errors in the course of conducting environmental studies. A QA/QC program has been developed by the U.S. Environmental Agency (USEPA) that has been published as a manual and as a software program (USEPA, 1991). This approach provides a framework for sampling design, sample handling, and laboratory procedures to minimize errors. To attain these goals, it adopts procedures for the use of duplicate, split, spiked, evaluation, and calibration samples. Furthermore, it can provide information to identify sources of error and determine the uncertainty of the measurements.

REFERENCES

Arshad M. C., B. Lowery, and B. Grossman. 1996. Physical tests for monitoring soil quality. p. 123–141. *In* Methods for assessing soil quality. SSSA Spec. Publ. 49. SSSA, Madison, WI.

Bates, T.E. 1993. Soil handling and preparation. p. 19–24. *In* M.R. Carter (ed.) Soil sampling and methods of analysis. Lewis Publ., Ann Arbor, MI.

Buol, S.E., F.D. Hole, and R.J. McCracken. 1973. Soil genesis and classification. Iowa State Univ. Press, Ames.

Crépin, J., and R.L. Johnson. 1993. Soil sampling for environmental assessment. p. 5–24. *In* M.R. Carter (ed.) Soil sampling and methods of analysis. Lewis Publ., Ann Arbor, MI.

Das, A.C. 1950. Two-dimensional systematic sampling and associated stratified and random sampling. Sankhya 10:95–108.

Dick, W.A., and T.C. Daniel. 1987. Soil chemical and biological properties as affected by conservation tillage: Environmental impacts. p. 125–147. *In* T.J. Logan et al. (ed.) Effects of conservation tillage on groundwater quality: Nitrates and pesticides. Lewis Publ., Chelsen, MI.

Ellert, B.H, and J.R. Bettany. 1995. Calculation of organic matter and nutrients stored in soils under contrasting management regimes. Can. J. Soil Sci. 75:529–538.

Fisher, R.A., and F. Yates. 1963. Statistical tables for biological, agricultural and medical research. Hafner Publ. Co., New York.

Freese, F. 1981. Elementary forest sampling. Agric. Handb. 232. USDA Oregon State University, Corvallis, OR.

Halvorson, J.J., J.L. Smith, and R.I. Papendick. 1997. Issues of scale for evaluating soil quality. J. Soil Water Conserv. (In press).

James, D.W., and K.L. Wells. 1990. Soil sample collection and handling: Technique based on source and degree of field variability. p. 25–44. *In* R.L. Westerman (ed.) Soil testing and plant analysis. 3rd ed. SSSA Book Ser. 3. SSSA, Madison WI.

Kempthorne, O., and R.R. Allmaras. 1986. Errors and variability of observations. p. 1–31. *In* A. Klute (ed.) Methods of soil analysis. Part 1. 2nd ed. Agron. Mongr. 9. ASA and SSSA, Madison WI.

Moncrief, J.F., W.E. Fenster, and G.W. Rehm. 1984. Effect of tillage on fertilizer management. p. 45–56. *In* Conservation tillage for Minnesota. Univ. of Minnesota. Agric. Ext. Serv. Publ. AG-B-2402. Univ. of Minnesota, St. Paul.

Oliver, M.A. 1987. Geostatistics and its application to soil science. Soil Use Manage. 3:8–20.

Peterson, R.G., and L.D. Calvin. 1986. Sampling. p. 33–51. *In* A. Klute (ed.) Methods of soil analysis. Part 1. 2nd ed. Agron. Monogr. 9. ASA and SSSA, Madison WI.

Sarrantonio, M., J.W. Doran, M.A. Liebig, and J.J. Halvorson. 1996. On-farm assessment of soil quality and health. p. 83–105. *In* Methods for assessing soil quality. SSSA Spec. Publ. 49. SSSA, Madison, WI.

Soil Survey Staff. 1975. Soil taxonomy. Agric. Handb. 436. U.S. Gov. Print. Office, Washington, DC.

USEPA. 1991. ASSESS users' guide. U.S. Environ. Protection Agency, Environmental Monitoring Syst. Lab. EPA/600/8-91001. USEPA, Las Vegas, NV.

Warrick, A.W., D.E. Myers, and D.E. Neilsen. 1986. Geostatistical methods applied to soil science. p. 53–82. *In* A. Klute (ed.) Methods of soil analysis. Part 1. 2nd ed. Agron. Monogr. 9. ASA and SSSA, Madison, WI.

Webster, R., and M.A. Oliver. 1990. Statistical methods in soil and land resource survey. Oxford Univ. Press, Oxford, UK.

Westerman, R.L. (ed.). 1990. Soil testing and plant analysis. 3rd ed. SSSA Book Ser. 3. SSSA, Madison, WI.

Wollum, A.G., II. 1994. Soil sampling for microbiological analysis. p. 1–14. *In* J.M. Bigham (ed.) Methods of soil analysis. Part 2. SSSA Book Ser. 5. SSSA, Madison, WI.

7 Physical Tests for Monitoring Soil Quality

M. A. (Charlie) Arshad
Agriculture & Agri-Food Canada
Beaverlodge, Alberta, Canada

Birl Lowery
University of Wisconsin
Madison, Wisconsin

Bob Grossman
USDA-NRCS
Lincoln, Nebraska

Physical condition of a soil has direct and indirect effects on crop production and environmental quality. Well aggregated soils, if managed properly, are able to maintain a balance of air and water so as to promote nutrient cycling and root development, while resisting erosion, surface sealing, and other degradative processes.

Key attributes of soil quality include physical, chemical, and biological properties. Physical and chemical criteria to characterize soil quality have been discussed in detail by Arshad and Coen (1992). This chapter describes relatively simplified versions of some physical tests that can be used to assess and monitor changes in soil quality. Other physical properties relating to soil water characteristics (water content, infiltration, hydraulic conductivity, and field water-holding capacity) are discussed in Lowery et al. (1996, this publication).

This chapter is divided into two parts. The first part describes standard soil properties that are relatively static and do not change readily with agricultural practices. These include parameters that are generally estimated or measured as a part of soil characterization such as the standard soil profile description, rooting depth, morphological features, texture, and others. The second part includes properties that are subject to measurable changes with different management practices. The sensitive properties in this category are bulk density, aggregate stability, penetration resistance, and others. This set of properties needs to be measured frequently to assess the impact of management practices on soil quality. These are the main indicators of changes in soil quality, and more importantly, they can indicate the direction of such changes. Such measurements, if made at selected time intervals, can indicate changes, positive or negative, in soil quality under a

---

Copyright © 1996 Soil Science Society of America, 677 S. Segoe Rd., Madison, WI 53711, USA. *Methods for Assessing Soil Quality*, SSSA Special Publication 49.

given set of agroclimatic and land-use conditions. Soil management techniques that lead to improved soil quality can thus be adopted and practices that result in decline of soil quality can be discouraged.

SAMPLING

A number of points should be considered for sampling to monitor soil quality parameters. Soil properties vary with location and with soil horizon or depth. Therefore we need to sample horizontally and vertically. For most properties, the variation declines with increasing depth; thus subsoil needs to be sampled less intensively than top soil to attain comparable accuracy. It is advisable to collect at least three composite samples from each site (subarea of uniform topography). These sites or subareas in the field may represent different soil types. Taking random samples from the subareas allows one to compare these sites and increases the precision of the estimates over the entire area. It does take extra time and add cost to the analyses. The number of samples required would depend upon the type of parameters to be monitored. Fewer samples are needed to estimate bulk density than other parameters such as water retention at field capacity or texture; hydraulic conductivity requires a much larger number of samples. For physical tests described in this chapter the following general guidelines are recommended: take 8 to 10 samples to represent the site and mix these samples to form one composite or bulk sample. Use this composite sample or subsample for analyses. Make sure that each sample contributes the same amount to the composite.

In order to assess changes in soil properties, it is essential that sampling and measurements are restricted to defined periods and locations so that they are reproducible and interpretation of results are not confounded by weather, season, management, and soil variability. For example, determine bulk density in the fall or spring before any soil disturbance due to tillage or frost; determine aggregation and penetration resistance in spring just before seeding at field capacity. Avoid sampling at times when soils are wet or very dry.

SITE CHARACTERIZATION

Record of morphological attributes of the soil profile can give useful information. Features that can be recorded without using sophisticated equipment include color, structure, root, and pore-size distribution. Characteristics of the site should include legal location, landform, climate, land use, drainage, and topography. (For details see Sarrantonio et al., 1996, this publication; the information can be obtained from the published Soil Survey reports).

Equipment and Supplies

Measuring tape, shovel and hand lens or other type of magnifying glass.

Procedure
1. Select a representative location in the field that considers the patterns in relief as associated with tillage and management. Dig a pit exposing

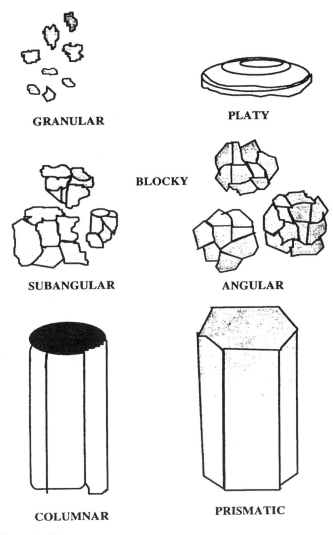

Fig. 7–1. Types of soil structure.

vertical face approximately 1 m (3 ft) across to an appropriate (2 m, 6 ft) depth. In general, the pit should extend 20 to 30 cm below the lowest layer to be described. Preselect two walls of the pit that will be exposed best to the sun light.
2. Mark layers to the depth of interest with nails, wooden, or other stakes. Measure and record thickness of each layer. Assign each layer to a structure subclass using Fig. 7–1 and Table 7–1.
3. Record visual features such as color, cracks, their widths and lengths, pattern, and continuity. Record soil surface features such as crusting or surface sealing and their extent.

Table 7–1. Size classes of soil structure.

Size classes	Shape of structure			
	Platy†	Prismatic and Columnar	Blocky	Granular
	mm			
Very fine	<1	<10	<5	<1
Fine	1–2	10–20	5–10	1–2
Medium	2–5	20–50	10–20	2–5
Coarse	5–10	50–100	20–50	5–10
Very coarse	>10	>100	>50	>10

† In describing plates, *thin* is used instead of *fine* and *thick* instead of *coarse*.

4. Hand texture each layer according to the method for soil texture by *feel* and record textural class.
5. Determine abundance and size of roots along with plant species and record appropriate classes using Table 7–2. This can be done by counting the number of roots in a 15 × 15 cm (6 × 6 in.) area at 30.5 cm (1 ft), 61 cm (2 ft), 91.5 cm (3 ft), etc.
6. Determine abundance and size of pores (as in Step 5) and record appropriate class using Table 7–3.
7. Examine soil profile to determine if there is a hard pan, stone layer, lime, salt, or other root restricting layer. Record its thickness, and its depth from the soil surface.

ROOTING DEPTH

Depth of soil to a layer that would restrict root growth strongly affects crop production. There is a relationship between crop yield and soil depth because the top-soil depth is important for water storage and nutrient supply for plant production. Pierce (1991) compiled a comprehensive list of publications relating to effect of erosion and top-soil depth on crop yields. He noted, however, that the results of such studies were site-specific. Recently, Malhi et al. (1994) reported that barley (*Hordeum vulgare* L.) yields on two Mollisols were substantially reduced by artificial removal of top-soil, which resulted in decreased soil depth in addition to loss of soil fertility. In their experiments, additions of N or P fertilizers to eroded topsoils improved yields, but the yields were less on noneroded soils under the same fertilizer treatment. Among many soil properties, topsoil

Table 7–2. USDA system of describing abundance and size classes of pores and roots.

No dm^{-2}†	Size		No dm^{-2}	Size		
	Very fine	Fine		Medium	Coarse	Very coarse
	<0.5 mm	1–2 mm		2–5 mm	5–10 mm	>10 mm
0–10	Few pores, few roots		<1	Few pores, few roots		
10–50	Common pores, common roots		1–5	Common pores, common roots		
>50	Many pores, many roots		>5	Many pores, many roots		

† Denotes number of pores or roots per square decimeter (10 × 10 cm).

Table 7–3. Changes in selected physical properties (0–30 cm) and wheat yields with long-term cultivation (Bowman et al., 1990).

Years of cultivation	Average depth to lime	Water content, MPa		Wheat yield
		–0.03	–1.5	
	cm	kg kg^1		kg ha^1
0	60	0.130	0.070	--
3	55	0.080	0.068	2700
20	44	0.098	0.065	1500
>60	30	0.086	0.055	1150

thickness has been identified as a major indicator of soil quality (Power et al., 1981). In a study conducted by Bowman et al. (1990), wheat (*Triticum aestivum* L.) yield declined by almost 57% with a 50% decrease in soil depth (Table 7–3).

Factors that influence rooting depth include depth to bedrock, stone layer, high salt contents, hard pan, frozen layer, and water table.

Core Method for Rooting Depth

Equipment and Supplies

1. Core cylinders (7.5 cm in diam. Same as described in section on bulk density).
2. Measuring tape or ruler.
3. Sharp and rigid knife or spatula.
4. Heavy duty leather gloves.
5. Labeled plastic freezer bags.
6. Top loading balance.
7. Refrigerator set at 5°C.
8. Oven to dry roots at 65°C.
9. Wire-screen cylinder (15 cm diam. × 30 cm long) with openings of 0.5 mm.
10. Bucket (20 L) to accommodate wire-screen cylinder (Item 9).
11. Weighing cans.

Procedure

1. Take soil cores near or after maximum crop plant development at 0- to 10- and 10- to 20-cm depths and at 20-cm intervals thereafter to 1 m on either side of the plant. Best time for sampling is when soil is at field capacity.
2. Slice the core with the knife or spatula at 10-cm intervals starting at the soil surface. Break individual core slices and record visual features such as color, abundance, and size of roots and pores as in Item 5 and 6 in the section on soil characterization (Table 7–2). Record orientation (horizontal or vertical) and continuity of pores, concretions, nodules, casts, pans, and any other features restricting root development. Transfer core segments into labeled plastic freezer bags. Seal the bags and store in the refrigerator set at 5°C for subsequent root separation.

Table 7–4. Terms for rock fragments (Soil Survey Staff, 1993).

Shape and size	Noun	Adjective
Spherical, cubelike, or equiaxial:		
2–75 mm diam.	Pebbles	Gravelly
2–5 mm diam.	Fine	Fine gravelly
5–20 mm diam.	Medium	Medium gravelly
20–75 mm diam.	Coarse	Coarse gravelly
75–250 mm diam.	Cobbles	Cobbly
250–600 mm diam.	Stones	Stony
600 mm diam.	Boulders	Bouldery
Flat:		
2–150 mm long	Channers	Channery
150–380 mm long	Flagstones	Flaggy
380–600 mm long	Stones	Stony
600 mm long	Boulders	Bouldery

3. For root separation, place soil from each core slice in wire-screen cylinder. Immerse the cylinder in water-filled bucket until soil becomes soft. For sandy soils, the time of immersion is short but it takes a longer time period for clay soils.
4. Move the cylinder up and down in the water to remove soil from the roots.
5. Washed roots from each depth are transferred into weighing cans, dried at 65°C and weighed.

Compare root distribution patterns in different crop and soil management systems. Record changes with time. Measurement of rooting depth is one of the most inexpensive methods to assess soil quality.

SOIL TEXTURE

Soil texture refers to the weight percentage of sand (0.05 to 2.0 mm), silt (0.002 to 0.05 mm) and clay (<0.002 mm) in which the total composition equals 100%. It is based on the soil sample that passes through a 2-mm sieve. If coarser material (>2 mm) is present in substantial amounts (>15%) an additional description such as gravelly or stony can be included (see Table 7–4). The dominant size fraction (<2 mm) is used to describe the *soil textural class,* such as sand, sandy clay, silty clay, and others. If no fraction is dominant the textural class is loam. Figure 7–2 shows the defined limits for various textural classes used by the soil scientists.

Texture of the soil is one of the most stable attributes of the soil, being modified only slightly by cultivation and other practices that causes mixing of different layers. Textural class indicates the ease with which a soil may be cultivated. Soils high in clay are often called *heavy* as they require greater draft power to cultivate compared with the sandy soils that are termed *light.* Soil texture also affects water and nutrient retention. For example, clay soils retain more water against gravity and more nutrients (especially positively charged ions) against leaching than sandy soils.

Several methods for determination of soil texture are available. Two methods chosen here are simple and less time consuming. The qualitative field method

PHYSICAL TESTS

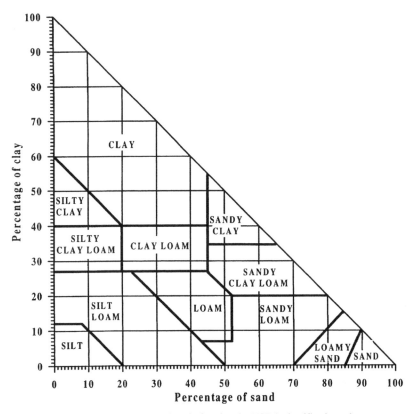

Fig. 7–2. Textural triangle for soil textural analysis using the USDA classification scheme.

is based on the feel of the soil material, which may be gritty (sand), smooth (silt), or sticky (clay) when kneaded or rubbed between the fingers. For a quantitative evaluation, the hydrometer method developed by G.J. Bouyoucos (1962) is proposed. It measures the density of the suspension at a given depth as a function of time. With time this density decreases as the largest particles, and then progressively smaller ones, settle out of the region of the suspension being measured. With the standardized Bouyoucos hydrometer, a settling time of 40 s is needed at 20°C (68°F) to measure the concentration of clay and silt (all the sand having settled through), and a time of about 2 h is needed to measure clay left in suspension because by now the smallest silt particles have fallen below the floating hydrometer. A temperature correction must be applied to the hydrometer reading for accurate results because it influences the density and viscosity of the water.

Soil Texture by Feel Method

Determination of soil texture by feel method is a tactile evaluation of the soil's resistance to deform (alter its shape by stress or by twisting). If a moist soil is easily deformed and forms a weak ribbon when rubbed between the fingers it is loam. Clay soils are difficult to deform and make strong ribbons. Sand feels gritty and silt gives a smooth, flour or talc-like feel. Soils high in organic matter may not form a ribbon. Clay soils are sticky in most cases but would depend on the type of clays. Soils dominated by montmorillonite (expanding) clays, for example, feel different from soils containing similar amounts of kaolinitic (non-expanding) clays. Most soils contain clays of mixed mineralogy. For soils high in kaolinitic clays, use the hydrometer method for particle size analysis.

Equipment and Supplies

Water

Procedure

1. Place 20 to 30 g of soil on the palm of your hand. Add water dropwise and knead to break aggregates; soil has the right moisture when it is moldable.
2. Squeeze the soil to form a ball. If soil does not form a ball it is *sand*. If it forms a ball, proceed to Step 3.
3. Hold ball of soil between thumb and forefingers gently pushing and squeezing it upward into a ribbon. If soil does not form a ribbon it is *loamy sand, silt or coarse silt loam* (use hydrometer method). If soil forms a ribbon proceed to Step 4.
4. If soil forms a ribbon 2.5 cm long before breaking, wet and rub the ribbon with forefingers. If soil feels gritty, it is *sandy loam* and if it is very smooth and no grit at all, it is *silt*. If it feels smooth, it is *silt loam* and if neither grittiness nor smoothness predominates then it is *loam*.
5. In Step 3, if soil forms a ribbon 2.5 to 5 cm long before breaking, wet and rub as in Step 4. If soil feels gritty, it is *sandy clay loam*; if no grit but smooth it is *silty clay loam*. If neither grittiness nor smoothness predominates the soil is *clay loam*.
6. In Step 3, if soil forms a ribbon longer than 5 cm before breaking, wet and rub as in Step 4. Soil texture is *sandy clay* if it feels gritty, *silty clay* if there is no grit but feels smooth and *clay* if neither grittiness nor smoothness predominates.

Comments

One needs to practice with a soil sample for which texture has been determined by the laboratory procedure. With experience it would be easier to make distinctions among different textural classes. The U.S. Soil Survey staff has texture estimates for most cultivated areas—use these estimates if available in your area but be advised that the textural class map unit component may not apply to a point or small area.

Soil Texture by Hydrometer Method

Equipment and Supplies

1. Standard hydrometer, ASTM 152H, with Bouyoucos scale in g L^{-1}.
2. Sedimentation cylinder with 1-L mark 36 ± 2 cm from the bottom of the inside.

3. Electric stirrer or a blender (milkshake machine).
4. Calgon solution (50 g L^{-1}, sodium hexametaphosphate).
5. Plunger, perforated disk fastened to a rod.
6. Thermometer.
7. Oven to dry samples at 105°C.
8. Tin cans.
9. Top loading balance.
10. Distilled water.
11. Timer.

Procedure

1. Calibration of hydrometer - Add 100 mL of Calgon solution to the cylinder and make the volume to 1 L with distilled water. Mix thoroughly with plunger and let stand until the temperature is constant (20–25°C). Lower the hydrometer into the solution carefully (avoid dropping hydrometer as it is fragile). Record the scale reading (R_o) at the top edge of the meniscus surrounding the stem.
2. Weigh 50.0 g of soil into a 0.5 L cup, add 100 mL of Calgon solution and about 300 mL of distilled water. Allow to soak for 20 to 30 min. While you wait, weigh another sample of the same soil (about 10 g) in a tin can, transfer into the oven at 105°C for 24 h (samples of clay soils may require longer period) and determine its oven-dried weight.
3. Quantitatively transfer the soil sample from the cup into a dispersing cup of the mixer and mix for 5 min.
4. Quantitatively transfer the suspension from the mixing cup into a cylinder and fill it to the 1 L mark with distilled water. Allow time for the contents to equilibrate with room temperature (20–25°C).
5. Insert the plunger into the suspension and move it up and down to mix the contents thoroughly. Make certain that no material is on the bottom of the cylinder. This can be done with strong upward strokes of the plunger and by spinning the plunger while the disk is just above the sediment. Finish with two to three smooth strokes. Record the exact time of completion of stirring.
6. Lower the hydrometer very carefully into the suspension and record the hydrometer reading after 40 s (R_{40s}) of the completion of stirring. Remove the hydrometer and repeat Steps 5 and 6 using the average of these two readings in the calculations.
7. Remove the hydrometer carefully after the second 40-s reading, rinse, and wipe it clean.
8. Similar to the 40-s reading, obtain a hydrometer reading at 2 h (R_{2h}).

Calculations

$$\% \text{ Silt} + \% \text{ Clay} = \frac{R_{40S} - R_0}{\text{oven dried weight in g}} \times 100$$

$$\% \text{ Clay} = \frac{R_{2h} - R_0}{\text{oven dried weight in g}} \times 100$$

$$\% \text{ Sand} = 100 - (\% \text{ silt} + \% \text{ clay})$$

Comments

Standard methods for particle-size analyses require that soil particles be dispersed in solution by both physical and chemical means. Therefore, some samples may require additional steps to disperse the particles. Some examples are soils containing gypsum, limestone, soluble salts, Fe, and Mn concretions, amorphous minerals, and high in organic matter (>2% organic C). The simplified method described in this section is not recommended for such soils. For special pretreatments and details please refer to Gee and Bauder (1986).

BULK DENSITY AND POROSITY

The bulk density is defined as the ratio of the mass (M) of oven-dried soil to its bulk volume (V), which includes the volume of the particles and the voids (pore space) between the particles. It is commonly designated as ρ_b but for consistency with other chapters we will use the abbreviation BD. Its value for most mineral soils is often about one-half of the density of the solid particles called particle density (PD). Particle density is expressed as the ratio of the total mass of the solid particles to their total volume, excluding the spaces between soil particles. The particle density usually ranges from 2.5 to 2.8 g cm^{-3}. The bulk density is a more dynamic property. It varies with structural condition of the soil; it is altered by cultivation, compression by animals and agricultural machinery, weather, and others. It generally increases with depth in the profile and ranges between 1.0 to 1.7 g cm^{-3}. In swelling soils, it varies with water content (Blake & Hartge, 1986). In such soils, the bulk density obtained should be accompanied by the water content of the soil at the time of sampling.

Bulk density is a widely used measurement. It is required to convert water content in percentage by weight (gravimetric) to percentage by volume (volumetric), to calculate porosity and to estimate soil weights that are too large to weigh directly such as the weight of a ha-30 cm or acre-foot. Bulk density values also are required to convert percentages of different nutrient elements to the weight of nutrients on a volume or area basis. Soil bulk density also can serve as an indicator of soil compaction and relative restrictions to root growth. As illustrated in Table 7–5, however, the relative significance of soil bulk density varies with soil textural class. Different methods for measuring bulk density are described by Blake and Hartge (1986). The most usual and simple method for bulk density is to cut out a cylindrical core of known volume and find the mass of the dried soil. This method is described below:

Method

Equipment and Supplies

1. Core cylinders: core samplers vary in design from a simple thin-walled metal can to a cylindrical sleeve with removable sample cylinders that

Table 7–5. Estimated soil bulk density thresholds for root restricting compacted conditions as determined by soil textural class.

Soil textural class	Minimum bulk density for root restrictions
	$g\ cm^{-3}$
Coarse, medium, and fine sands and loamy sands	1.80
Very fine sand, loamy very fine sand	1.77
Sandy loams	1.75
Loam, sandy clay loam	1.70
Clay loam	1.65
Sandy clay	1.60
Silt, silt loam	1.55
Silty clay loam	1.50
Silty clay	1.45
Clay	1.40

     fit inside. Core sampler can be inexpensively made by cutting the required length of a stainless steel pipe (of about 7.5 cm or larger i.d.) and sharpening one of the edges with the bevel to the outside.
2. Labeled plastic freezer bags. Numbered weighing cans, large enough to hold core samples.
3. Sharp and rigid knife or spatula.
4. Heavy leather gloves.
5. Measuring tape or ruler.
6. Top loading balance.
7. Oven to dry sample at 105°C.

Procedure

1. Place the core sampler on a smooth and even soil surface. Press the sampler into the soil using heavy leather gloves. Use the premarked line on the sampler to determine the depth to push the sampler. With the sampler still in the soil, remove soil from around the container to facilitate the removal of the sampler without loosing any of its contents. Remove the sampler and trim off any excess soil flush with the end of the cylinder using a sharp knife or spatula.
2. Empty the soil contents from the sampler into a preweighed can (quart size paint can work well and they are reusable). Close and weigh the can and the soil. Place the cores in a labeled plastic bag and seal.
3. Dry the soil samples in the oven set at 105°C to a constant weight.

Calculations

Calculate bulk density (BD), water content (G_W, $g\ g^{-1}$), soil porosity (P), and water-filled pore space (WFPS) as follows:

$$BD\ (g\ cm^{-3}) = \frac{\text{Weight of oven-dry soil in grams}}{\text{Volume of soil in core sampler in } cm^3}$$

Weight of oven-dry soil = (Weight of can + oven dry soil) − weight of can.

Volume of soil core = $\pi r^2 h$, where r is the radius of the core and h is the height or depth of the core. A 7 cm long and a 7 cm wide core has a volume of $3.14 \times 3.5 \times 3.5 \times 7 = 269.5$ cm^3

Water content (G_W in %, g g^{-1}) =

$$\frac{\text{Weight of moist soil - weight of oven dry soil}}{\text{Weight of oven dry soil}} \times 100$$

Soil porosity (P in %, cm^3 cm^{-3}) =

$[1 - (BD/PD)] \times 100$ where PD, the particle density is usually 2.65 g cm^{-3}

$$\% \text{ WFPS} = \frac{\% \text{ Volumetric water content}}{\% \text{ Soil porosity}} \times 100$$

where % volumetric water content = % gravimetric water content × soil bulk density

Comments

Cores should be obtained at or near field capacity to avoid soil shattering or compaction during sampling. If soil is compacted and dry, carefully empty about 20 L (5 gallons) of tap water on the area that is to be sampled and allow it to drain for about 2 to 4 h before sampling. This bulk density method is not suitable for loose soil material.

SOIL STRUCTURE AND AGGREGATION

Clay, silt, and sand (called primary particles or soil separates) generally do not occur as discrete particles, but are frequently combined together to form secondary particles (peds, clods, aggregates, or concretions). An aggregate can be defined as a group of two or more primary particles that are bound together forming a stronger unit than the surrounding mass. An individual natural soil aggregate is called a ped, and should not be confused with a clod, a fragment or a concretion. A clod is formed as a result of physical disturbance such as tillage where the soil is molded or broken into a transient unstable mass. A fragment is formed by rupture of a soil mass across natural surfaces or planes of weakness. A concretion is the result of local concentration of compounds that irreversibly cement the soil particles together.

The arrangement and organization of both primary particles and aggregates under field conditions into certain shapes or patterns is referred to as soil structure. Classification of soil structure is based on type, size, and grade. Common structure types include granular, platy subangular and angular blocky, prismatic, and columnar (Fig. 7–1; Table 7–1). Granular structure is typically found in most surface horizons, platy often in E horizons and other types commonly occur in B horizons. Soils vary in their degree of aggregation and have four grades: structureless (no aggregation), weak (structure barely observable), moderate (moderately durable), and strong (durable peds). Description of soils according to the foregoing nomenclature is used for soil classification.

Aspects of soil structure are quantitatively characterized by determining the size distribution and stability of aggregates. The amount and quality of soil organic matter, types of clays, wetting and drying, freezing and thawing, type and amount of electrolytes affecting colloidal dispersion, biological activity, cropping, and tillage systems affect soil aggregation. Improved aggregation increases porosity, especially the macropores, which favor high infiltration rate, good tilth and adequate aeration for plant growth. Aggregate instability or their breakdown by rainfall results in plugging of pores by fine aggregates or particles and restricts water flow–infiltration rate and may cause surface sealing (soil crusting) and anaerobic conditions. Maintenance of crop residues as is the case with conservation tillage systems, protect the soil surface against raindrop impact, thus reducing aggregate breakdown and surface sealing. Also, microorganisms decompose the residue and produce compounds that stabilize aggregates.

Methods

Size Distribution of Aggregates

Various indices have been proposed for expressing the distribution of aggregate sizes. A widely used index is the *mean weight diameter* (MWD) based on weighting the masses of aggregates of the various size classes according to their respective sizes. The most common procedure for testing the water stability of soil aggregates is the *wet sieving method*. This method simulates the action of flowing water. The *dry sieving method*, that pertains to wind erosion prediction is not described in this chapter. These disruptive forces are generally related to water and wind erosion; however, the disintegrating forces occurring during sample taking, preparation, and analysis do not duplicate the field phenomena. Consequently, the relationship between aggregate-size distribution obtained in the laboratory and that existing in the field is somewhat empirical. For detailed discussion, refer to Kemper and Rosenau (1986).

Equipment and Supplies

1. A sieve (20 cm in diam.) with openings of 8 mm and a pan.
2. Sieves (15 cm in diam.) with openings of 4, 2, 1, and 0.5 mm nested together with 4-mm sieve at the top, 0.5-mm sieve at the bottom and others in the middle. The necessary hardware is required to hold the sieves together with a handle to carry them.
3. Bucket, 20 L.
4. Weighing cans or pint size paint or some other cans.
5. Top-loading balance.
6. Oven for drying sample at 105°C.

Procedure

1. Choose a representative area of the field, being careful to avoid tire tracks. Take samples with a shovel or coring tube as described in pro-

cedure under the section on bulk density. Best time to sample is when the soil is not too wet or too dry (for example, just prior to seeding).
2. Transfer the soil sample on the 8-mm sieve and shake gently, allowing soil to collect in the pan. If any aggregates do not pass through the sieve, gently break them up until the entire sample passes through the sieve.
3. Weigh exactly 40 g of the sieved soil and spread it evenly over the top sieve of the sieve nest. Take another exact 40-g subsample in a weighing can, transfer into oven at 105°C for 24 h and determine its oven-dried weight.
4. Place the nest of sieves, including the sample, in the bucket so that it is suspended 5 to 7 cm above the bottom of the bucket. Fill the bucket with water until the water just begins to wet the soil in the top sieve. Allow the soil to wet by absorbing water for 5 min.
5. Move the nested sieves up and down in the water at a rate of 30 oscillations per minute (one oscillation is an up and down stroke of 3.7 cm in length) for 3 min. Other specifications can be used but it is important that they be consistent throughout these measurements.
6. Remove the nest of sieves from the water carefully. Dismantle the nest of sieves and transfer contents from each sieve to weighing or other cans (be sure to label them so you know which sieve they came from!). Place the cans in the oven to determine oven-dried weight of each aggregate size fraction.

Calculations

A = weight of oven dry soil (g) = (weight of can + oven dry soil) – weight of can

Determine weight of oven dry aggregate size fractions by this formula

B = % of sample in 4 to 8 mm size aggregates = weight of oven dry 4 to 8 mm aggregates (g) × 100/A

C = % of sample in 2 to 4 mm size aggregates = weight of oven dry 2 to 4 mm aggregates (g) × 100/A

D = % of sample in 1 to 2 mm size aggregates = weight of oven dry 1 to 2 mm aggregates (g) × 100/A

E = % of sample in 0.5 to 1 mm size aggregates = weight of oven dry 0.5 to 1 mm aggregates (g) × 100/A

F = % of sample in <0.5 mm size aggregates = weight of oven dry <0.5 mm aggregates (g) × 100/A

$$\text{MWD} = \left[\frac{B}{6} + \frac{C}{3} + \frac{D}{1.5} + \frac{E}{0.75} + \frac{F}{0.25}\right] \times 100$$

Where values 6, 3, 1.5, 0.75, and 0.25 are the mean diameter (intersieve) in mm of 4- to 8-, 2- to 4-, 1- to 2-, 0.5- to 1-, and <0.5-mm size fractions, respectively.

Aggregate Stability

Aggregate stability refers to the resistance of soil aggregates to breakdown by water and mechanical manipulation. The most commonly used procedure for

PHYSICAL TESTS

testing the water stability of soil aggregates is the wet sieving method (Kemper & Rosenau, 1986).

Equipment and Supplies

1. Sieves with openings of 2 and 1 mm. The necessary hardware is required to hold the sieves with a handle to carry them.
2. Bucket, 20 L.
3. Weighing cans.
4. Top loading balance.
5. Oven to dry samples at 105°C.

Procedure

1. Take about 50 g of air-dried or field-moist soil and transfer to 2-mm sieve nested with 1-mm sieve. Shake the sieves gently to obtain 1- to 2-mm size aggregates (that pass through the 2-mm sieve but are retained on 1-mm sieve). Take a larger soil sample if you cannot get at least 20 g of 1- to 2-mm aggregates.
2. Weigh 10 g of 1- to 2-mm aggregates and spread it evenly over the 1-mm sieve. Take another 10 g of 1- to 2-mm aggregate subsample in a weighing can and determine its oven-dried weight.
3. Place the sieve with aggregate sample in the bucket filled with water so that the water surface just touches the aggregates. Allow the sample to wet by absorbing water for 5 min.
4. Move the sieve up and down in the water at the rate of 30 oscillations per minute (one oscillation is an up and down stroke of 3.7 cm in length) for 3 min. Other specifications can be used but it is important that they be consistent throughout these measurements.
5. Remove the sieve and transfer the aggregates to weighing cans. Dry the aggregates at 105°C and determine oven-dried weight of aggregates.

Calculations

A = weight of oven dry aggregate sample = (weight of can + weight of oven dry aggregate sample) − weight of can

B = weight of oven dry aggregates remaining on 1-mm sieve (in Step 5) = (weight of can + weight of oven-dry aggregates on 1-mm sieve)

$$\% \text{ water aggregate stability (WAS)} = \frac{B}{A} \times 100$$

Comments

No allowance is made for correcting the weights of aggregates for the coarse primary particles retained on the 1-mm sieve. This is done in standard procedures to avoid designating them falsely as aggregates. Disperse the material collected from each sieve by using a mechanical stirrer and a sodic dispersing agent, then wash the material back through the same sieve. The weight of the sand retained after the second sieving is then subtracted from the total weight of undispersed material retained after the first sieving. The following formula is used to determine the percentage of stable aggregates.

% stable aggregates

$$\frac{\text{weight of 1- to 2-mm aggregates retained} - \text{weight of sand}}{\text{total sample weight} - \text{weight of sand}} \times 100$$

Alternative (Water-Drop) Method for Determining Aggregate Stability

Water-drop method (McCalla, 1944) is based on the number or quantity of water drops required to disintegrate the structure of a unit of soil. It is simple and inexpensive.

Equipment and Supplies

1. Burette (50 mL) with stand and clamps.
2. 1-mm screen.
3. Spatula.
4. Distilled water.

Procedure

1. Place an air-dried soil lump weighing 0.15 g on a 1-mm screen.
2. Calibrate burette so that the distilled water is delivered in the form of drops of uniform size (4 to 5 mm in diam.) at a rate of 1 drop per 5 s.
3. Place and adjust the burette directly on top of the soil lump at a height of exactly 30 cm.
4. Allow the water drops to strike the soil lump.
5. Record the number of drops required to break down the soil lump and wash it through the screen.

Comments

Structural stability of different soils have been studied by this method to determine the effects of microbiological and organic matter treatments and other such factors (Wei et al., 1985). The influence of various cultural practices on different soil types can be assessed and compared with this method. It may be pointed out that for well-aggregated organic matter-rich top soils, the end point for soil lump breakdown is difficult to determine. This is because the lump frequently breaks into several small aggregates which are difficult to keep together (McCalla, 1944).

PENETRATION RESISTANCE

Penetration resistance (PR) is the capacity of the soil in its confined state to resist penetration by a rigid object. Shape and size of the penetrating object must

Table 7–6. Penetration resistance classes (adapted from Soil Survey Staff, 1993).

Classes	Penetration resistance
	Mpa
Extremely low	<0.01
Very low	0.01–0.1
Low	0.1–1
Moderate	1–2
High	2–4
Very high	4–8
Extremely high	8

be defined. Any device designed to measure resistance to penetration is called a penetrometer. Penetration resistance is expressed in units of pressure, typically megapascals (MPa, 10 bars). Penetration resistance depends strongly on the soil water content, which should be specified when reporting these data.

Determination of PR at the field capacity is a useful strategy for evaluation of root limitations; however, a penetrometer has to exert greater pressure than a root tip in penetrating a soil. This is because a penetrometer, unlike a root, cannot diverge from its direct line of advance when a resistant aggregate is in its way. As an example, Ehlers et al. (1983) found that while root growth was severely limited at a penetrometer pressure of 3.6 MPa in conventionally tilled soil, the corresponding limit in untilled soil was higher at about 5 MPa. In the untilled soil, the roots by-passed the resistant barriers using continuous channels left by earthworms (e.g., *Lumbricus* or *Aporrectodea* sp.) and decayed roots that were not preserved in the tilled soil.

As a guideline, a range of PR values in relation to root restriction is shown in Table 7–6. The classes in this table pertain to the pressure required to push the flat end of a cylinderical rod with a diameter of 6.4 mm a distance of 6.4 mm into the soil in about 1 s (Bradford, 1986). A significant restriction of root growth for many important annual crops is encountered at about 2 MPa. Below 1 MPa, root restriction may be assumed to be small. The type of root system (fine seminal roots vs. tap roots) is an important factor that determines the ability of a root to penetrate the soil.

Method

Direct-reading penetrometers in different models and sizes are commercially available and are described in detail by Bradford (1986). The standard instrument is the pocket penetrometer shown in Fig. 7–3. They weigh from 170 to 200 g and are 160 to 180 mm (6–7 in.) long.

Fig. 7–3. A pocket penetrometer.

Equipment and Supplies

1. Pocket penetrometer.
2. Top loading balance

Procedure

1. Select the test location. The location must be defined in relation to horizontal nonuniformities such as tillage relief, wheel tracks, planting rows, and others. Suggested location is the midway between the rows in both wheel-tracked and nontracked interrows. Penetration resistance measurement at the field capacity is recommended because this water content is repeatable from season to season. The insertion may be horizontal or vertical.
2. Move the indicator sleeve to the lowest reading (zero) on the penetration scale.
3. Grip the handle and push the piston needle into the soil at a constant rate of about 3 cm^{-1} until the engraved line 6 mm from the blunt tip is flush with the ground surface
4. Remove the penetrometer from the soil and read the scale.
5. Clean the piston and return the sliding indicator to its zero position. Repeat the test at least five times across a distance of 50 to 60 cm. The sample variance and mean values with standard deviations should be reported.

Comments

For the 6.4-mm diameter tip, the measured force in kilograms is multiplied by 0.31 to obtain the unconfined compressive strength in MPa. To extend the range of the instrument, weaker, and stronger springs may be substituted. For each spring the scale should be calibrated. This can be done by pressing the instrument downward vertically while the tip rests on a top-loading balance.

REFERENCES

Arshad, M.A., and G.M. Coen. 1992. Characterization of soil quality: Physical and chemical criteria. Am. J. Altern. Agric. 7:25–31.

Blake, G.R., and K.H. Hartge. 1986. Bulk density. p. 363–375. *In* A. Klute (ed.) Methods of soil analysis. Part 1. 2nd ed. Agron. Monogr. 9. ASA and SSSA, Madison, WI.

Bouyoucos, G.J. 1962. Hydrometer method improved for making particle size analyses of soils. Agron. J. 54:464–465.

Bowman, R.A., J.D. Reeder, and G.E. Schuman. 1990. Evaluation of selected soil physical, chemical and biological parameters as indicators of soil productivity. p. 64–70. *In* J.W.B. Stewart (ed.) Proc. of the Int. Conf. on Soil Quality in Semi-Arid Agriculture, Univ. of Saskatchewan, Saskatoon. 11–16 June 1990. Saskatchewan Institute of Pedology, Saskatoon, Canada.

Bradford, J.M. 1986. Penetrability. p. 463–477. *In* A. Klute (ed.) Methods of soil analysis. Part 1. 2nd ed. Agron. Monogr. 9. ASA and SSSA, Madison, WI.

Ehlers, W., V. Kopke, F. Hesse, and W. Böhm. 1983. Penetration resistance and root growth of oats in tilled and untilled loess soil. Soil Tillage Res. 3:261–275.

Gee, G.W., and J.W. Bauder. 1986. Particle size analysis. p. 383–409. *In* A. Klute (ed.) Methods of soil analysis. Part 1. 2nd ed. Agron. Monogr. 9. ASA and SSSA, Madison, WI.

Kemper, W.D., and R.C. Rosenau. 1986. Aggregate stability and size distribution. p. 425–441. *In* A. Klute (ed.) Methods of soil analysis. Part 1. 2nd ed. Agron. Monogr. 9. ASA and SSSA, Madison, WI.

Lowery, B. M.A. Arshad, L. Lal, and W.J. Hickey. 1996. Soil Water parameters and soil quality. p. 143–155. *In* J.W. Doran and A.J. Jones (ed.) Methods for assessing soil quality. SSSA Spec. Publ. 49. SSSA, Madison, WI.

Malhi, S.S., R.C. Izaurralde, M. Nyborg, and E.D. Solberg. 1994. Influence of topsoil removal on soil fertility and barley growth. J. Soil Water Conserv. 49:96–101.

McCalla, T.M. 1944. Water-drop method of determining stability of soil structure. Soil Sci. 58:117–121.

Pierce, F.J. 1991. Erosion productivity impact prediction. p. 35–52. *In* R. Lal and F.J. Pierce (ed.) Soil management for sustainability. Soil Water Conserv. Soc., Ankeny, IA.

Power, J.F., F.M. Sandoval, R.E. Ries, and S.D. Merrill. 1981. Effects of topsoil and subsoil thickness on soil water content and crop production on a disturbed soil. Soil Sci. Soc. Am. J. 45:124–129.

Sarrantonio, M., J. Halvorson, and J.W. Doran. 1996. On-farm assessment of soil health. p. 83–105. *In* J.W. Doran and A.J. Jones (ed.) Methods for assessing soil quality. SSSA Spec. Publ. 49. SSSA, Madison, WI.

Soil Survey Staff. 1993. Soil survey manual. USDA-SCS. U.S. Gov. Print. Office, Washington, DC.

Wei, Q.P., B. Lowery, and A.E. Peterson. 1985. Effect of sludge application on physical properties of a silty clay loam soil. J. Environ. Qual. 14:178–180.

8 Soil Water Parameters and Soil Quality

Birl Lowery and William J. Hickey
University of Wisconsin
Madison, Wisconsin

M. A. (Charlie) Arshad
Agric and Agri-Food Canada
Beaverlodge, Alberta, Canada

Rattan Lal
Ohio State University
Columbus, Ohio

Water, comprising two-thirds of the Earth's surface, is fundamental to life on this planet, and soil water is essential to soil organisms and plant life. Soil organisms such as ants, beetles, and earthworms reside in the soil matrix but much of the microfauna, which includes protozoa, rotifers, and nematodes, reside in soil water. Soil is a porous media in which soil water and air are found in the pores. Consequently, the nonlimiting water range for plant and microbial activity is greatly influenced by soil aeration and mechanical resistance, particularly so in dense, poorly structured soils (Letey, 1985). When soil is water-saturated and air is excluded for long periods of time, many soil organisms suffer from a lack of O_2. Plant and soil organisms require optimum levels of both water and O_2, so the ratio of water- to air-filled pores is critical. Linn and Doran (1984) noted that soil microbial activity under different tillage systems appears to be closely related to water-filled pores (percentage of saturation) of soil but aeration drives the system.

Relationships or interactions between soil and water are complex. Not only does soil water contain various chemicals that influence its behavior, but flow and retention of water in soils are keys to our understanding of soil processes. In addition to its considerable impact on soil physical, chemical, and biological properties, water plays a major role in soil formation and its changes over time. As water moves through soil, it leaches chemicals and small soil particles from the upper soil and displaces them within the lower soil profile. Water flows through soil under the force of gravity and the influence of soil matrix and other potentials. On the other hand, water is retained in the soil by soil adhesive forces. The

---

Copyright © 1996 Soil Science Society of America, 677 S. Segoe Rd., Madison, WI 53711, USA.
Methods for Assessing Soil Quality, SSSA Special Publication 49.

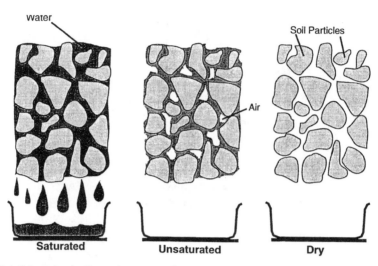

Fig. 8–1. Schematic of soil at various stages of saturation.

amount of water retained by a given soil affects that soils ability to support biological activities, including plant growth since plants obtain water from the soil.

Soil is a major reservoir for water. If one-half of the total porosity (which is 50% in an ideal soil) of soil that is 1 m deep is full of water to a depth of 1 m, this represents a 25-cm depth of water, and a hectare-meter of soil contains 2.5×10^3 m^{-3} of water. It is this water that is retained in the root zone that is the lifeline of soil organisms and plants. Soil water retention and flow characteristics are dynamic properties, undergoing tremendous changes over time and space. These properties are controlled by soil physical properties such as texture and structure.

In this chapter, we have outlined some alternative and very simple, but reliable methods for measuring some key properties of soil water essential for comparing the impact of different soil management techniques; however, a more complete set of scientifically accepted methods for measuring soil physical properties and soil–water relations have been presented by Klute (1986a). These methods include a range of procedures from very complicated to some simple ones.

SOIL WATER CONTENT

Soil consists of a three-phase system composed of solid (soil particles), liquid (water and solutes), and gas (air; Fig. 8–1). The solid part of soil is not a continuous mass of solid material, but it is broken into individual grains or aggregates with open space (pores) between the solid matrix that can contain the other two phases (water and air). When soil is void of water all the pores are filled with air; however, if the air is replaced with water, the soil is said to be saturated (Fig. 8–1). Soil water content at saturation is an indication of the soils total porosity. A 1 m^3 volume of soil with 50% pores has 0.5 m^3 of pores. When this soil is saturated, it will contain 0.5 m^3 of water.

Soil water content is a basic soil parameter and a quantitative evaluation of it is needed for almost every aspect of soil and related sciences from those dealing with soil organisms and plant growth to environmental concerns; however, methods of measuring water content, other than the oldest and standard of oven drying, are often not widely accepted. Some people feel new methods are too expensive and difficult to be used and/or adopted by a wide range of scientists and others. There have been many devices developed for indirectly measuring soil water content, including very expensive ones such as time domain reflectometry (TDR), based on the dielectric constant of a soil (Topp & Davis, 1980) and neutron moisture gauges, based on thermalization of fast neutrons by H atoms in the soil, and low-cost but not as accurate ones such as moisture resistance cells, based on electrical conductivity of fixed electrodes in soil (gypsum blocks, fiberglass cells, and water mark sensors; Gardner, 1986). The TDR method of measuring water content has become widely accepted among research scientists and engineers. Excellent reviews of the development and use of TDR have been presented by Zegelin et al. (1992) and Dalton (1992). For most practical uses, however, where an occasional sample is needed, oven drying a sample of soil is a good method. It should be noted that repeated sampling will result in soil disturbance. The oven-drying method can be done in one of two ways. The simplest is to collect a soil sample, determine the wet weight, then dry it, and weigh it again. Water content is expressed as the mass of water per unit mass of soil and this is known as the *gravimetric method*. The second, and most useful, method is to take a known volume of soil and weigh it, then dry the soil, and weigh it again. In this case, water content is expressed as the volume of water per unit volume of soil. This is known as the *volumetric method*. From this latter sampling, one can obtain soil water content as well as bulk density (the weight or mass of a given volume of dry soil) of the soil. Using the bulk density, the total porosity can be obtained and used to determine the percentage of saturation (Hillel, 1982) or water-filled pore space (Linn & Doran, 1984).

Equipment

1. Metal cylinder (irrigation pipe or food can);
2. Balance for weighing samples;
3. Oven for drying samples at 105°C;
4. Shovel to extract soil samples and a block of wood and hammer to insert cylinder into soil;
5. Paint cans and plastic bags;
6. Pie or pizza pan.

Method

1. In order to measure volumetric water content, it is necessary to obtain a known volume of soil. This can be done using a cylinder of known volume. A cylinder can be constructed from a used food or coffee can or an aluminum cylinder cut from a section of irrigation pipe (as discussed in

Sarrantonio et al., 1996, this publication). The volume is obtained by carefully measuring the volume of water required to fill the can or by the difference in weight from the empty can and the weight when the can is filled with water. In the metric system, 1 cm^3 (which is 1 mL) of water is equal to 1 g of water at 4°C (Nobel, 1974). The volume also can be obtained by measuring the inside diameter and length of the can/cylinder. The volume (V) can be obtained from the following equation

$$V = \pi r^2 h,$$

where r is the radius and h is the height of the can/cylinder. When the numerical value of π (3.14) is inserted and the radius is replaced with the diameter (d) of the cylinder, the above equation becomes:

$$V = 0.785 \, d^2 \, h$$

2. Take a soil sample by slowly pushing the cylinder into the soil. To do this, it might be necessary to place a board on top of the cylinder. The cylinder should be completely filled with soil. This can be accomplished by placing a short ring the size of the cylinder above it during sampling. Dig the cylinder full of soil out and cut off excess soil with a knife (the knife blade should be longer than the diameter of the cylinder) or steel spatula. Collect a minimum of three samples for each treatment.
3. Place the cylinder full of soil into a plastic bag. If more than one sample is to be taken, remove the soil from the cylinder and store it in a bag with a label indicating where the sample was taken. To minimize water loss, the sample should be double-bagged. Another, and perhaps better, storage container would be an unused paint can (paint cans can be purchased from local paint companies). Paint cans provide a good seal against water loss from sample by evaporation.
4. Weigh the sample (including the bag or paint can). Remove the sample from the bag and place it on a pie or pizza pan. Dry the sample at 105°C for 24 h or until the weight of dry soil is stable. If a paint can is used to store the sample and transport it from the field, the sample can be dried in the can. The can with lid removed can be placed directly into the oven. If the sample is not intact and contains little water, it will take less time to dry than if the sample is saturated and intact. After 24 h of drying, let sample cool, then obtain the weight of dry soil plus container. After this has been done, remove (all) the soil and get a weight of the empty container. The difference between the two weights is the weight of the soil (Note: the wet soil also must be corrected for container weight).

Calculations

The volumetric water content (V_w), the volume of water per unit volume of soil, can be expressed in any units such as m^3 m^{-3} (cm^3 cm^{-3}) or percentage as indicated in the following equation.

$$\% \text{ water on a volume basis} = \frac{(\text{wet soil weight}) - (\text{dry soil weight})}{(\text{volume of sample})(\text{density of water})^*} \times 100$$

* If you are using metric units, the density of water is 1000 kg m^{-3} (1 g cm^{-3}).

The gravimetric water content (G_w), mass of water per unit mass of soil, can be expressed in any units such as kg kg^{-1} (g g^{-1}) or percentage as indicated in the following equation:

$$\% \text{ water on a weight basis} = \frac{\text{wet soil weight} - \text{dry soil weight}}{\text{dry soil weight}} \times 100.$$

Additional Parameters Obtained from Water Content Sampling

The depth of water (D_w) in soil is a common means of expressing soil water, especially with respect to the amount needed to grow a crop. This can be determined from volumetric water content for a given depth of soil using the following equation:

Depth of water (D_w) = (volumetric water content)(depth of soil).

The resulting value of D_w can be expressed in meters, centimeters, or any length units.

The bulk density (BD) (mass of soil for a given volume) of soil also can be determined from a sample collected for volumetric water content determination. This can be obtained from the following formula.

$$\text{Bulk density} = \frac{\text{dry soil weight}}{\text{volume of soil}}.$$

Bulk density is expressed in g cm^{-3} or kg m^{-3}. If the bulk density is known, volumetric water content can be determined from gravimetric water content using the following formula

$$V_w = [(BD)(G_w)]/DW,$$

where DW is the density of water and assumed to be 1 g cm^{-3}.

The total porosity (P) of a soil can be obtained from bulk density by

$$P = 1 - (BD/PD),$$

where PD is the particle density. The particle density of most soils is 2.65 g cm^{-3} (2650 kg m^{-3}).

Air-filled porosity (P_a) is needed for many soil-related investigations and has been found to be a good indicator of soil biological and chemical activities. This can be obtained by the difference between the P and the volumetric water content (i.e., $P_a = P - V_w$). The percentage saturation, (amount of pores filled with water relative to total volume of pores) S given by

$$S = V_w/P.$$

Another good indicator of aeration is water-filled pore space (WFPS), which was reported by Linn and Doran (1984) as WFPS = V_w/P. You will note that this is the same as S. They suggested that this is a good indicator of aeration-dependent microbial activity. They reported maximum relative aerobic microbial activity at 60% saturation (WFPS). A diagrammatic illustration of the relationship between WFPS and microbial activities important to C and N cycling in soil is given by Parkin et al. (1996, this publication). For temperate-region soils, this relationship is consistent across a wide range of soil textures (Doran et al., 1990).

WATER INFILTRATION

Water supplied by rainfall or irrigation must enter the soil before it can be of value to soil organisms and plants. The process of water entering the soil is called infiltration. The rate at which water infiltrates (infiltration rate) into soil is a function of the type of soil, its physical condition such as aggregation and stability of aggregates, and the water content. The initial water content (the water content when the infiltration process first starts) influences the soils ability to pull water into small pores. Water infiltrates into soil because of gravity and matrix potentials. Gravity provides a downward force on water and it is pulled into soil (at the wetting front) under a negative pressure (matrix potential), which is greater in small pores than large pores and is a function of soil water content. Although the negative pressure that pulls water into a clay soil is greater than that for a sand, the infiltration rates for the two soils are often vastly different. The rate for the sand is much greater, particularly if the clay has little or no structure. Infiltration rate for most sand and well-structured soil is much greater than that of a massive clayey soil. A crust tends to form on some soils, especially those with a large surface silt content and low organic matter content. Surface crusts severely reduce water infiltration. Low infiltration rate could lead to soil erosion and limit the amount of soil profile water recharge (Johnson et al., 1984; Andraski et al., 1985; Andraski & Lowery, 1992).

As previously noted, the infiltration rate of a particular soil is a function of the soils initial water content. When soil is wet, the initial infiltration rate will be smaller than that of a dry soil. Thus, for comparing infiltration rates of different soils, the soils should be at a similar moisture content when infiltration measurements are made. We suggest that infiltration measurements be made when soils are at field moisture capacity.

The two common methods of assessing infiltration are (i) sprinkler, where water is sprayed on the soil to simulate rainfall and (ii) flooded infiltration, where water is flooded on the soil surface and water intake rate is obtained. The sprinkler method is not simple but it can be done if one is willing to construct the proper device, as described by Peterson and Bubenzer (1986). The flooded/ponded infiltration method is simple, and we will suggest the use of a food or coffee can or metal cylinder similar to that described in Sarrrantonio et al. (1996, this publication) for this method. Ponded infiltration can be done with a single- or double-ring infiltrometer. For the relative comparisons necessary for soil quality assessment, we suggest the single ring.

Equipment

1. Cylinder (irrigation pipe or food can);
2. Bentonite or well-drilling mud;
3. Container of water (3 to 4 L).

Methods

Single-Ring (Ponded Infiltration) Infiltrometer

1. A large-diameter cylinder (i.e., 15 cm or greater) can be used as the single-ring infiltrometer. Thus, the single-ring infiltrometer also can be constructed from aluminum irrigation pipe or a large can or bucket. Cut the bottom (assuming the top has been removed) from the can or bucket, or cut off a section of irrigation pipe about 30 cm long. The volume of the cylinder should be obtained. This can be done by measuring the diameter and height of the cylinder and calculating its volume. Make a reference (zero) mark near the bottom of the cylinder (10 cm from bottom). Attach a ruler or make 0.5-cm marks along the upper part of the cylinder with a permanent marker.
2. Place a wooden block on the cylinder and push it into the soil to the reference (zero) mark.
3. One to two days prior to measuring the infiltration, fill the cylinder with water and let it drain. This will bring the soil to field moisture capacity. As discussed later in this chapter, to obtain field moisture capacity the soil is saturated and allowed to drain for 24 h or more depending on soil texture.
4. After the soil has equilibrated to field moisture capacity, infiltration can be measured. Make a paste of clay (bentonite or well-drilling mud) and apply it along the edge of the cylinder (along the inside of cylinder) to prevent water from channeling along cylinder walls.
5. Quickly fill the cylinder with water and record the time required for water to move from one 0.5- or 1-cm mark to the next. This will provide estimates of change in infiltration rate over time.
6. Calculate the infiltration rate (i) using the following equation:

$$i = (VW/t)(1/A),$$

where VW is the volume of water that infiltrates, t is the time required for water to infiltrate, and A is the area of cylinder ($A = \pi r^2$). From this equation, the infiltration rate can be obtained in any length per unit time units such as cm/min. Using these data, one can compare the infiltration rate obtained for different management systems to assess their impact on soil quality. We anticipate that improved soil quality will result in greater and more stable aggregates, thus producing higher infiltration rates for a given soil. It is important to note that in some cases high infiltrations rates have been associated with rapid leaching of certain chemicals such as NO_3^- to groundwater. The specific relevance of soil infiltration rates to soil quality, however, is largely determined by soil landscape position and local climate characteristics which define the seasonal distribution and intensity of rainfall. There is a potential for excessive soil erosion for those soils with low infiltration rates, located on steep landscape positions under high-intensity rainfall.

ASSESSING WATER FLOW RATE (HYDRAULIC CONDUCTIVITY)

The rate at which water flows through soil affects many properties of the soil, including infiltration, drainage, solute (nutrient) flow within the soil, and soil erosion. Water flows rapidly through sandy soil and soil with stable aggregate and structure, which reduces the potential for soil erosion and the need for artificial drainage, assuming there is no restrictive layer close to the soil surface (Lowery et al., 1981).

We present a method for assessing water flow via the hydraulic conductivity of saturated soil cores removed from the field. Hydraulic conductivity (K) values may provide a potential means for comparing the impact of different management practices on water flow. For example, large macropores created by earthworms will result in rapid water movement under saturated conditions (Linden et al., 1994). Abundant earthworm activity has been attributed to good soil quality. Because it is difficult to collect a representative sample of the many field conditions, such as macropores, it should be noted that in situ measurements of K are preferred; however, we are not aware of a low-cost, simple, in situ technique for measuring saturated hydraulic conductivity (K_s) in unsaturated soil. The most commonly accepted technique for field measuring K_s is the constant head/borehole permeameter. Recent reviews of current devices and methods have been provided by Amoozegar (1992), Elrick and Reynolds (1992), and Stephens (1992).

For the simple method described in this chapter, a section of irrigation or clear plastic pipe is recommended as a sampling device. Water will be ponded in the pipe similar to what is done in the single-ring infiltrometer, but unlike infiltration, the sample will be removed from the field for this experiment. The rate at which water flows from the pipe through the soil is an indication of the hydraulic conductivity of a given soil under saturated conditions. This procedure is called the falling head permeameter (Spangler & Handy, 1982; Klute & Dirksen, 1986).

Equipment

1. Clear plastic or other pipe, approximately 1 m long;
2. Shovel;

3. Cloth (nylon) and rubber bands or tape;
4. Ruler (1 m long).

Methods

1. A clear plastic pipe is best suited as a sampling device for this experiment. A 10-cm diam. or larger clear plastic or an irrigation pipe is recommended, but a 7.6-cm diam. pipe could be used if a larger pipe is not available. The pipe should be at least 100 cm long.
2. The area (A) of soil sampled depends on the diameter of the pipe. The area of the pipe is determined by:

$$A = \pi r^2 = 0.785 \, d^2,$$

where r is the radius and d is the diameter. This is the area of both the soil and the stand pipe. The K_s can be determined from the following equation:

$$K_s = (L/t) \ln(h_1/h_2),$$

where, L is the length of sample, t is time it takes for a column of water to move from some height, h_1, above the top of the soil to a second position, h_2, above the soil surface. Multiple readings are recommended.
3. If plastic pipe is used, one end should be sharpened to make it easy to push it into the soil. The pipe can be sharpened by turning it in a lathe or with a file. Make a mark on the pipe about 20 cm (this can vary from 5 to 30 cm) from (the bottom) the sharp end. This will be the length of sample (L in the equation in Step 2).
4. Saturate the area from which the sample is to be collected at least 1 or 2 h before collecting a sample. This can be accomplished by ponding water in the area. Cut the bottom from a large bucket (note this bucket should be several cm larger than the sample pipe). Place the bucket in the area that is to be sampled and fill it with water. Allow several hours for the water to infiltrate and distribute through the profile.
5. Before collecting a sample, push the unsharpened end of the pipe into the soil to about 2 cm and carefully remove it without removing the soil. Fill the ring depression created by the pipe with a clay slurry (or drilling mud slurry). Slurry will prevent water channeling along the side wall of the pipe. Values of K_s have been shown to vary considerably over a short distance; thus, to minimize variability, the area sampled and the number of samples should be large. It is recommended that six or more samples per treatment be evaluated.
6. Push the sharpened end of pipe into the soil, up to the mark (line that indicates sample length). Using a shovel, remove the tube with the soil sample and place a nylon cloth over the end to hold the soil in place. Secure the cloth with a large rubber band or tape.
7. Take the sample to the laboratory or workshop to make measurements. Place it on a ring stand or suitable support device. Attach a ruler to the

outside of the pipe with the zero mark located at the top of the soil. Fill the tube with tap water and measure the time required for the water to flow at 10-cm intervals. If a 1-m ruler is used, one can obtain six to seven readings (depending on length of sample). Using the equation in Step 2, calculate the hydraulic conductivity of the saturated soil. Using the values obtained for h_1, h_2, and t, calculate the hydraulic conductivity for each set of values then average these for a mean K_s value.

Large values of hydraulic conductivity from this experiment indicate rapid water movement under saturated conditions. Such values are expected for good quality soil. Except for very sandy soils, large average values obtained from some management systems may be indicative of soils that have improved aggregation and greater macroporosity, which may be related to greater soil biological activity. Well-aggregated soils will exhibit similar behavior.

FIELD WATER-HOLDING CAPACITY

Field water-holding capacity (FWHC) is the amount of water retained by the soil after it has been saturated in the field and allowed to drain for 12 to 48 h. This assumes that drainage will be negligible after this time. The time required to reach FWHC depends on soil texture. Most sandy soils (without layers) will take 12 to 24 h, while clay soils require 48 h or more before the drainage becomes negligible. For practical purposes, the FWHC is the amount of water retained by the soil against a tension of 10 to 33 kPa. The amount of water retained by the soil depends on the number and size of pores that are primarily governed by the texture, structure, organic matter, and mineralogy of the soil. Water from large pores (*macropores*) will drain under the influence of gravity or under a low suction or pressure. High negative pressure or large suction needs to be applied to drain the water from small pores (*micropores* or *capillary pores*). The *permanent wilting point* (PWP) is the soil water content below which plants cannot extract sufficient water from the soil to meet their transpiration demand and they are unable to recover from wilting when provided sufficient water. The PWP has been technically defined as the water content held against a tension of 1500 kPa or the amount of water that remains in a soil sample that has been equilibrated at 1500 kPa. The water that remains after a soil reaches FWHC and before it dries to the PWP can readily be absorbed by plant roots and is known as the *available water holding capacity* (AWC) of the soil. Thus,

AWC = (soil water content at FWHC) − (soil water content at PWP).

Equipment

1. Cylinder (irrigation pipe);
2. Nylon cloth and rubber bands or tape;
3. Wire mesh;
4. Shovel;

5. Container of water;
6. Oven and balance for drying and weighing sample.

Method

1. Take a soil core sample using a section of irrigation pipe or a food can. Three to six replicates are suggested. Cores should be 5 to 8 cm in diam. and about 2 to 3 cm in height. Select a flat area or level the soil for collecting samples, push the cores into the soil completely filling the cylinder. Remove the cylinder with the soil. Cut or trim the ends of the core flat with the ends of the cylinder. Fasten a piece of cheesecloth or nylon cloth to the lower end of the core secured by a rubber band or tape.
2. Place the cores in a pan of water to obtain maximum or saturated water contents. Soak the cores overnight in water at a level just below the top of the core. Soaking overnight should result in saturation. Time required for saturation varies with the type of soil, sandy soil may take less than an hour, while clay soil takes much longer.
3. Remove the saturated soil cores from the water and place them on a screen or other wire mesh (rack) with sufficient space below for water to drain from the sample. At this stage, the soil is near its maximum water retention capacity.
4. Cover the top of the core with a sheet of plastic to prevent evaporation. Let it drain for 48 to 72 h.
5. Determine water content as previously described in the section on soil water content. This volume of water in the soil at the end of 48 to 72 h is the FWHC.

Water holding capacity could also be estimated after the hydraulic conductivity is measured by allowing the soil to drain for 2 to 3 d and measure the water content as noted in Step 5. An approach for estimating soil water-holding capacity in the field is given by Sarrantonio et al. (1996, this publication).

WATER RETENTION

The amount of water retained by a given soil under a range of tensions is another important indicator of soil quality. Since there are no simple methods for assessing this the reader is referred to Klute (1986b) for laboratory methods and Bruce and Luxmoore (1986) for field methods. Soil water retention characteristics can be estimated, however, from components of the minimum data set given in Doran and Parkin (1996, this publication), such as particle-size distribution, organic matter content, and bulk density (Gupta & Larson, 1979). Management that results in increased soil organic matter is thought to improve soil quality. Soil aggregation improves with increasing organic matter resulting in more macropores. With the increase in macropores, total porosity increases and water retention at low tensions is less but the AWC increases.

REFERENCES

Amoozegar, A. 1992. Compact constant head permeameter: A convenient device for measuring hydraulic conductivity. p. 31–42. *In* G.C. Topp et al. (ed.) Advances in measurement of soil physical properties: Bringing theory into practice. SSSA Spec. Publ. 30. SSSA, Madison, WI.

Andraski, B.J., T.C. Daniel, B. Lowery, and D.H. Mueller. 1985. Runoff results from natural and simulated rainfall for four tillage systems. Trans. ASAE 28:1219–1225.

Andraski, B.J., and B. Lowery. 1992. Erosion effects on soil water storage, plant water uptake and corn growth. Soil Sci. Soc. Am. J. 56:1911–1919.

Bruce, R.R., and R.J. Luxmoore. 1986. Water retention: Field methods. p. 635–662. *In* A. Klute (ed.) Methods of soil analysis. Part 1. 2nd ed. Agron. Monogr. 9. ASA and SSSA, Madison, WI.

Dalton, F.N. 1992. Development of time-domain reflectometry for measuring soil water content and bulk soil electrical conductivity. p. 143–168. *In* G.C. Topp et al. (ed.) Advances in measurement of soil physical properties: Bringing theory into practice. SSSA Spec. Publ. 30. SSSA, Madison, WI.

Doran, J.W., L.N. Mielke, and J.F. Power. 1990. Microbial activity as regulated by soil water-filled pore space. p. 94–99. *In*Trans. 14th Congr. Int. Soil Sci. Soc., Kyoto, Japan. 12–18 Aug. 1990. ISSS, Wagenegen, the Netherlands.

Doran, J.W., and T.B. Parkin. 1996. Quantitative indicators of soil quality: A minimum data set. p. 25–37. *In* J.W. Doran and A.J. Jones (ed.) Methods for assessing soil quality. SSSA Spec. Publ. 49. SSSA, Madison, WI.

Elrick, D.E., and W.D. Reynolds. 1992. Infiltration from constant-head well permeameters and infiltrometers. p. 1–24. *In* G.C. Topp et al. (ed.) Advances in measurement of soil physical properties: Bringing theory into practice. SSSA Spec. Publ. 30. SSSA, Madison, WI.

Gardner, W.H. 1986. Water content. p. 493–544. *In* A. Klute (ed.) Methods of soil analysis. Part 1. 2nd ed. Agron. Monogr. 9. ASA and SSSA, Madison, WI.

Gupta, S.C., and W.E. Larson. 1979. Estimating soil water retention characteristics from particle size distribution, organic matter percent, and bulk density. Water Resour. Res. 15:1633–1635.

Hillel, D. 1982. Introduction to soil physics. Academic Press, New York.

Johnson, M.D., B. Lowery, and T.C. Daniel. 1984. Soil moisture regimes of three conservation tillage systems. Trans. ASAE 27:1385–1390.

Klute, A. (ed.). 1986. Methods of soil analysis. Part 1. 2nd ed. ASA and SSSA, Madison, WI.

Klute, A. 1986. Water retention: Laboratory methods. p. 635–662. *In* A. Klute (ed.) Methods of soil analysis. Part 1. 2nd ed. Agron. Monogr. 9. ASA and SSSA, Madison, WI.

Klute, A., and C. Dirksen. 1986. Hydraulic conductivity and diffusivity: Laboratory methods. p. 687–732. *In* A. Klute (ed.) Methods of soil analysis. Part 1. 2nd ed. Agron. Monogr. 9. ASA and SSSA, Madison, WI.

Letey, J. 1985. Relationship between soil physical properties and crop production. Adv. Soil Sci. 1:277–294.

Linden, D.R., P.F. Hendrix, D.C. Coleman, and P.C.J. van Vliet. 1994. Faunal indicators of soil quality. p. 91–106. *In* J.W. Doran et al. (ed.) Defining soil quality for a sustainable environment. SSSA Spec. Publ. 35. ASA, Madison, WI.

Linn, D.M., and J.W. Doran. 1984. Effect of water-filed pore space on carbon dioxide and nitrous oxide production in tilled and untilled soils. Soil Sci. Soc. Am. J. 48:1267–1272.

Lowery, B., G.F. Kling, and J.A. Vomocil. 1981. Overland flow from sloping land: Effects of perched water tables and subsurface drains. Soil Sci. Soc. Am. J. 46:93–99.

Nobel, P.S. 1974. Biophysical plant physiology. W.H. Freeman & Co., San Francisco, CA.

Parkin, T.B., J.W. Doran, and E. Franco-Vizcaino. 1996. Field and laboratory tests of soil respiration. p. 231–245. *In* J.W. Doran and A.J. Jones (ed.) Methods for assessing soil quality. SSSA Spec. Publ. 49. SSSA, Madison, WI.

Peterson, A.E., and G.D. Bubenzer. 1986. Intake rate: Sprinkler infiltrometer. p. 845–867. *In* A. Klute (ed.) Methods of soil analysis. Part 1. 2nd ed. Agron. Monogr. 9. ASA and SSSA, Madison, WI.

Sarrantonio, M., J.W. Doran, M.A. Liebig, and J.J. Halvorson. 1996. On-farm assessment of soil quality and health. p. 83–105. *In* J.W. Doran and A.J. Jones (ed.) Methods for assessing soil quality. SSSA Spec. Publ. 49. SSSA, Madison, WI.

Spangler, M.G., and R.L. Handy. 1982. Soil engineering. 4th ed. Harper & Row Publ., New York.

Stephens, D.B. 1992. Application of the borehole permeameter. p. 43–68. *In* G.C. Topp et al. (ed.) Advances in measurement of soil physical properties: Bringing theory into practice. SSSA Spec. Publ. 30. SSSA, Madison, WI.

Topp, G.C., and J.L. Davis. 1980. Electromagnetic determination of soil water content: Measurements in coaxial transmission lines. Water Resour. Res. 16:574–582.

Zegelin, S.J., I. White, and G.F. Russell. 1992. A critique of the time domain reflectometry technique for determining field soil-water content. p. 187–208. *In* G.C. Topp et al. (ed.) Advances in measurement of soil physical properties: Bringing theory into practice. SSSA Spec. Publ. 30. SSSA, Madison, WI.

9 Soil Organic Carbon and Nitrogen

Lawrence J. Sikora

USDA-ARS
Beltsville Agricultural Research Center
Beltsville, Maryland

D. E. Stott

USDA-ARS
Purdue University
West Lafayette, Indiana

The Soil Science Society of America defines soil organic matter (SOM) as the organic fraction of soil exclusive of undecayed plant and animal residue (SSSA, 1987). Most analytical procedures for SOM, however, do not distinguish between decomposed and undecomposed plant and animal residues that pass a 2-mm sieve during the soil preparation process. In the broadest sense, SOM consists of diverse components such as living organisms, slightly altered plant and animal organic residues, and well-decomposed organic residues that vary considerably in their stability and susceptibility to further decomposition (Magdoff, 1992).

Soil organic matter has long been considered the key quality factor of soil. Soil organic matter is a source of and a sink for plant nutrients in soils and is important in maintaining soil tilth, aiding infiltration of air and water, promoting water retention, reducing erosion, and controlling the efficacy and fate of applied pesticides (Gregorich et al., 1993). Soil organic matter impacts soil productivity in a number of ways (Stott & Martin, 1990). It serves as a storehouse for plant nutrients that are released slowly, especially N. Organic matter aids in the solubilization of plant nutrients from insoluble minerals present in the soil, has a high adsorptive or exchange capacity for nutrient cations, and aids in trace element nutrition of plants through chelation reactions. Increases in a soil's organic matter content lead to a greater and more varied soil population, thus increasing biological control of plant diseases and pests. Soil organic matter's dark pigment favors absorption of heat by the soil, thus speeding the warming of the soil in the spring.

Soil organic matter also impacts the partitioning of precipitation that affects soil productivity, soil erosion by water, and water conservation (Stevenson, 1994). Organic matter increases the water holding capacity of soil, thereby decreasing the potential for saturated soil conditions and runoff events. Increases

---

Copyright © 1996 Soil Science Society of America, 677 S. Segoe Rd., Madison, WI 53711, USA.
Methods for Assessing Soil Quality, SSSA Special Publication 49.

in SOM have long been associated with increased soil aggregate stability (Tisdall & Oades, 1982) and mean aggregate diameter. Larger, denser aggregates are less susceptible to wind erosion than smaller sized aggregates. The breakdown of aggregates at the soil surface is considered to be the first step in crust formation and surface sealing (Luk et al., 1990). Surface sealing impedes water infiltration and increases erosion.

Soil organic matter has been shown to adsorb both natural and anthropogenic toxic substances thereby reducing their toxicity (Stott & Martin, 1989). Upchurch et al. (1966) found that phytotoxicity of aromatic herbicides was inversely related to SOM level. Application rates for herbicides are often tied to the SOM content.

The fractions of soil organic C can be divided by biological, physical or chemical basis. Based on physical characteristics, soil can be fractionated by dry or wet mechanical sieving that will separate soil particles into sand, silt and clay-sized fractions. Each fraction has OM associated with it and levels in each fraction are affected by management practices. Dalal and Mayer (1986) found that after cultivation the organic C associated with sand-sized particles declined rapidly; that associated with silt-sized particles declined the least; and that associated with clay-sized particles increased from 48 to 61%. These data suggest that clay particles protect SOM. Research efforts have attempted to characterize fractions of OM below 2 mm in size. Soil organic matter between 0.20 and 2.0 mm is considered macroorganic matter (MOM). Between 0.050 and 0.20 mm, the SOM is called particulate OM or POM. Chemically, SOM can be divided by solubility in alkali, acids, or organic solvents. The classical alkaline extraction followed by mineral acid precipitation separates and measures the amount of fulvic acid, humic acid, and humins (Stevenson, 1994). Biologically, SOM has been classified as readily available to degradation, moderately available and slowly available or recalcitrant. Bell (1993) referred to the biological division of SOM, namely soils having *active* and *passive* C fractions. In general, these fractions are pools of readily decomposable OM and plant nutrients with a short half life (Gregorich et al., 1993). The role of the biological and physical fractions of SOM in determining soil quality is a subject covered in other chapters in this book.

The ratio of C to N in OM provides information on the capacity of the soil to store and recycle nutrients. The C/N ratio of soils is approximately 10 to 1. Higher ratios may indicate recent additions of manure or plant residues. Rasmussen et al. (1980) found increasing ratios with long-term burning of wheat (*Triticum aestivum* L.) and decreasing ratios in soils receiving manure. Kononova et al., (1975) noted that in land of the same soil type but used differently, the C/N ratios varied widely and depended on the care taken in removing semidecomposed plant residues during preparation of the soil samples for analysis.

Virgin grassland soils traditionally lose OM rapidly after they are first cultivated (Allison, 1973). The loss of OM is usually exponential, declining rapidly during the first 10 to 20 yr, then continuing more slowly until a new equilibrium is reached after 50 to 60 yr (Jenny, 1941). The SOM level declines to a new steady-state level regulated by newly established abiotic and biotic parameters of the cropped ecosystem (Tate, 1987). The reasons for OM decline are stimulation of decomposition after cultivation or reduced plant residue additions to soil as a result of decreased plant yields. Therefore, any factor that increases the plant biomass or slows the decomposition of OM will tend to increase the equilibrium level of SOM.

Farming practices can influence SOM content significantly but, the changes are slow in temperate climates (Johnston, 1993). In Missouri, 50 yr of cropping to continuous corn (*Zea mays* L.) caused a 56% decline in SOM as compared with SOM levels of virgin timber or mixed grasses (USDA, 1980). Continuous cropping to timothy (*Phleum pratense* L.) caused the least decline in SOM (19%) and 3- to 6-yr crop rotations were intermediate in SOM loss. The major influence of crop management practice on SOM levels was probably related to differences in the amount of crop biomass produced above and below ground as reflected by abundance and nature of crop rooting characteristics; however, even with continuous corn, where SOM losses were greatest, it would take 3 to 5 yr before the decline in total SOM levels was detectable by conventional analytical procedures. This study illustrates the importance of understanding the potential equilibrium levels of SOM that are attainable under different cropping systems.

Soil texture and soil environmental conditions affect SOM decomposition rates. Spain (1990) and Feller et al. (1991) found significant positive correlations between soil texture and soil organic C content. The protection of SOM from decomposition is attributed to adsorption of OM onto clay surfaces (Oades, 1989), encapsulation by clay particles (Tisdall & Oades, 1982) or entrapment of OM in small pores of aggregates that makes it inaccessible to microorganisms (Elliot & Coleman, 1988). Both temperature and moisture regimes affect the equilibrium OM content of soils (Buckman & Brady, 1969). Increased temperature decreases OM content while increased moisture increases OM. Within zones of uniform moisture conditions and comparable vegetation, average total OM and N contents increase two to three times for each 10°C decline in mean annual temperature.

Recent concern for increasing atmospheric CO_2 levels and global warming has resulted in increased interest in the sequestration or tie-up of plant C by SOM. Long-term crop rotations and optimum N fertilizer practices can result in higher equilibrium SOM contents due to greater residue additions and/or lower decomposition rates. Results of an 8-yr study in the Western Corn Belt demonstrated that from 100 to 200 kg of C ha^{-1} yr^{-1} can be sequestered by SOM, even in some continuous monocropping systems that receive sufficient levels of N fertilizer (Varvel, 1994). Although statistically significant, these increases in SOM represent < 3% of the total C present in soil. This again illustrates that increasing the SOM content of temperate area soils is a slow process. Increasing the stable organic C content of soil is generally achieved over a long time. Large amendments can increase the OM content quickly. Tester (1990) added 134 Mg of beef manure per hectare for four consecutive years and increased the OM content from five- to seven-fold based on a weight measurement; however, because the soil bulk density decreased 40%, the actual OM increase on a volume basis was three- to four-fold. The equilibrium SOM level for the amended soils was not determined, but it is likely to take several years before equilibrium is reached. These examples demonstrate that, unless large amendments are made to soils, total organic C increases from changed farming practices require several years to significantly increase stable SOM levels.

Interpretation and prediction of the effects of soil management on organic N mineralization and availability to plants, or loss to the environment, depend on understanding the unique roles played by living and nonliving components of SOM. The majority of SOM is contained in plant and animal debris and soil humus. These nonliving components of SOM play an important role in determining the soil physical and chemical environment within which living organisms function. Heterotrophic soil microorganisms and fauna, a relatively small proportion of total OM (1–5%), function as important catalysts for transformation and cycling of N and other nutrients. The importance of soil microbial biomass as a significant sink and source of N and C is discussed by Jenkinson (1988).

Because of its influence on so many factors, Larson and Pierce (1991) suggested that SOM is the single most important indicator of soil quality and productivity. With all these factors taken into consideration, does a standard exist by which OM content would designate a level of soil quality? Gregorich et al. (1994) recommended accumulation of data from a minimum set of SOM component analyses that describe quantitatively how SOM influences soil functions such as nutrient storage, physical structure, and organism population and activity. Because SOM is a key to defining quality but by itself will not provide sufficient information for overall assessment, measurement must be made in light of other quality indicators discussed elsewhere in this book.

ORGANIC CARBON DETERMINATION

No attempt will be made to present a chronicle of OM or OC analyses currently used or available for use because several text cover this aspect (Stevenson, 1994). Our intention is to present a variety of methods presently used or are published, along with some of their benefits and shortcomings. Recommendations are presented for determination of organic C based on availability to most users, practicality and accuracy.

Organic C makes up approximately 58% of SOM by weight. This constant allows for the determination of SOM by direct measurement of OC. The methods for SOM range from a simple color determination to an automated method which records precise levels of C in the sample of soil (Table 9–1). Color determination is limited because SOM and soil color are related to soils within a landscape and this relationship is affected by moisture. The automated methods, while precise, require capital investment and technical expertise. The loss-on-ignition (LOI) method requires a balance and a high temperature oven. Dichromate oxidation is commonly used in research laboratories and in some state testing laboratories for C analyses, but hazards in handling chemicals and cost of disposing of wastes are drawbacks to this method. Spectrophotometric analyses of soil extracts require considerable expertise in interpretation of results and is considered an experimental procedure. Therefore, we recommend the automated dry combustion method and direct determination of CO_2 by chromatography method for determination of OC in soil (Sheldrick, 1986). Alternative method is the LOI method

Table 9–1. Examples of soil organic C determinations.

Method	Principle	Accuracy	Comment	Reference
Walkley–Black	Wet oxidation with $K_2Cr_2O_7$ in H_2SO_4, titration against indicator	Incomplete oxidation, correction factor of approximately 1.3 required	Correction factor varies with soils. Interference from chlorides most troublesome.	Nelson & Sommers, 1982; Walkley & Black, 1934
Modified Tinsley	Wet oxidation with $K_2Cr_2O_7$ in H_2SO_4, plus heat, titration against indicator.	95% recovery of C as compared with dry combustion	No correction required.	Tinsley, 1950
Modified Mebius	Wet dichromate digestion, 170 C, 30 min., titration.	95–100% recovery of C compared with dry combustion	No correction required.	Yeomans & Bremner, 1988
Spectrophotometric estimate of wet digestion	Wet dichromate digestions, absorbance of resulting solution at 590 nm.	Sucrose or glucose used as standards. 0.99 correlation with titrimetric method.	Temperature and duration of heat important. No correction required.	Walinga et al., 1992
Spectrophotometric estimate of soil extract	KOH–NaEDTA extract and absorbance at 260 nm correlated to OC.	$r^2 = 0.89$ compared with dry combustion.	More of a research tool than soil analysis method.	Bowman et al., 1991
Loss-on-ignition (LOI)	Loss of dry weight after 2 h at 360°C. Preliminary heating at 105°C to remove moisture.	Based on regression analyses comparing LOI to Walkley Black.	Gypsum-containing soils retain moisture after 105°C heating. Additional 2 h heating at 150°C required.	Schulte et al., 1991
Soil color	Munsell chroma chart reading versus organic matter.	Correlation of 0.92 for dry soils, 0.94 for moist soils.	Correlation is better within a soil landscape.	Fernandez et al., 1988
Reflectance	Correlation of reflectance at 630 nm and organic matter.	Good estimate within a landscape.	Sophisticated equipment and training. Possible use in prescription farming.	Schreier et al., 1988
Automated dry combustion	Detection of CO_2 loss or weight loss after heat ≥ 950°C.	1–2% coefficient of variation	Small sample size, must be uniform and representative and pass 0.05-mm sieve. Carbonates interfere.	Sheldrick, 1986.

using a temperature of 360°C. All methods should include a companion analysis of a standard compound to determine efficiency of recovery.

Carbonates and, in some cases, gypsum will interfere with accurate determination of OC. The methods that determine OC content by analyzing for evolved CO_2 would be affected by the presence of carbonate. Carbonate level in soils varies with soil-forming processes. In general, moderate to highly weathered soils with moderate to low leaching potential may contain significant carbonate concentrations. Soils with a pH > 7.4 probably contain appreciable levels of carbonates. Testing for carbonate in soils with dilute acid would determine the presence of carbonates that must be removed by acid treatment prior to OC determination. If carbonates are not removed prior to analysis by dry combustion, their concentration must be determined by acid treatment and subtracted from the total C value to obtain the organic C content of the soil. Carbonates do not interfere with the LOI method because temperature (360°C) is below that where CO_2 forms from carbonate, but gypsum will interfere because moisture bound to gypsum is not removed at 105°C, the temperature used to determine soil moisture content. An additional heating step at 150°C prior to the heating at 360°C will remove gypsum-bound water. Soils that contain gypsum are moderately to highly weathered and occur in areas with moderate to low leaching potential, conditions similar to those where carbonates are present.

ORGANIC N DETERMINATION

Organic N analyses are less varied than organic C (Table 9–2). Because there is no direct method for determining organic soil N alone, methods determine total soil N and from this value is subtracted the amount of inorganic N determined by other methods. Total N in soils can be determined by the Kjeldahl method or the Dumas method (Bartholomew & Clark, 1965). The Kjeldahl method is a wet oxidation method whereby the N is converted to ammonium by digestion with concentrated sulfuric acid and catalysts that promote the conversion organic N to ammonium. The Dumas method involves dry-oxidation at high temperature in the presence of Cu and O_2, which reduce the N gases to N_2 that is

Table 9–2. Total organic N determinations.

Method	Principle	Accuracy	Comment	Reference
Kjeldahl N	Acid plus catalyst digestion followed by determination of ammonium.	Complete digestion is time and temperature dependent.	Separate analysis of inorganic N and subtraction required. Fixed ammonium is not recovered.	Bremner & Mulvaney, 1982
Automated dry combustion	Dumas method, oxidation to N_2 and chromatographic detection.	Near 100% recovery	Sample size is small, needs to be uniform and representative and pass 0.05 mm sieve. Determination and subtraction of inorganic N required. fixed ammonium is recovered in varying amounts depending on temperature.	Yeomans and Bremner, 1988; Bremner & Mulvaney, 1982

detected by gas chromatography. Soils with appreciable amounts of fixed ammonium will compromise the accuracy of either the Kjeldahl or Dumas method to determine total N. Fixed ammonium is that ammonium found between clay lattices. It is difficult to remove and, hence, detect unless extreme measures are taken to destroy or expand the lattice. Therefore total N analysis of soils known to contain significant amounts of fixed ammonium will result in lower N contents than actually present. The agronomic significance of fixed ammonium is unclear because this ammonium is highly unreactive and generally does not contribute significantly to the N cycle within a cropping year.

We recommend the automated dry combustion (Dumas) method and direct analysis of N_2 method for determination of organic N (Bremner & Mulvaney, 1982). The inorganic N content of the soil and, possibly, the fixed ammonium content in those soils known to have appreciable quantities of fixed ammonium must be subtracted from the total N concentration resulting from the dry combustion method to obtain the organic N content. Several manufacturers provide equipment for total N determination. Many laboratories have semiautomated micro-Kjeldahl apparatus that are suitable for analyses, but the use and disposal of chemicals used in the Kjeldahl method are problematic.

SAMPLING OF SOILS FOR ORGANIC CARBON OR NITROGEN DETERMINATION

The technique of sampling soils for analyses is as critical as the method used for analyses. Soils are highly variable even within a field and the OC or ON content are affected by when the field is sampled, how the field is sampled and how the sample is prepared for analysis.

Sampling of soils for the determination of organic C or N should be consistent from year to year. The time of sampling should be a time when the most recent amendments would have the least effect on biasing the organic C determination. Sampling after manure application or after legume plow down may bias the organic C or N data. Soils should be screened through 2-mm or smaller sieve to remove plant materials and sieve size reported along with the OC or ON concentration. Screening through even smaller sieves (<1 mm) is better for dry combustion method that requires small, uniform samples. Stones removed by sieving should be accounted for in calculating the OC or ON concentration based on volume or area basis. Gregorich at al. (1994) suggested that the depth of the Ap horizon and bulk density of the soil also be measured because management practices may induce changes in OC content based on volume as well as based on mass content. A moisture determination on the soil sample allows correction for varying soil moisture contents.

DISCUSSION

If a land manager had a choice to make one or two measurements to determine the quality of the soil, a first estimate could be gained through the determination of organic C or N content using the dry combustion method. The soil C and N content are highly correlated with soil productivity, erodibility, water infiltration, and the capacity of the soil to act as an environmental buffer by absorbing potential pollutants. The OM content however, does not provide a complete soil quality determination and must be considered within the context of the soil physical characteristics, as well as other chemical and biological characteristics. Also, comparisons of total organic C and N contents are inappropriate across regional boundaries as climatic conditions and landscape characteristics greatly influence the amount of C an N that can accumulate within a given soil.

In analyzing a soil for quality, what standard can be used to determine if OM is of good quality? Can quantity be equated to quality? When measuring SOM, what values or changes are important to consider in relation to soil quality? Larson and Pierce (1994) suggest monitoring the dynamics of SOM change. Sustainable agriculture principles suggest that change be minimized as long as the management practice is productive and there is a balance of inputs and outputs in the system. If the SOM changes little from 1 yr to the next, and the system is productive and competitive, the SOM content may be considered optimum. Buckman and Brady (1969) suggest that SOM should be maintained at an economic maximum consistent with a suitable physical condition of the soil, satisfactory biochemical activity, adequate availability of nutrients, and profitable crop yields. Essentially, there is not an established quantity of SOM that can be defined as good or less than optimum quality because SOM evaluation must include farming practice and climate as covariables.

The determination of a target level for organic C and N for a good quality soil is highly dependent on the climatic region and soil texture. A fertile, high quality soil in the southwestern region of the USA may have only 1% organic C content, yet a Midwestern USA. soil with the same organic C content may be considered a poor quality soil. Land managers must take a look at the various soils in their region and determine the range of OM contents that are related to soils considered to be of high quality. Another indicator of an appropriate level might be the OM content of a field with the same or similar soil type in the owner's field that has not been intensely cultivated that stimulates SOM breakdown. A third source for determining a target level may be historical records of the same field.

It is difficult to recommend a specific amount of an amendment that when added to a soil would significantly increase the OM level. Generally, it is thought that organic residues (plant or manure) left on the surface of a soil will decompose more slowly than those that are incorporated into the soil profile. This is thought to result from surface residues being subject to greater shifts in temperature and more rapid drying after precipitation (Stott & Martin, 1989). Similar transformations occur in forests or prairie where residues predominate on the surface and are not mixed physically with soil. Incorporated residues decompose more rapidly and are less likely assimilated into SOM. The quality of residue also affects decomposition rate and the amount retained within the soil. Legume residues will decompose quickly, adding little to the OM reserve, while high lignin residues will decompose slowly with a greater proportion of the residue C remaining in organic matter (Stott & Martin, 1990).

Development of site specific management strategies are often based on the variability of SOM contents within a single field or area. Soil productivity estimates coupled with appropriate seeding and N fertilizer rates are based on SOM content as are herbicide and insecticide application rates. Real time sensors that can be mounted on a tractor have been developed for estimating OM contents as the tractor travels over open fields. These sensors use soil color changes within a landscape to estimate the changes in OM content (Fernandez et al., 1988; Schulze et al., l992). With proper standardization and proper equipment, instrumental soil color evaluation might provide a simple, economical way for land users to monitor changes in OM content of a field over time. When a plant canopy is present, such as in forests or grasslands, color of undisturbed soil can not be obtained.

CONCLUSION

The recommended method for accurate determination of both organic C and N is the automated dry combustion method, which detects total C and N directly. The inorganic N content of soils must be subtracted from the total N to obtain organic N concentration. Soils with appreciable carbonate or fixed ammonium content require separate treatments or analyses to determine values that must be subtracted from the total C or N value to reflect accurate OC or ON content. The dry combustion method should be available in each state, either at the State testing laboratories or at a commercial laboratories at a moderate price. The LOI (Loss on Ignition) determination for OC is recommended as an alternate low technology, moderate expense method. The wet chemical oxidation methods, such as the Walkley–Black or Kjeldahl, are less costly in materials, but higher in labor costs, require knowledge of analytical chemistry techniques, and use hazardous chemicals that pose safety and waste disposal problems.

More accurate and informative OC and ON values are obtained when soil samples are taken in the same fashion every year. To adjust for any variation, we recommend that results include date of sampling, depth of sampling, sieve size used to screen soil and whether results were corrected to include stones removed by sieving, and bulk density. As discussed in Doran and Parkin (1996, this publication), results expressed on a volume basis are preferred because they take into account the depth of soil sampled and any changes in bulk density that may result from changes in farming practices.

REFERENCES

Allison, F.E. 1973. Soil organic matter and its role in crop production. Dev. Soil Sci. 3. Elsevier, Amsterdam.

Bartholomew, W.V., and F.E. Clark (ed.). 1965. Soil nitrogen. Agron. Monogr. 10. ASA, SSSA, and CSSA, Madison, WI.

Bell, M.A. 1993. Organic matter, soil properties and wheat production in the high valley of Mexico. Soil Sci. 156:86–93.

Bowman, R.A., W.D. Guenzi, and D.J. Savory. 1991. Spectroscopic method for estimation of soil organic carbon. Soil Sci. Soc. Am. J. 55:563–566.

Bremner, J.M., and C.S Mulvaney. 1982. Nitrogen: Total. p. 595–624. *In* A.L. Page et al. (ed.) Methods of soil analysis. Part 2. 2nd ed. Agron. Monogr. 9. ASA and SSSA, Madison, WI.

Buckman, H.O., and N.C. Brady. 1969. The nature and properties of soils. Macmillan Co., London.

Dalal, R.C., and R.J. Mayer 1986. Long-term trends in fertility of soils under continuous cultivation and cereal cropping in southern Queensland: III. Distribution and kinetics of soil organic carbon in particle-size fractions. Aust. J. Soil Res. 24:293–300.

Doran, J.W., and T.B. Parkin. 1996. Quantitative indicators of soil quality: A minimum data set. p. 25–37. *In* J.W. Doran and A.J. Jones (ed.) Methods for assessing soil quality. SSSA Spec. Publ. 49. SSSA, Madison, WI.

Elliot, E.T., and D.C. Coleman. 1988. Let the soil work for us. p. 23–32. *In* H. Eij Sackers and A. Quispel (ed.) Ecological implications of contemporary agriculture. Proc. 4th. European Ecology Symp., Wageningen, the Netherlands. 7–12 Sept. Munksgaard Int., Copenhagen.

Feller, C., E. Fritsch, R. Ross, and C. Valentin. 1991. Effects of the texture on the storage and dynamics of organic matter in some low activity clay soils (West Africa, particularly). Cah. ORSTOM, Ser. Pedol.26:25–36.

Fernandez, R.N., D.G. Schulze, D.L. Coffin, and G.E. Van Scoyoc. 1988. Color, organic matter, and pesticide adsorption relationships in a soil landscape. Soil Sci. Soc. Am. J. 52:1023–1026.

Gregorich, E.G., M.R. Carter, D.A. Angers, C.M. Monreal, and B.H. Ellert. 1994. Towards a minimum data set to assess soil organic matter quality in agricultural soils. Can. J. Soil Sci. 74:367–385.

Gregorich, E.G., C.M. Monreal, B.H. Ellert, D.A. Angers, and M.R. Carter. 1993. Evaluating changes in soil organic matter. p. 10-1–10-17. *In*. D.F. Acton. (ed.) A program to assess and monitor soil quality in Canada: Soil quality evaluation program summary (interim). Center Land and Biol. Res. Contr. 93-49. Agric. Res. Branch, Agric. Canada, Ottawa.

Jenkinson, D.S. 1988. Determination of microbial biomass carbon and nitrogen in soil. p. 368–386. *In* J.R. Wilson (ed.) Advances in nitrogen cycling in agricultural ecosystems. CAB Int., Oxon, England.

Jenny, H. 1941. Factors of soil formation. McGraw-Hill, New York.

Johnston, A.E. 1993. Significance of organic matter in agricultural soils. p. 1–18. *In* Organic substances in soil and water: Natural constituents and their influences on contaminant behavior. Spec. Publ. 135. The Royal Soc. of Chem., Cambridge, England.

Kononova, M.M., T.Z. Nowakowski, and N. Walker. 1975. Humus of virgin and cultivated soils. p. 475–526. *In* J.E. Gieseking (ed.) Soil components. Vol. 1. Organic components. Springer-Verlag, New York.

Larson, W.E., and F.J. Pierce. 1991. Conservation and enhancement of soil quality. p. 175–203. *In* Evaluation of sustainable management in the developing world. Vol 2. Tech. Papers. IBSRAM Proc. 12(2). Int. Board for Soil Res. and Manage., Bangkok, Thailand.

Larson, W.E., and F.J. Pierce. 1994. The dynamics of soil quality as a measure of sustainable management. p. 37–51 *In* J.W. Doran et al. (ed.) Defining soil quality for a sustainable environment. Soil Sci. Soc. of Am. Spec. Publ. 35. SSSA, Madison, WI.

Luk, S.H., W.E. Dubbin, and A.R. Mermut. 1990. Fabric analysis of surface crusts developed under simulated rainfall in loess soils, China. p. 29–40. *In* R.B. Bryan (ed.) Soil erosion: Experiments and models. Catena Suppl. 17. Cantena Verlag, Cremlingen-Destedt.

Magdoff, F. 1992. Building soils for better crops: Organic matter management. Univ. of Nebraska Press, Lincoln.

Nelson, D.W., and L.E. Sommers. 1982. Total carbon, organic carbon and organic matter. p. 539–579. *In* A.L. Page et al. (ed.). Methods of soil analysis. Part 2. 2nd ed. Agron. Monogr. 9. ASA and SSSA., Madison, WI.

Oades, J.M. 1989. An introduction to organic matter in mineral soils. p. 89–159. *In* J.B. Dixon and S.B. Weed (ed.) Minerals in soil environments. 2nd ed. SSSA Book Ser. 1. SSSA, Madison, WI.

Rasmussen, P.E., R.R. Allmaras, C.R. Rohde, and N.C. Roager. 1980. Crop residue influences on soil carbon and nitrogen in a wheat–fallow system. Soil Sci. Soc. Am. J. 44:596–600.

Schreier, H., R. Wiart, and S. Smith. 1988. Quantifying organic matter degradation in agricultural fields using PC-based image analysis. J. Soil Water Conserv. 43:421–424.

Schulte, E.E., C. Kaufman, and J.B. Peter. 1991. The influence of sample size and heating time on soil weight loss-on-ignition. Commun. Soil Sci. Plant Anal. 22:159–166.

Schulze, D.G., J.L. Nagel, G.E. Van Scoyoc, T.L. Henderson, M.F. Baumgardner, and D.E. Stott. 1992. Significance of organic matter in determining soil colors. p. 71–90. *In* J.M. Bigham and E.J. Ciolkosz (ed.) Soil color. SSSA Spec. Publ. 31. SSSA, Madison, WI.

Sheldrick, B.H. 1986. Test of the Leco CHN-600 determinator for soil carbon and nitrogen analysis. Can. J. Soil Sci. 66:543–545.

Soil Science Society of America. 1987. Glossary of soil science terms. SSSA, Madison, WI.

Spain, A. 1990. Influence of environmental conditions and some soil chemical properties on the carbon and nitrogen contents of some tropical Australian rain-forest soils. Aust. J. Soil Sci. Res. 28:825–839.

Stevenson, F.J. 1994. Humus chemistry: Genesis, composition, reactions. 2nd ed. John Wiley & Sons, New York.

Stott, D.E., and J.P. Martin. 1989. Organic matter decomposition and retention in arid soils. Arid Soil Res. Rehabil. 3:115–148.

Stott, D.E., and J.P. Martin. 1990. Synthesis and degradation of natural and synthetic humus material. p. 37–64. *In* P. MacCarthy et al. (ed.) Humic substances in soil and crop sciences: Selected readings. ASA and SSSA, Madison, WI.

Tate, R.L., III. 1987. Soil organic matter: Biological and ecological effects. John Wiley & Sons, New York.

Tester, C.F. 1990. Organic amendment effects on physical and chemical properties of a sandy soil. Soil Sci. Soc. Am. J. 54:827–831.

Tinsley, J. 1950. The determination of organic carbon in soils by dichromate mixtures. p. 161–164. *In* Trans. 4th Int. Cong. Soil Sci., Amsterdam. 24 July–1 Aug. 1950. ISSS, Amsterdam.

Tisdall, J.M., and J.M. Oades. 1982. Organic matter and water stable aggregates in soil. J. Soil Sci. 33:141–163.

U.S. Department of Agriculture. 1980. Report and recommendations on organic farming. p. 29–35. U.S. Gov. Print. Office, Washington, DC.

Upchurch, R.P., F.L. Selman, D.D. Mason, and E.J. Kamprath. 1966. The correlation of herbicide activity with soil and climatic factors. Weed Sci. 14:42–49.

Varvel, G.E. 1994. Rotation and N fertilizer effects on changes in soil carbon and nitrogen. Agron. J. 86:319–325.

Walkley, A., and T.A. Black. 1934. An examination of the Degtjareff method for determining soil organic matter, and a proposed modification of the chromic acid titration method. Soil Sci. 37:29–38.

Walinga, I., M. Kithome, I. Novozamsky, V. J. G. Houba, and J. J. van der Lee. 1992. Spectrophotometric determination of organic carbon in soils. Commun. Soil Sci. Plant Anal. 23:1935–1944.

Yeomans, J.C., and J.M. Bremner. 1988. A rapid and precise method for routine determination of organic carbon in soil. Commun. Soil Sci. Plant Anal. 19:1467–1476.

10 Measurement and Use of pH and Electrical Conductivity for Soil Quality Analysis

Jeffrey L. Smith

USDA-ARS
Washington State University
Pullman, Washington

John W. Doran

USDA-ARS
University of Nebraska
Lincoln, Nebraska

Soil quality is important for long-term sustainability of the earth's biosphere and land base. Historically, soil quality has been associated mainly with productivity for food and fiber. More recent concepts of soil quality have emphasized the degradation of soil by mechanisms such as erosion, over grazing, over application of waste materials, and toxic chemical additions. Most recently, however, we have realized that soil quality is a holistic term describing the many functions of soil as related to ecosystem health, crop nutrition, and quality, the filtering or buffering of environmental pollutants, and maintaining food and fiber production. Thus soil quality has become the umbrella term for overall soil health, taking numerous factors into consideration, as opposed to focusing only on specific degrading or agrading processes. For example, a soil that has been treated with sewage sludge for a number of years may be highly productive and contain superior levels of organic matter; however, if the Cd level in the soil is increasing due to sludge addition, the crops produced may be unsuitable for direct human or food-animal consumption and the soil will be of low quality due to its' potential hazard to health. The potential uptake of Cd by field crops, however, is to a large extent determined by soil properties that influence the mobility of metals such as soil pH and soil electrical conductivity (McLaughlin et al., 1994).

The analysis of soil quality is complex due to the many factors that make up a good soil. For the example given above, sound judgment will have to be made on the benefits of sludge amendment, it's chemical composition, it's long-term effect on soil chemical and biological properties, and any detrimental effects from the amendment or it's management. For this simple example of amending a soil with sewage sludge for several years, a number of factors need to be mea-

---

Copyright © 1996 Soil Science Society of America, 677 S. Segoe Rd., Madison, WI 53711, USA. *Methods for Assessing Soil Quality,* SSSA Special Publication 49.

sured and judgments made to estimate the overall effects on soil quality. It is clear that more complicated examples could be envisioned making the task of estimating soil quality even more difficult.

Numerous meetings have been held in the last few years to determine what soil properties should be measured to describe soil quality. Many of the lists are long and would take a significant amount of resources to measure all of the soil properties included. In addition, there has been significant emphasis on biological measurements for determining soil quality that are usually complex both in measurement and interpretation. Therefore, our research efforts have focused on determining which chemical, biological, and physical soil quality indicators would respond most quickly to soil management changes. In addition, we have made extensive analysis of spatial and temporal variability of the proposed soil quality indicators. In this chapter we report on the measurement and use of pH, as a measure of soil acidity, and electrical conductivity (EC), an estimate of salt content or chemical supply, to evaluate soil quality. Unlike some of the other soil quality indicators included in the minimum data set presented in Doran and Parkin (1996, this publication), the in-field measurement and interpretation of soil pH and EC are straightforward, more accessible to producers and specialists alike, and important to interpretation of management related changes in other soil chemical, physical, and biological properties. For example, the knowledge that the pH of a soil is above 7.2 indicates a need to correct for carbonate C when dry combustion methods are used for organic C analyses (see Sikora, 1996, this publication) and also represents a potential for N loss via ammonia volatilization where ammoniacal fertilizers are applied. On-farm measurements of pH and EC also are important for monitoring the effects of agricultural management practices on the efficiency of N use and related environmental impacts (Patriquin et al., 1993).

SOIL ACIDITY AND ELECTRICAL CONDUCTIVITY

Soil acidity or pH is a measure of hydrogen ion (H^+) activity in the soil solution and is specifically defined as the $-\log_{10}$ of the [H^+] concentration. In the natural state, the H^+ activity in soil is a function of the parent material, time of weathering, vegetation, climate, and topography. In addition to these soil forming factors, soil pH is influenced by the season of the year, cropping and soil management practices, use of ammoniacal fertilizers, acid precipitation, sludge and manure applications, soil organic matter, and biological activity. Soil management and biological activity are of most interest since they can be manipulated in the short term to enhance soil quality.

Most soils of the humid forest and subhumid to semiarid grassland categories have pH values between 4 and 8 (Fig. 10–1). Values above and below this range are usually due to Ca and Na salts or H^+ and Al ions, respectively, in soil solution. In general, within the pH range of 6 to 7.5, pH itself has little direct effect on plant roots and microorganisms. Since pH is a logarithmic function, a pH of 6 is 10 times more acid than a pH of 7 and 100 times more acid than a pH of 8. In the pH range of 6 to 7.5 the H^+ concentration decreases by more than 31-fold, thus there is a wide H^+ concentration range that does not directly effect roots

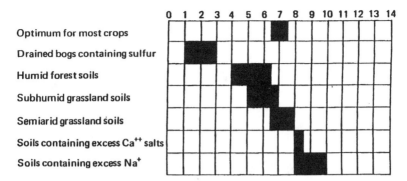

Fig. 10–1. Typical pH ranges for various types of soils (Troeh & Thompson, 1993).

or microorganisms. The main pH effect, in the normal soil pH range of 4 to 8 is on the availability and toxicity of elements such as Fe, Al, Mn, Mb, B, Cu, Cd, and others to plants and microorganisms. Figure 10–2 shows the relative availability of macro and micronutrients over the pH range of 4 to 10. Most plant nutrients are readily available within the pH range of 6.0 to 7.5. This nutrient effect is manifested in the increased yield at more neutral pHs as shown in Table 10–1 where higher yields were maintained in the pH range of 5.7 to 7.5. The use of data from Ohio, where soil acidity is often more uniform throughout the soil profile, is probably an over-simplification because many soils of the Great Plains contain moderately acid surface layers that are underlain by neutral to calcareous subsoils of higher pH.

The optimum pH range for most soil microorganisms is from 5 to 8 (Table 10–2). Soil pH can be an important determinant of the relative predominance and activity of different microbial groups as related to microbially mediated processes such as nutrient cycling (nitrification, denitrification, and others), plant disease, decomposition of natural and synthetic organic chemicals, and microbial transformation of atmospherically important trace gases such as CH_4. The process of microbial nitrification in soils is largely the result of aerobic bacteria that obtain energy from the oxidation of NH_4^- and NO_2^- to NO_3^-. The optimum pH for nitrification is generally between 6.5 to 8. Above 8, NH_4^+ is present mainly as NH_3 gas, which can be lost from soil and also inhibits the activity of NO_2^- oxidizing bacteria. Below pH of 5.5 to 6 bacterial nitrification is greatly reduced in most soils. Control of soil pH can influence the incidence of certain plant diseases and the degradation of organic chemicals. For example, potato scab, a disease caused by the Actinomycete *Streptomyces scabies* in neutral to alkaline soils, can be controlled by the acidification of soil through application of elemental S. As illustrated in Fig. 10–3, the degradation and efficacy of certain agricultural chemicals also can be greatly influenced by soil pH.

Another significant factor affecting soil pH is the buffering capacity of the soil. Soil clay particles are negatively charged and thus attract positive basic ions such as calcium Ca^{++}, Mg^{++}, Na^+, and K^+ and positive acidic ions such as H^+ and Al^{+++} to neutralize this charge. These ion reserves held on the soil particles help buffer the soil at a constant pH. When an acid material is added to soil, the added

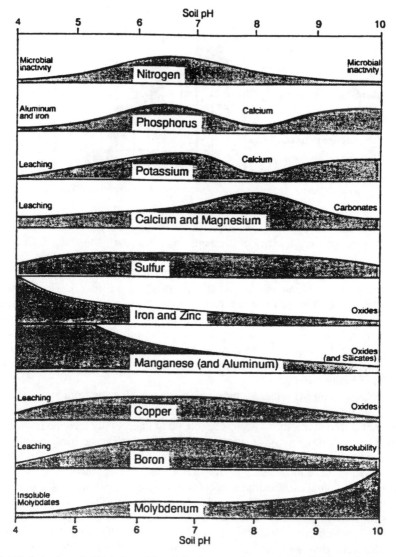

Fig. 10–2. A schematic illustration of the relation between plant nutrient availability and soil reaction, pH (Troeh & Thompson, 1993).

H^+ ions exchange with cations such as Ca^{++}, Mg^{++}, K^+ on the soil exchange sites (the basic cation goes into solution) and the effect of the acid is neutralized. The opposite occurs when excess base is added. The base exchanges with H^+ ions and the soil solution is buffered from the added base. These reactions will occur until the buffering capacity of the soil is nearly exhausted and then the pH steadily changes.

A good example of soil buffering capacity is the effect of long-term addition of NH_4^+ based fertilizers on the pH of soil. In most soils ammonium is rapid-

Table 10–1. Yield of crops grown in corn, small grain, legume or Timothy rotation at different pH values (Ohio Agricultural Experiment Station, 1938).

Crop	Relative average yield at pH indicated				
	4.7	5.0	5.7	6.8	7.5
Corn	34	73	83	100	85
Wheat	68	76	89	100	99
Oats	77	93	99	98	100
Barley	0	23	80	95	100
Alfalfa	2	9	42	100	100
Sweet clover	0	2	49	89	100
Red clover	12	21	53	98	100
Alsike clover	13	27	72	100	95
Mammoth clover	16	29	69	100	99
Soybean	65	79	80	100	93
Timothy	31	47	66	100	95

Table 10–2. Minimum, maximum and optimum pH values for different microbial groups and biochemical processes (after Bazin & Prosser, 1988; Bender & Conrad, 1995; Killham, 1994; Paul & Clark, 1989).

Group–Process	Range	Optimum
Bacteria	5–9	7
Nitrification–denitrification	6–8	6.5–8
NH_3 inhibition of NO_2 oxidation	>8	--
S-oxidizers	1–8	2–6
CH_4-oxidizers	4–9	6.6–7.5
Actinomycetes	6.5–9.5	8
Fungi	2–7	5
Blue green bacteria	6–9	>7
Protozoa	5–8	7

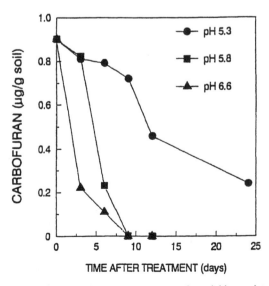

Fig. 10–3. Effects of soil pH on Carbofuran, a corn rootworm insecticide, persistence in a Sharpsburg silty clay loam soil (after Petersen, 1990).

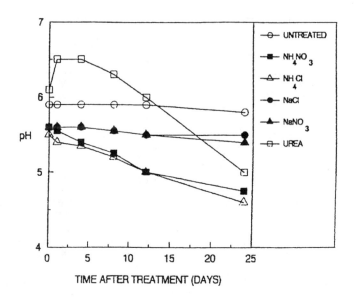

Fig. 10–4. Relationship between fertilizer N type and salt content to soil pH with time in a Sharpsburg silty clay loam soil. Fertilizer N and NaCl were applied at a rate of 200 mg/kg soil. Soil pH before treatment was 5.8. (after Petersen, 1990).

ly converted to NO_3^- with the release of H^+ ions that acidify soil. In the case of Pacific Northwest dry land wheat soils, the pH has decreased less than 1 pH unit, despite more than 30 yr of fertilizing with ammoniacal N sources. This demonstrates the large buffering capacity of these soils; however, this long-term change is serious because, as the soil buffering capacity decreases, more rapid changes in pH will occur in shorter time periods. The continued application of high rates of ammonium fertilizer is an example of a practice that is directly affecting soil quality and is not sustainable in the long term, unless lime is eventually added.

Several years may be required to identify the sustainability of agricultural practices through measuring soil quality attributes such as pH, however, the seasonal measurement of pH on field soils provides a real time indication of changes in nutrient availability and the acid-base reaction of the soil. This measurement can be useful for field comparisons, identifying the effects of management practices on short-term changes in soil pH and the potential for cumulative effects on soil quality as related to soil productivity and environmental quality. The importance of soil pH in understanding the short-term effects of fertilizer management is illustrated in Fig. 10–4. Addition of high levels of ammoniacal fertilizers to slightly acid soils can result in a 1 unit drop in pH, due to microbial nitrification, within a period of 3 to 4 wk. This can result in impaired crop growth, decreased herbicide effectiveness, or reduced microbial decomposition. Note that the addition of any salt to soil results in an apparent decline in soil pH due to a change in H^+ activity. Also, the hydrolysis of urea can result in higher soil pH the first week or two after application to soil. In neutral to slightly alkaline soils urea hydrolysis could temporarily raise soil pH levels above 8, which could restrict nitrification, result in significant loss of N through ammonia volatilization, and lead to NO_2^- accumulation.

Another soil quality indicator that is useful and easily measured is electrical conductivity. The electrical conductivity of a solution is related to the total cations or anions in the solution. These ions can be either cations (Ca^{++}, Mg^{++}, K^+, Na^+, H^+) or anions (NO_3^-, SO_4^{2-}, Cl^-, HCO_3^-, CO_3^{2-}, OH^-), which are soluble in the soil solution. Electrical conductivity has generally been associated with determining soil salinity, however, electrical conductivity also can serve as a measure of soluble nutrients—both cations and anions. Thus within a specific range, EC would indicate good nutrient availability for plants, with the low end indicating nutrient poor soil that is structurally unstable and disperses readily and the high end salinity problems. Both pH and EC need to be measured since a good range for electrical conductivity could be found in an acid soil that was unsuitable for plant growth.

Since EC is a measurement of how well a solution conducts electricity the units are the reciprocal of the resistance to conduct electricity, which is measured in ohms, thus EC is measured as $ohms^{-1}$, commonly written as mhos. Electrical conductivity is measured in a cell, most of which contain two electrodes spaced at a distance of 1 cm, and thus the units for EC are mhos/cm. Normal values in soil solutions are small and are reported as mmhos/cm or 1/1000 mhos/cm. The international unit for conductivity is Siemens per meter (S/m) or deciSiemens per meter (dS/m). The relationship between these two units is mmhos/cm = dS/m.

Although only approximate (Rhoades, 1982), other useful relationships are:

Total cation (or anion concentration), meq/L $\cong$ 10 × EC (in dS/m).

Salt concentration, mgrams/L $\cong$ 640 × EC (in dS/m).

Osmotic pressure, Bars at 25°C $\cong$ 0.39 × EC (in dS/m).

Where NO_3^- is the predominant ion in the soil solution, a very useful relationship derived from the above is:

EC (1:1 soil to water mixture) in dS/m ×

10 (mg NO_3^-–N per kg soil/14 mg/meq)

or

$EC_{1:1}$ (dS/m) × 140 $\geq$ mg NO_3^-–N/kg of soil

This relationship assumes complete extractability of NO_3^- in water, a valid assumption for the majority of soils with low anion-exchange capacity, and also that NO_3^- is the major anion in the soil solution; however, the analyst should be warned that while an EC of 0.1 dS/m is indicative of a soil NO_3^-–N concentration <14 mg/kg it is not known how much of the soil EC value results from the contribution of other anions in soil solution such as SO_4^{2-}, Cl^-, HCO_3^-, CO_3^{2-}, and others. The background contribution of such other ions, however, can often be determined through comparison with the EC reading of control soils that have either

Table 10–3. The relationship between conductivity and degree of salinity for the 1:1 method and the saturated paste method (after Dahnke & Whitney, 1988).

Texture	Nonsaline	Degree of salinity				Ratio $EC_{1:1}/EC_e$
		Slightly saline	Moderately saline	Strongly saline	Very saline	
		mmhos/cm (dS/m)				
		1:1 Method ($EC_{1:1}$)†				
Coarse to loamy sand.	0–1.1	1.2–2.4	2.5–4.4	4.5–8.9	9.0+	~0.56
Loamy fine sand to loam	0–1.2	1.3–2.4	2.5–4.7	4.8–9.4	9.5+	~0.59
Silt loam to clay loam	0–1.3	1.4–2.5	2.6–5.0	5.1–10.0	10.1+	~0.63
Silty clay loam to clay	0–1.4	1.5–2.8	2.9–5.7	5.8–11.4	11.5+	~0.71
		Saturated paste (EC_e)†				
All textures	0–2.0	2.1–4.0	4.1–8.0	8.1–16.0	16.1+	

† $EC_{1:1}$ and EC_E, electrical conductivity of soil water in a 1:1 soil/water mixture and saturated paste extract, respectively.

not received or have received lower applications of fertilizer N. Therefore the following relationship should be used to estimate soil NO_3^-–N levels from measurements of soil $EC_{1:1}$:

$$[EC_{1:1} - \text{Background}] \, (dS/m) \times 140 \approx \text{mg } NO_3^-\text{–N per kg soil}$$

Applications of this relationship are presented later in this chapter and in Sarrantonio et al. (1996, this publication) where comparisons of EC and NO_3^-–N levels for organic and conventionally managed farms in North Dakota are used as examples.

Soil salinity can greatly influence physical, chemical, and biological properties and processes in soil. Most soils are considered slightly saline if the EC of a saturated paste extract (EC_e) exceeds 2 dS/m (mmhos/cm), which is equivalent to an EC for a 1:1 soil-to-water mixture ($EC_{1:1}$) of 1.0 to 1.4 dS/m for coarse and fine textured soils, respectively (Table 10–3). As illustrated in Table 10–4, the salt tolerance of agricultural crops varies considerably and ranges from soil EC_e values of 1.0 to 3.2 ($EC_{1:1}$ of 0.6 to 2.0) dS/m for salt sensitive species and 2.7 to 8.0 ($EC_{1:1}$ of 1.7 to 5.1) dS/m for salt tolerant species. Microorganisms vary considerably in their tolerance to salt, with actinomycetes and fungi being more tolerant of higher soil EC than most bacteria that are more sensitive; however, some halophilic bacteria have a high tolerance to salt. As demonstrated by the experimental findings presented in Table 10–5, soil ECs within the range encountered in most intensively managed agricultural soils (i.e., $EC_{1:1}$ values of 0.6 to 5 dS/m) can have pronounced effects on microbial processes such as respiration, decomposition of organics, ammonification, nitrification and denitrification, and chemical processes such as ammonia volatilization. These data also illustrate that EC thresholds above which processes such as ammonification, nitrification, or pesticide degradation are impaired by increased salinity can vary as a result of soil amendment with decomposable organics (such as alfalfa, *Medicago sativa* L.) or soil pH (see example for carbofuran degradation). It is not surprising that bacterial processes such as nitrification and denitrification appear most sensitive to soil EC and can be greatly affected by soil $EC_{1:1}$ values within the range of 0.6

Table 10–4. Salt tolerance of agricultural crops: Threshold EC_e or $EC_{1:1}$ (25°C) at the point of initial yield decline and yield decrease per unit EC beyond threshold.†

Crop species	Threshold (T)		Yield decrease per unit EC beyond threshold EC (D)
	EC_e	$EC_{1:1}$‡	
	dS/m or mmhos/cm		%
Barley (*Hordeum vulgare* L.)	8.0	4.5–5.7	5.0
Cotton (*Gossypium hirsutum* L.)	7.7	4.3–5.5	5.2
Sugar beet (*Beta vulgaris* L.)	7.0	3.9–5.0	5.9
Wheat (*Triticum aestivum* L.)	6.0	3.4–4.3	7.1
Ryegrass, perennial (*Lolium perenne* L.)	5.6	3.1–4.0	7.6
Soybean [*Glycine max* (L.) Merr.]	5.0	2.8–3.6	20.0
Fescue, tall [*Festuca arundinacea* (Schreber)]	3.9	2.2–2.8	5.3
Wheatgrass, crested (*Agropyron desertorum* Schultes)	3.5	2.0–2.5	4.0
Peanut (*Arachis hypogea* L.)	3.2	1.8–2.3	29.0
Rice, paddy (*Oryza sativa* L.)	3.0	1.7–2.1	12.0
Vetch, common (*Vicia sativa* L.)	3.0	1.7–2.1	11.0
Tomato [*Lycopersicon lycopersicum* (L. Karsten)]	2.5	1.4–1.8	9.9
Alfalfa (*Medicago sativa* L.)	2.0	1.1–1.4	7.3
Potato (*Solanum tuberosum* L.)	1.7	1.0–1.2	12.0
Maize (*Zea mays* L.)	1.7	1.0–1.2	12.0
Clover, berseem (*Trifolium alexandrium* L.)	1.5	0.8–1.1	5.7
Orchardgrass (*Dactylis glomerata* L.)	1.5	0.8–1.1	6.2
Pepper (*Capsicum annuum* L.)	1.5	0.8–1.1	14.0
Grape (*Vitis vinifera* L.)	1.5	0.8–1.1	9.6
Lettuce (*Lactuca sativa* L.)	1.3	0.7–0.9	13.0
Cowpea (*Vigna unguiculata* L.)	1.3	0.7–0.9	14.0
Bean (*Phaseolus vulgaris* L.)	1.0	0.6–0.7	19.0

† Based on Maas and Hoffman (1977), relative yield = 100 – [D (EC – T)].
‡ $EC_{1:1}$ = electrical conductivity of a 1:1 soil/water mixture relative to that of a saturated paste extract (EC_e) for coarse to fine textured soils.

to 2.5 dS/m. Aulakh and Singh (1995) found, for a range of coarse to fine textured alkaline soils of low organic C content from the Punjab of India, that denitrification losses of NO_3^-–N under waterlogged conditions were inversely related to soil $EC_{1:1}$ above a threshold value of 0.6 to 0.7 dS/m. Weier et al. (1993) found that soil NO_3^-–N levels resulting in soil $EC_{1:1}$ values above 1.5 to 2.0 dS/m resulted in a 16 to 28% reduction in the NO_3^- lost through denitrification and also a greater percentage of the gaseous products being lost as N_2O, over 91% in one case (Table 10–6). This finding has major potential implications with regard to the relationships between soil quality and the quality of both air and water environments.

MEASUREMENT OF pH AND ELECTRICAL CONDUCTIVITY

The measurement of both pH and EC provides a more complete indication of the chemical nature of the soil. When measured together, the effects of management practices and cropping systems on soil quality will be detected more easily.

There are several acceptable methods for measuring pH and EC, some more complicated than others. The measurement of pH is usually performed with a combination electrode and calibrated millivolt meter (so called pH meter) or by

Table 10–5. Influence of soil electrical conductivity on microbial and chemical processes in soils amended with salts or N fertilizers.

Microbial–Chemical Process	Soil amendment	Relative change	Electrical conductivity (1:1)		Reference (Soil)
			Range tested	Threshold	
			dS/m		
Respiration	NaCl	17–47% decrease	0.7–2.9	0.7	McCormick & Wolf, 1980 (Sandy loam, pH 5.6–5.8 init. EC = 0.4–0.6)
Decomposition	NaCl + alfalfa	2–25% decrease	0.7–3.4	0.7	
Ammonification	NaCl	10–32% decrease	0.7–2.9	0.7	
	NaCl + alfalfa	No change	0.7–3.4	–	
	NaCl + alfalfa	39–71% decrease	4.4–175	9.3	
Nitrification	NaCl	10–37% decrease	0.7–2.9	0.7	
	NaCl + alfalfa	No change	0.7–3.4	–	
	NaCl + alfalfa	43–91% decrease	4.4–175	9.3	
Nitrification	NaCl, CaCl$_2$, Na$_2$SO$_4$ + urea & NH$_4$ fertilizers	8–83% decrease	3–12	6	McClung & Frankenberger, 1985, 1987 (Sandy to clay loam, pH 5.8–8; EC 0.18–3)
NO$_2$ accumulation		up to 3.4% of N applied	3–12	6	
NH$_3$ volatilization		up to 370% increase	3–12	3	
Mineralization	NaCl, CaCl$_2$, Na$_2$SO$_4$ + urea & NH$_4$ fertilizers	12–34% decrease	3–12	2–3	
Nitrification		16–98% decrease	3–12	3–6	
Carbofuran decomposition	NaCl, pH 6.7, N fertilizers, pH 5.8	17–40% reduction	0.9–2	2	Petersen, 1990 (silty clay loam; EC 0.6)
		9–82% reduction	0.9–2	0.9	
Denitrification	NO$_3$–N	4–32% reduction	0.4–1	0.6	Aulakh & Singh, 1995 (sands to loams; pH 7.5–8.4; EC 0.2–4)
		32–88% reduction	1–1.8	1.0	
Denitrification	NO$_3$–N	up to 16–28% reduction	0.02–2.8	1.5–2	Weier et al., 1993 (soils-see Table 10–6)
		N$_2$O major product	0.02–2.8	1.5–2.5	

Table 10–6. Relationship between soil NO_3–N level and soil electrical conductivity, microbial denitrification, and proportion gaseous N lost as N_2O for soils of varying texture (pH range 6.5 to 7.4) amended with glucose (0.5 mg C/g soil) and incubated at 90% Water-filled Pore Space for 5 d at 25°C (after Weier et al., 1993).

| Soil | Particle size | | | NO_3–N | EC† (1:1) | Denitrification | | $N_2O/(N_2O + N_2)$ × 100‡ |
	Sand	Silt	Clay			Rate	NO_3 lost	
	%			mg/kg	dS/m	kg N/ha/d		%
Valentine sand	90	7	3	2.5	0.02	0.1	37	10
				142	1.13	5.4	61	21
				280	2.52	11.3	71	67
Hord silt loam	34	46	20	5.8	0.04	0.2	51	2
				145	1.03	6.3	75	20
				284	2.08	14.2	90	38
Yolo silt loam	22	54	24	43	0.18	1.4	41	10
				182	1.00	9.0	70	37
				321	2.23	6.5	10	53
Sharpsburg silty clay loam	3	63	34	89	0.74	3.8	56	14
				228	1.50	12.6	80	63
				367	2.75	10.6	42	>91

† Electrical conductivity of 1:1 soil to water mixture; can be estimated by $\frac{\text{mg/kg } NO_3\text{-N}}{140}$.

‡ N_2O as a % of $(N_2O + N_2)$ produced at time of maximum denitrification.

colorimetric methods where the color of the soil solution is compared with a standard colored indicator solution based on pH. Electrical conductivity is measured with a conductivity meter that is standardized with a salt solution (KCl). For field analysis we use a hand-held portable combination pH and EC meter or separate pocket meters as described in Sarrantonio et al. (1996, this publication). The methods described below use standard meter equipment rather than colorimetric methods.

To measure soil pH and EC the soil must be mixed with some amount of distilled water since we want to measure ions in solution. There are a number of procedures that have been developed to obtain a soil–solution mixture for measuring pH and EC including saturating the soil, mixing the soil with an equal volume or weight of water, and for pH, by mixing the soil with a dilute salt solution such as $CaCl_2$. The method discussed here will use a soil/water ratio of 1:1, which is the easiest method for field measurements. This is a volume to volume ratio where a level scoop of water is added to a firmed and level scoop of soil. For the field assessments described in Sarrantonio et al. (1996, this publication), a standard scoop (29.5 mL volume) is used, which is a common coffee measure available in most grocery stores in the USA. The conductivity of a 1:1 mixture has been shown to be dependent upon soil texture but is related to EC_e in a very consistent manner (Table 10–3); however, it is easier and quicker than the saturated paste method that can be used for more rigorous analysis if the 1:1 method indicates a possible problem. The use of 0.01 M $CaCl_2$ has been suggested as an extractant instead of distilled water to mask small differences in salt concentration between samples that would affect apparent pH; however, where soil EC also is measured, it does not seem worth the effort of obtaining a separate $CaCl_2$ extract for pH since the correlation between $CaCl_2$–pH and water–pH is greater than 0.90 and the presence of salts usually only decreases the pH reading by 0.2 to 0.3 pH units (McLean, 1982).

In the following procedure EC will be measured first followed by pH on the same sample. Where separate meters are used for EC and pH measurements it is important to take the EC measurement first because salts diffusing out of commercial pocket pH meters can result in erroneous EC readings.

Procedure: Soil Electrical Conductivity and pH

1. Measure two firmed and level scoops of soil into a 240 mL (half-pint) container (jelly jar).
2. Add two scoops of distilled water*, then mix or shake the suspension at least 20 times in up and down motion to uniformly disperse the soil, let stand 5 to 10 min.
3. During this time calibrate the EC meter by placing the conductivity cell into a 0.01 M KCl solution. The conductivity meter should read 1.41 dS/m (mmhos/cm) at 25°C, if not, adjust meter to read this value (Note: meters with an expanded scale may read 14.1 or 141, which indicates readings must be divided by 10 or 100 to obtain units of dS/m).
4. Insert the conductivity cell into the soil–water suspension and read the conductivity in dS/m (mmhos/cm).
5. Calibrate the pH meter with standard pH solutions (pH 7 and pH 4 or 10 as needed) then rinse the electrode with distilled water. Place the calibrated electrode into the soil–water suspension and record the pH.

   * Test water supply for EC and pH, especially when alternative tap or rain water sources are used where distilled water is not available. Where necessary, subtract water supply EC value from sample values.

This method appears to be the most useful for quick field measurements. If field measurements for either EC or pH indicate potential problems, then it is recommended that a laboratory based measurement be made for confirmation. It also should be noted that if a soil is saturated, or nearly so, the amount of water added may need to be reduced to obtain a near 1:1 ratio of soil to water on a weight basis (Note: approaches to adjusting results for soil water content are given in more detail in Sarrantonio et al., 1996, this publication). Generally at moisture contents of field capacity or below this can be ignored.

In measuring soil EC and pH on a field scale it is recommended that a number of samples be taken at different parts of the field. Be careful to ensure that the depth of soil sampling is consistent. Since these parameters vary across the landscape it is more important to take samples in different areas than several in one area. Thus taking several individual samples across a field gives more information than taking repeated subsamples in fewer locations. In this way problem areas may be defined, isolated from better areas of the field, and treated separately–resulting in more economic management. More detailed discussions of sampling approaches to account for spacial variability in the field are given in Sarrantonio et al. (1996, this publication) and Dick et al. (1996, this publication).

INTERPRETATION OF RESULTS

On-farm measurements of pH and EC are important for monitoring effects of agricultural management practices on the interaction of nutrient and proton cycles in soil that can result in the decoupling of soil–plant N cycling and the balance of protons and basic ions resulting in soil acidification due, in part, to leaching of basic cations and NO_3^-–N (Patriquin et al., 1993). The acidity produced during formation of NO_3^- from ammoniacal fertilizers is neutralized by OH^- released when plants take up NO_3^-–N; however, where nitrified NO_3^- is lost due to leaching before being taken up by plants, the soil can be acidified. Bouman et al. (1995) reported that long-term application of high rates of anhydrous ammonia and urea (90 to 180 kg N/ha) to cereal crops resulted in a marked acidification of Canadian prairie soils; however, where NO_3^- leaching losses were minimized by matching fertilizer application rates to the N requirement of the crop (45 kg N/ha), little acidification occurred. These are examples of how simple measures, such as soil pH and EC, can be useful indicators of the efficiency and environmental friendliness of fertilizer and cropping management practices.

In general, $EC_{1:1}$ values between 0 and 1.5 dS/m (mmhos/cm) and pH values between 6 and 7.5 are acceptable for general plant growth and microbial activity; however, site-specific interpretation for soil quality will depend on specific land use and crop tolerances that vary considerably for pH and EC (Tables 10–1 and 10–4). For example, a soil with an $EC_{1:1}$ of 3.0 would represent poor soil quality with respect to growing beans (*Phaseolus* sp.), some grasses, most legumes, maize, pepper (*Capsicum* sp.), potatoes (*Solanum tuberosum* L.), rice (*Oryza sativa* L.), or tomatoes [*Lycopersicon lycopersicum* (L.) Karsten], but would represent acceptable soil quality for growing sugar beets (*Beta vulgaris* L.), barley (*Hordeum vulgare* L.), cotton (*Gossypium hirsutum* L.), and wheat (*Triticum aestivum* L.). The interpretation of EC, or pH, in this case is based on crops and crop production; however, a soil $EC_{1:1}$ of 1.5 dS/m (mmhos/cm) associated with a good level of quality for crop production could be associated with poor soil quality if the conductivity resulted mostly from soil nitrates (~210 mg N/kg soil) or a small amount from potentially toxic elements such as Mn, Se, Cd, or others. This could result in poor soil quality because of the increased potential for environmental degradation due to nitrate leaching, or crop products with excessively high levels of nitrates, or toxic elements that pose a threat to human and animal health. Thus soil quality needs to be interpreted based on many soil quality factors, but EC and pH are good initial indicators for changes in soil quality.

We will present a short discussion on our measurements of pH and $EC_{1:1}$ consisting of 220 samples taken in a 0.5 ha plot in southeastern Washington State. The field is in the 50 cm rainfall zone with a crop rotation of winter wheat–dry peas. We sampled 220 locations centered on a regularly spaced grid pattern The samples were analyzed for biological and chemical parameters including EC and pH. We found significant negative correlations between soil EC and pH for samples taken in late fall and early summer with correlation coefficients of 0.34 and 0.49, respectively. Patriquin et al. (1993) attributed similar correlations between soil EC and pH to the acidity associated with microbial mineralization and nitrification processes on farms in eastern Canada. They supported this assumption by the fact that they also observed a high positive correlation between EC and soil NO_3^- content. We also found a significant positive correlation between EC and NO_3^-–N at both sampling times with correlation coefficients of 0.84 for fall and 0.74 for the summer samplings (Fig. 10–5). It appears that the EC at these sampling times was largely influenced by the amount of soluble NO_3^- in the soil. Nitrate was negatively correlated to pH to a similar extent as EC (data not shown).

The spatial distribution of EC and pH values across a field can be both dramatic and problematic. Figure 10–6 shows a smoothed contour map of pH, EC (dS/m), and NO_3^-–N (mg/kg) across a 100 × 50 m field (0.5 ha). The values of each parameter range from low (light color) to high (dark color). It is visually apparent that pH and EC are negatively correlated since light and dark areas match. In addition, as stated above, EC and NO_3^-–N are correlated across the field as shown by the light and dark areas matching up. It is obvious that a good sampling scheme must be developed to avoid classifying the quality of a field as acceptable or nonacceptable based on a few areas that do not represent the overall condition. For example, concentrating our sampling in the 70 to 100 m zone (Fig. 10–6) would indicate low NO_3^-–N and EC and high pH values that are not representative of the entire field. In addition, it can be surmised that a mean value of the whole field may not be a good indicator of the field especially if one is trying to increase soil quality. The mean values for $EC_{1:1}$, pH, and NO_3^-–N in this field are 0.21 dS/m, 5.14, and 20.0 mg/kg, respectively. An $EC_{1:1}$ of 0.21 dS/m (mmhos/cm) is acceptable, however, a large area of this field has EC values between 0.3 and 0.4, which could indicate a potential future problem if EC levels continue to rise appreciably. The average pH level of 5.14 represents a lower level of soil quality and a need for remediative action, particularly in more acidic spots in the field. The average NO_3^-–N content of 20.0 mg/kg soil (ppm) is reasonable for wheat production but, as shown in Fig. 10–6, at least one-third of the field has NO_3^-–N contents below 10 mg/kg. Thus caution must be used in sampling fields and interpreting the results.

In conclusion, the parameters of EC and pH are easily measured in the field and can provide valuable information for assessing the soil condition for plant growth and nutrient cycling as related to both production of food and fiber and the associated quality of water and air environments. These parameters also can be indicators of effects on biological activity where certain microbial mediated processes are affected by shifts in pH or EC. Soil and crop management practices can have a significant effect on pH and EC in relatively short time periods making these parameters good indicators of change. Elevated levels of soil EC can indicate the potential for loss of soluble components such as NO_3^-–N and resultant contamination of surface and subsurface water supplies. A trend for soil acidification where ammoniacal fertilizers are used, can be an indicator of the inefficient use of those fertilizers and increased N loss due to leaching from the plant rooting zone. Within the context of soil quality, the measurement of direction and rate of change are of greatest importance.

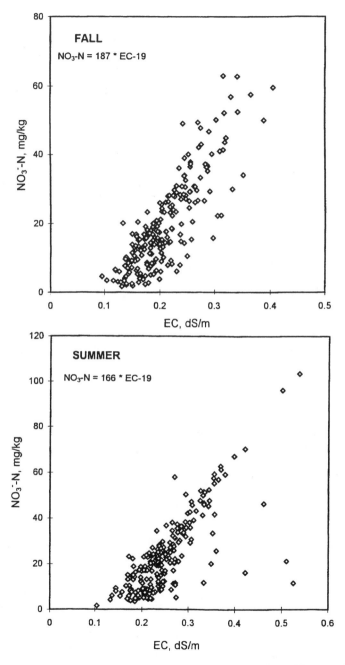

Fig. 10–5. The relationship between $EC_{1:1}$ (dS/m) and NO_3^-–N sampled in the fall (top) and summer (bottom) periods, $n = 220$, for a wheat field in western Washington State.

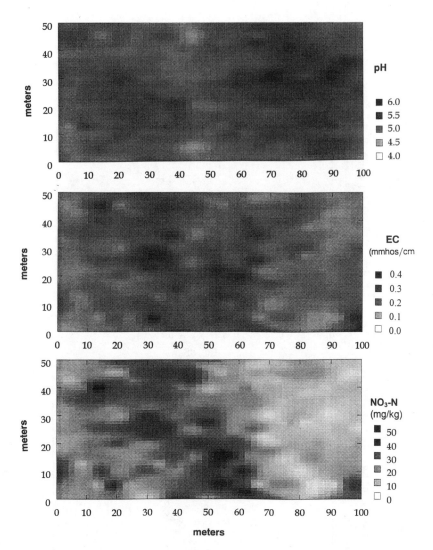

Fig. 10–6. Spatial variability of pH, $EC_{1:1}$(dS/m) and NO_3^-–N sampled in the fall over a 50 by 100 m area (0.5 ha) for a wheat field in western Washington State.

REFERENCES

Aulakh, M.S., and Bijay Singh. 1995. Gaseous losses of N through denitrification from soils under different cropping systems. Progress Rep. IN-ARS-526. Sept. 1993–Aug. 1994. Punjab Agric. Univ., Ludhiana, India.

Bazin, M.J., and J.I. Prosser. 1988. Physiological models in microbiology. Vol. I. CRC Ser. in Mathematical Models in Microbiology. CRC Press, Boca Raton, FL.

Bender, M., and R. Conrad. 1995. Effect of CH_4 concentrations and soil conditions on the induction of CH_4 oxidation activity. Soil Biol. Biochem. 27:1517–1527.

Bouman, O.T., D. Curtin, C.A. Campbell, V.O. Biederbeck, and H. Ukrainetz. 1995. Soil acidification from long-term use of anhydrous ammonia. Soil Sci. Soc. Am. J. 59:1488–1494.

Dahnke, W.C., and D.A. Whitney. 1988. Measurement of soil salinity. p. 32–34. *In* Recommended chemical soil test procedures for the North Central Region. North Central Reg. Publ. 221. Revised. North Dakota Agric. Exp. Stn. Bull. 499. Fargo, ND.

Dick, R.P., D.R. Thomas, and J.J. Halvorson. 1996. Standardized methods, sampling, and sample pretreatment. p. 107–121. *In* J.W. Doran and A.J. Jones (ed.) Methods for assessing soil quality. SSSA Spec. Publ. 49. SSSA, Madison, WI.

Doran, J.W., and T.B. Parkin. 1996. Quantitative indicators of soil quality: A minimum data set. p. 25–37. *In* J.W. Doran and A.J. Jones (ed.) Methods for assessing soil quality. SSSA Spec. Publ. 49. SSSA, Madison, WI.

Killham, K. 1994. Soil ecology. p. 24–28. Cambridge Univ. Press, Cambridge, England.

Maas, E.V., and G.J. Hoffman. 1977. Crop salt tolerance-current assessment. J. Irrig. Drain. Div. Am. Soc. Civ. Eng. 103:115–134.

McClung, G., and W.T. Frankenberger, Jr. 1985. Soil nitrogen transformations as affected by salinity. Soil Sci. 139:405–411.

McClung, G., and W.T. Frankenberger, Jr. 1987. Nitrogen mineralization rates in saline versus salt-amended soils. Plant Soil. 104:13-21.

McCormick, R.W., and D.C. Wolf. 1980. Effect of sodium chloride on CO_2 evolution, ammonification, and nitrification in a Sassafras sandy loam. Soil Biol. Biochem. 12:153–157.

McLaughlin, M.J., L.T. Palmer, K.G. Tiller, T.A. Beech, and M.K. Smart. 1994. Increased soil salinity causes elevated cadmium concentrations in field-grown potato tubers. J. Environ. Qual. 23:1013–1018.

McLean, E.O. 1982. Soil pH and lime requirement. p. 199–224. *In* A.L. Page (ed.) Methods of soil analysis. Part 2. 2nd ed. Agron. Monogr. 9. ASA and SSSA, Madison, WI.

Ohio Agricultural Experiment Station. 1938. Handbook of experiments in agronomy. Ohio Agric. Exp. Stn. Spec. Cir. 53. Ohio Agric. Exp. Stn., Columbus.

Paul, E.A., and F.E. Clark. 1989. Soil microbiology and biochemistry. Academic Press, New York.

Patriquin, D.G., H. Blaikie, M.J. Patriquin, and C. Yang. 1993. On-farm measurements of pH, electrical conductivity, and nitrate in soil extracts for monitoring coupling and decoupling of nutrient cycles. Biol. Agric. Hortic. 9:231–272.

Petersen, B. 1990. Interactions between agricultural chemicals and their effects on pesticide persistence in soil. Ph.D. diss. Univ. of Nebraska, Lincoln.

Rhoades, J.D. 1982. Soluble salts. p.167–179. *In* A.L. Page (ed.) Methods of soil analysis. Part 2. 2nd ed. Agron. Monogr. 9. ASA and SSSA, Madison, WI.

Sarrantonio, M., J.W. Doran, M.A. Liebig, and J.J. Halvorson. 1996. On-farm assessment of soil quality and health. p. 000–000. *In* J.W. Doran and A.J. Jones (ed.) Methods for assessing soil quality. SSSA Spec. Publ. 49. SSSA, Madison, WI.

Sikora, L.J., and D.E. Stott. 1996. Soil organic carbon and nitrogen. p. 157–167. *In* J.W. Doran and A.J. Jones (ed.) Methods for assessing soil quality. SSSA Spec. Publ. 49. SSSA, Madison, WI.

Troeh, F.R., and L.M. Thompson. 1993. Soils and soil fertility. Oxford Univ. Press, New York.

Weier, K.L., J.W. Doran, J.F. Power, and D.T. Walters. 1993. Denitrification and the dinitrogen/nitrous oxide ratio as affected by soil water, available carbon, and nitrate. Soil Sci. Soc. Am. J. 57: 66–72.

11 Assessing Soil Nitrogen, Phosphorus, and Potassium for Crop Nutrition and Environmental Risk

Deborah L. Allan
University of Minnesota
St. Paul, Minnesota

Randy Killorn
Iowa State University
Ames, Iowa

Measures of soil mineral N, P, and K are important for predicting potential plant productivity. In addition, N and P are important environmental components of soil quality. For soils in many regions it is possible to evaluate soil test results to determine the soil's capacity to supply the nutrient requirements of major crops. There is less information available to set upper limits for pollution risk. The goal for land managers should be to maintain nutrient levels in the range desirable for optimal plant growth but low enough to avoid environmental contamination.

While a large body of literature exists regarding the standard laboratory testing for mineral N, P, and K, there is little information in print about so-called "quick tests" or field tests of nutrient status. Thorough reviews for soil tests for these three nutrients appeared recently (Dahnke & Johnson, 1990; Fixen & Grove, 1990; Haby et al., 1990) and detailed descriptions of appropriate analytical procedures appear in Page et al. (1982). Methods for N analysis recently have been updated by Bundy and Meisinger (1994).

Nitrogen

Use of the residual (fall or early spring) NO_3^- test is the most common procedure for determination of mineral N in western Canada and the Great Plains region, where NO_3^- is less likely to be leached below the root zone compared with the eastern USA (Dahnke & Johnson, 1990; Bundy & Meisinger, 1994). In areas of higher precipitation, the in-season NO_3^- test (residual plus recently mineralized NO_3^-) has become a common management practice (Magdoff, 1991; Bundy & Meisinger, 1994). For the residual NO_3^- test, the soil is often sampled to a depth

---

Copyright © 1996 Soil Science Society of America, 677 S. Segoe Rd., Madison, WI 53711, USA.
Methods for Assessing Soil Quality, SSSA Special Publication 49.

of 2 ft (61 cm) or more, whereas in-season NO_3^- is measured in the top 1 ft (30.5 cm) of soil. Because NO_3^- is water-soluble, it can be extracted with water and then measured with a NO_3^- electrode; this method is the easiest but is less accurate than others (Keeney & Nelson, 1982). Detailed procedures for the NO_3^- electrode method are given in Gelderman and Fixen (1988). Ion chromatography is a very accurate, easy technique for NO_3^- detection, but requires special equipment. Nitrate–N measurement also can be accomplished by Cd reduction to NO_2^- (Barnes & Folkard, 1951) or by the reduction of NO_3^- to ammonia with Cu-coated Zn (Carlson, 1986) or steam distillation with Devarda alloy (Keeney & Nelson, 1982).

Both manual and automated methods for Cd reduction are described by Bundy and Meisinger (1994). In the automated method, NO_3^- is quantitatively reduced to NO_2^- by passing the sample through a copperized Cd column. The reduced NO_3^- and original NO_2^- are then determined by diazotizing with sulfanilamide followed by coupling with N-(1-naphthyl)ethylenediamine dihydrochloride. This reaction yields a magenta-colored soluble dye, which is read at 520 nm (Bundy & Meisinger, 1994). In this chapter, we have detailed the extraction procedure only, assuming that automated instrumentation to quantify soil NO_3^- will be available to the user.

Because ammonium levels are low under most favorable growing conditions, NH_4–N is rarely measured for soil test purposes. Both exchangeable NH_4^+ and NO_3^- can be extracted with 2 M KCl, and most automated ion analyzers are equipped to detect both ions. On a number of auto analyzers, soil extracts are analyzed for ammonia by the salicylate method. Ammonium also can be detected as ammonia by specific ion electrode (in solutions made alkaline to convert NH_4^+ to NH_3), colorimetrically by indophenol blue, by steam distillation with MgO and titration, or by microdiffusion (Keeney & Nelson, 1982; Dahnke & Johnson, 1990; Bundy & Meisinger, 1994).

Phosphorus

Recommended soil analysis methods for P in various regions of the USA are given in Table 11–1. For simplicity, we will present the two standard extraction procedures for P determination in the North Central and Western Regions: Bray P-1 (Bray & Kurtz, 1945) and the Olsen sodium bicarbonate methods (Olsen et al., 1954). Bray P-1 is better suited to acid and neutral soils while Olsen's method is preferable for calcareous soils (Knudsen & Beegle, 1988). In Minnesota, Bray P-1 is recommended for use in soils below pH 7.4 and Olsen's

Table 11–1. Recommended soil analysis methods.†

Region	Phosphorus	Potassium
Northeast	Morgan; Modified Morgan	Morgan; Modified Morgan
South	Mehlich-1	Mehlich-1
North Central	Bray P-1, Olsen	Ammonium Acetate, pH 7
West	Bray P-1, Olsen, AB-DTPA	Ammonium Acetate, pH 7
All regions	Mehlich-3	Mehlich-3

† Soil and Plant Analysis Council (1992).

for pH greater than 7.4 (Soil Testing Laboratory, University of Minnesota, St. Paul). An acid bottle test also can be used to determine which test to apply: if there is no or weak reaction, Bray P-1 should be used; for a strong reaction Olsen's method is preferred. The problem with using Bray P-1 for calcareous soils is that calcium carbonate can neutralize the acidity and precipitate the fluoride in the Bray P-1 extractant, resulting in low readings (Olsen & Sommers, 1982).

The two tests extract similar amounts of P where P concentrations are in the low range, but Bray P-1 extracts more at higher levels in noncalcareous soils (Knudsen & Beegle, 1988). Thus, separate calibrations are required for each method. Reducing methods for the Bray P-1 test include the ascorbic acid method or Fiske-Subbarow reducing agent (amino-naphthol-sulfonic acid). Since the ascorbic acid method is well adapted for both the Bray P-1 and Olsen methods, it is detailed in the recommended procedure below.

Two other extractants that may prove useful for soil quality measurements because of the large number of elements they extract are Mehlich 3 and ammonium bicarbonate-DTPA (AB-DTPA). Unfortunately, there are few correlation and calibration data available for field grown crops; only for correlations with other extractants. The Mehlich 3 procedure (Mehlich, 1984) is recommended for acid and neutral soils. It can be used to test for P, K, Ca, Mg, Na, Fe, Mn, Zn, and Cu and approximates the Bray P-1 test for P, ammonium acetate for K, Ca, and Mg, and EDTA for micronutrients. The AB-DTPA procedure is given in Soltanpour and Schwab (1977) and is recommended for alkaline soils. It can be used to test for P, K, Na, Fe, Mn, Cu, and Zn and approximates the Olsen method for P, neutral ammonium acetate for K, and Lindsay's DTPA extraction for micronutrients (Lindsay & Norvell, 1978).

Potassium

For the north central and western USA, the standard extractant for exchangeable K (which includes soil solution K) has been 1.0 M ammonium acetate, pH 7.0, for more than 50 yr (Table 11–1; Haby et al., 1990). After NH_4^+ displacement of exchangeable K, K in the extractant is commonly determined using the emission mode on an atomic absorption–emission spectrometer. A K-ion-selective electrode also can be used to detect K^+, but not when NH_4^+ is the replacement cation because of its interference with the electrode. A possible alternative is to use Ca acetate extractant, with K determined by ion-specific electrode; however, K values obtained in this manner have not been calibrated for crop response as for ammonium acetate.

Other possible extractants are Mehlich 3 and AB-DTPA as detailed above. Mehlich 3 and ammonium acetate yield highly correlated results, with Mehlich 3 extracting approximately 20% more K than ammonium acetate (Haby et al., 1990).

RELEVANCE TO OTHER MEASURES

While extensive calibration for specific crops in specific regions can predict fertilizer requirements for P and K, this is much more difficult for N because of the uncertainties of N losses and transformations, including mineralization. Thus the availability of mineral N relates to other measures of available N including soil organic N (Sikora & Stott, 1996, this publication) and tests of potentially mineralizable N (Drinkwater et al., 1996, this publication). For the anaerobic N test for potentially mineralizable N, 2 M KCl extract is used on field moist samples to obtain initial values for NH_4^+ and NO_3^- and then again on soils incubated anaerobically for 10 d, using the same procedure. Thus, this method is readily compatible with inorganic NO_3^- and NH_4^+ extraction. While a portion of soluble organic P is extracted by the sodium bicarbonate (Olsen's) method, some nonresponsive soils may supply a significant portion of P from organic sources that may not be reflected in measures detailed here. Nutrient availability to certain plants may be dominated by rhizobial (N) and mycorrhizal (P) symbioses (Dick et al., 1996a, this publication).

RECOMMENDED PROCEDURES

Soil sampling methods and concerns are detailed in Dick et al. (1996b, this publication).

Ammonium and Nitrate–Nitrogen

These methods are adapted from Keeney and Nelson (1982) and QuikChem Methods 12-107-04-1-b and 12-107-06-2-A (Lachat Instruments, 1993). They also are given in Bundy and Meisinger (1994).

Equipment

1. Balance to weigh samples
2. 100-mL Erlenmeyer flasks and volumetric flasks for standards
3. Automatic extracting solution dispenser (50 mL)
4. Rotating or reciprocal shaker
5. Autoanalyzer

Reagents

1. Extracting solution (2 M KCl)
   Dissolve 1.5 kg of reagent grade KCl in 8 L of distilled water and make to 10 L volume with deionized water.
2. Stock standard ammonium solution [100.0 ppm (mg L^{-1}) NH_4–N]:
   In a 1 L volumetric flask, dissolve 150 g KCl and 0.3819 g of reagent grade NH_4Cl (dried for 2 h at 110°C) in about 800 mL of water. Dilute to volume with water and mix.
3. Working standard solutions for NH_4–N:
   To eight 250-mL volumetric flasks add, respectively, 50.00, 20.00, 10.00, 5.00, 2.50, 1.25, 0.50, and 0.25 mL of the stock standard. This makes 20.00, 8.00, 4.00, 2.00, 1.00, 0.50, 0.20 and 0.10 ppm NH_4–N

standards, respectively. Dilute each to volume with 2 M KCl and mix. Standards should be stored in a refrigerator to inhibit microbial growth.
4. Stock standard nitrate solution [200.0 ppm (mg L^{-1}) NO$_3$–N]:
In a 1-L volumetric flask, dissolve 149 g KCl and 1.444 g of reagent grade KNO$_3$ in about 600 mL of water. Add about 1 mL of chloroform to inhibit microbial growth. Dilute to volume and mix.
5. Working standard solutions for NO$_3$–N:
First, make a working stock standard of 20 ppm NO$_3$–N. Add 100.0 mL stock solution (see no. 4 above) to a 1 L volumetric flask. Dilute to volume with 2 M KCl and mix. This solution should be made fresh weekly. To five 250-mL volumetric flasks add, respectively, 125.0, 25.0, 10.0, 1.0 and 0 mL of the working stock standard (20 ppm). This makes 10.00, 2.00, 0.80, 0.08, and 0 ppm NO$_3$–N standards, respectively. Dilute each to volume with 2 M KCl and mix.

Procedure

1. Weigh 5 g of air-dried, sieved soil (<2 mm) and place in a 100-mL Erlenmeyer flask.
2. Add 50 mL of extracting solution to each flask and shake for 1 hr. For NO$_3^-$ only, 30 min shaking time is sufficient.
3. Filter extracts through acid-washed Whatman no. 42 filter paper. If samples cannot be run within 24 h of extraction, they should be adjusted to pH 3-5 with dilute sulfuric acid and refrigerated. This requires that standards also be adjusted with sulfuric acid.
4. Perform analyses on samples and standards as specified for the autoanalyzer used.
5. Convert ppm concentration in filtrate to concentration in the soil:

ppm (μg g^{-1}) NO$_3$–N or NH$_4$–N in soil = ppm N in filtrate × 10

kg ha^{-1} NO$_3$–N or NH$_4$–N in soil =

$$\text{ppm soil N } (\mu g\ g^{-1} \text{ or mg kg})^{-1} \times \frac{\text{depth of sample (cm)}}{10}$$

× soil bulk density (g cm^{-3}).

lb acre^{-1} NO$_3$–N or NH$_4$–N in soil = kg ha^{-1} NO$_3$–N or NH$_4$–N × 0.89

Phosphorus

Bray P-1 Soil Test

Compiled from Knudsen and Beegle (1988) and Olsen and Sommers (1982).

Equipment

1. Standard 1-g soil scoop available from Soil Chem, (Rossville, IL) or balance to weigh samples.
2. 50-mL Erlenmeyer extracting flasks and volumetric flasks for standards.
3. Automatic solution dispensers (8 mL and 10 mL).
4. Spectrophotometer or colorimeter capable of determining absorbance at 882 nm.
5. Rotating or reciprocal shaker capable of at least 200 excursions per minute (epm).

Reagents

1. Extracting solution (0.025 M HCl in 0.03 M NH_4F):
   Dissolve 11.11 g of reagent grade NH_4F in about 9 L of distilled water. Add 250 mL of 1.00 M HCl (previously standardized) and make to 10 L volume with distilled water. Mix thoroughly. The pH of the solution should be 2.6 ± 0.05. Adjust pH if necessary. Store in polyethylene.
2. Stock standard P solution [50 ppm (mg L^{-1}) P]:
   Dissolve 0.02197 g of reagent grade KH_2PO_4, (dried in a desiccator), in about 25 mL of distilled water. Dilute to a final volume of 100 mL with extracting solution. Under refrigeration, this standard should be stable for 6 mo to 1 yr.
3. Working standard solutions:
   Pipette appropriate volumes of stock solution into volumetric flasks to make up working standard solutions of 0.2, 0.5, 1.0, 2.0, 3.0, 4.0, and 5.0 ppm (mg L^{-1}) P (respectively 1 mL stock in 250 mL, 1 mL in 100 mL, 2 mL in 100 mL, 4 mL in 100 mL, 6 mL in 100 mL, 8 mL in 100 mL, and 10 mL in 100 mL). Bring each flask to volume with extracting solution (See no. 1 above).
4. Acid molybdate stock solution:
   Dissolve 60 g ammonium molybdate, $(NH_4)_6Mo_7O_{24} \cdot 4\ H_2O$, in 200 mL of distilled water. If necessary, heat to about 60°C until solution is clear and allow to cool. Dissolve 1.455 g of antimony potassium tartrate in the molybdate solution. Under a fume hood, slowly add 700 mL of concentrated sulfuric acid. Cool and dilute to a final volume of 1.00 L. This solution may be blue, but will clear when diluted for use. Store in the dark under refrigeration. Bring to room temperature before use.
5. Ascorbic acid stock solution:
   Dissolve 132 g of ascorbic acid in distilled water and dilute to a final volume of 1.00 L. Store in the dark under refrigeration. Bring to room temperature before use.
6. Working solution:
   Prepare fresh each day by adding 25 mL of acid molybdate stock solution to about 800 mL of distilled water, mixing, adding 10 mL of ascorbic acid stock solution and making to a final volume of 1.00 L.

Procedure

1. Weigh or scoop 1 g of air-dried and sieved soil into an extraction flask.
2. Add 10 mL of extracting solution to each flask and shake at 200 or more epm for 5 min at 24 to 27°C.
3. Filter extracts through Whatman no. 2 filter paper or a similar grade of paper. Refilter if extracts are not clear.
4. Transfer a 2 mL aliquot to a test tube.
5. Add 8 mL of working solution so that thorough agitation and mixing occurs.
6. Allow 10 min for color development. Read percentage of transmittance or optical density on a colorimeter or spectrophotometer set at 882 nm. Color is stable for about 2 h.
7. Prepare a standard curve by aliquoting 2 mL of each working standard, and 8 mL of working solution, developing color, and reading intensity as for soil extracts. Plot intensity against concentration of working standards. Use to determine concentration in soil extracts.
8. Convert ppm (mg L^{-1}) concentration in filtrate to concentration in the soil:

$$\text{ppm } (\mu g\ g^{-1}) \text{ P in soil} = \text{ppm P in filtrate} \times 10$$

$$\text{kg ha}^{-1} \text{ P in soil} = \text{ppm soil P } (\mu g\ g^{-1} \text{ or mg kg}^{-1}) \times \frac{\text{depth of sample (cm)}}{10} \times \text{soil bulk density (g cm}^{-3}).$$

$$\text{lb acre}^{-1} \text{ P in soil} = \text{kg ha}^{-1} \text{ P} \times 0.89$$

Olsen (NaHCO$_3$) Soil Test

Compiled from Knudsen and Beegle (1988) and Olsen and Sommers (1982).

Equipment

Same as for Bray P-1 test, with the exception of a standard 2-g scoop, and 125-mL Erlenmeyer extracting flasks.

Reagents

1. Extracting solution, 0.5 M NaHCO$_3$, pH 8.5:
   Dissolve 420 g commercial grade NaHCO$_3$ in distilled water and make to a volume of 10 L. Adjust to pH 8.5 with 50% (weight: volume) NaOH. A magnetic stirrer or electric mixer is required to dissolve the salt.
2. Acid molybdate stock solution (Reagent A):
   Dissolve 60 g of ammonium molybdate $(NH_4)_6Mo_7O_{24} \cdot 4\ H_2O$, in 1250 mL of distilled water. Dissolve 1.455 g of antimony potassium

tartrate in 500 mL of distilled water. Add both of these solutions to 5.0 L of 5 M H_2SO_4 (Add the acid to the water! 148 mL of concentrated H_2SO_4 per L of water), mix and dilute to 10.0 L with distilled water. Store in a pyrex glass bottle in a dark cool place. Bring to room temperature before use.
3. Reagent B:
Dissolve 2.639 g of ascorbic acid in 500 mL of reagent A. This reagent must be prepared each day as needed since it will not keep for more than 24 h.
4. Stock standard phosphorus solution, 50 ppm (mg L^{-1}). Same as for Bray P-1 test.
5. Working standards. Same as for Bray P-1 test, but bring to volume with 0.5 M $NaHCO_3$.

Procedure

1. Scoop or weigh 2 g of air-dried and sieved soil into an extraction flask.
2. Add 40 mL of extracting solution to each flask and shake at 200 or more epm for 30 min at 24 to 27°C.
3. Filter extracts through Whatman no. 2 filter paper. Refilter if extracts are not clear.
4. Transfer 5 mL to a beaker or Erlenmeyer flask (50-mL or larger).
5. Add 15 mL of deionized water.
6. Add 5 mL of Reagent B and agitate flask so that thorough mixing occurs.
7. After 10 min for color development, read percent transmittance or optical density on a colorimeter set at 882 nm. Color is stable for at least 2 h.
8. Prepare a standard curve by aliquoting 5 mL of each working standard, 15 mL of deionized water and 5 mL of Reagent B, developing color and reading intensity as for soil extracts. Plot intensity against concentration of working standards. Use to determine concentration in soil extracts.
9. Convert ppm (mg L^{-1}) concentration in filtrate to concentration in the soil:

$$\text{ppm } (\mu g\ g^{-1}) \text{ P in soil} = \text{ppm P in filtrate} \times 20$$

$$\text{kg ha}^{-1} \text{ P in soil} = \text{ppm soil P } (\mu g\ g^{-1} \text{ or mg kg}^{-1}) \times \frac{\text{depth of sample (cm)}}{10} \times \text{soil bulk density (g cm}^{-3}).$$

$$\text{lb acre}^{-1} \text{ P in soil} = \text{kg ha}^{-1} \text{ P} \times 0.89$$

Potassium

Compiled from Brown and Warncke (1988) and Knudsen et al. (1982).

Equipment

1. Standard 1-g scoop or scale to weigh samples
2. Automatic extracting solution dispenser (10 mL)
3. Erlenmeyer extracting flasks (50-mL) and volumetric flasks (100-mL) for standard preparation.
4. Funnels or filter holding devices
5. Receiving receptacle (20 to 30 mL beakers or test tubes)
6. Rotating or reciprocating shaker capable of 200 epm
7. Atomic absorption/emission spectrometer (set in the emission mode for K)

Reagents

1. Extracting solution (1 M NH$_4$OAc at pH 7.0)
   Add 77.1 g of reagent-grade NH$_4$OAc to 900 mL of distilled water. After dissolution of the salt, adjust the pH to 7.0 with 3 M NH$_4$OH or 3 M acetic acid. Dilute to a final volume of 1 L. Store in refrigerator to inhibit microbial growth.
2. Stock standard solution [1000 ppm (mg L^{-1}) K]:
   Dissolve 1.9073 g oven dry reagent grade KCl in 1 M NH$_4$OAc at pH 7.0. Bring to a volume of 1000 mL with the extracting solution and mix well. Store in refrigerator to inhibit microbial growth.
3. Working standard solutions:
   Prepare a 100 ppm (mg L^{-1}) standard by diluting 100 mL of the 1000 ppm K stock solution to 1 L with extracting solution. Pipette 10, 20, 30, 40, and 50 mL of the 100 ppm K solution into 100-mL volumetric flasks and bring each to volume with extracting solution. These solutions will contain 10, 20, 30, 40, and 50 ppm K, respectively. The extracting solution serves as the 0 ppm standard.

Procedure

1. Weigh or scoop 1 g of air-dried and sieved soil into an extraction flask.
2. Add 10 mL of extracting solution.
3. Shake for 5 min on the shaker at 200 epm.
4. Filter the suspensions through Whatman no. 2 filter paper. Refilter if the extract is cloudy.
5. Set up the atomic absorption–emission spectrometer for K by emission. Determine the standard curve using the standards and obtain the concentrations of K in the soil extracts.
6. Convert ppm concentration in filtrate to concentration in the soil:

$$\text{ppm } (\mu g \text{ g}^{-1}) \text{ K in soil} = \text{ppm K in filtrate} \times 10$$

$$\text{kg ha}^{-1} \text{ K in soil} = \text{ppm soil K } (\mu g \text{ g}^{-1} \text{ or mg kg}^{-1}) \times \frac{\text{depth of sample (cm)}}{10} \times \text{soil bulk density (g cm}^{-3}\text{)}.$$

$$\text{lb acre}^{-1} \text{ K in soil} = \text{kg ha}^{-1} \text{ K} \times 0.89$$

INTERPRETATION OF RESULTS

Interpretation of soil tests, based on local calibration data, are available for the standard N, P, and K tests in each state or region. For example, in western Minnesota, the residual N credit amounts for corn (*Zea mays* L.) on soil samples obtained in early spring are 0 lb N/A for 0 to 6 ppm (μg g^{-1} or mg kg^{-1}) NO$_3$–N in a 60 cm sample, 35 lb N/A for 6 to 9 ppm, 65 lb N/A for 9 to 12 ppm, 95 lb N/A for 12 to 15 ppm, and 125 lb N/A for 15 to 19 ppm. At 19 ppm or greater NO$_3$–N, no fertilizer N would be recommended (Schmitt & Randall, 1994a). These N credits apply only when using Minnesota recommendations that include factors such as yield goal, soil organic matter, and previous crop. In Iowa, if the NO$_3$–N exceeds 25 ppm (μg g^{-1}) in a soil sample collected in early June (0–30 cm; 6–12 in corn height), no fertilizer is recommended. If the test indicates less than 25 ppm (μg g^{-1}), the fertilizer recommendation is calculated from: (25 − soil test) × 8, but not to exceed 160 lb N/A (Blackmer et al., 1993). Other presidedress NO$_3$ test critical levels have consistently been in the range of 20 to 25 ppm NO$_3$–N (Bundy & Meisinger, 1994). These interpretations may need to be altered for situations where N is primarily supplied from organic sources such as manure (Practical Farmers of Iowa, 1996).

Interpretations for correlation and calibration data for Bray P-1 or Olsen extractable P and ammonium acetate extractable K are available for each state in the Midwest and Western regions. Table 11–2 shows relative levels for soil test P and K in Minnesota soils. At high or very high levels, fertilizers are not recommended because there is little probability of crop yield increases.

While traditional soil tests for N and P have been interpreted according to plant response, there is increased interest in soil test interpretations to achieve environmental goals. These efforts are hampered by the lack of a research base and difficulties balancing production and environmental needs in soil test interpretation. An estimate of the upper limits of soil test NO$_3^-$ might be based on background soil levels (which will vary depending on soil type and time of sample collection) plus needs for the current year's crop. In Minnesota, for example, background preplant NO$_3$–N levels in the top 60 cm might range from 4 to 13 ppm (μg g^{-1}; Schmitt & Randall, 1994a). Because crop sufficiency can be measured at 19 ppm (μg g^{-1}), any NO$_3$–N above this amount may potentially be an environmental liability. In-season test results in Iowa from 90 randomly selected sites in 1988 (a drought year) showed that 53% had NO$_3$–N levels of 40 ppm (μg

Table 11–1. Relative levels for soil test P and K in Minnesota.†

Relative level	Phosphorus		Potassium (NH$_4$OAc)
	Bray P-1	Olsen P	
	ppm (μg g^1) in soil		
Very low	0–5	0–3	0–40
Low	6–10	4–7	41–80
Medium	11–15	8–11	81–120
High	16–20	12–15	121–160
Very high	21+	16+	161+

† Rehm et al., 1994.

g^{-1}) or more in the surface 30 cm, and in 1989, 52% had 43 ppm ($\mu g\ g^{-1}$) or more (Blackmer et al., 1992), indicating widespread potential for groundwater contamination.

Residual NO_3^- tests may be more suitable indicators of potential for environmental risk. In the state of Baden-Württemberg in Germany, farmers are paid a bonus to limit the NO_3^- content of their field drainage water. The surface 90 cm of soil in the late fall must contain <45 kg $NH_4^+ + NO_3$–N ha^{-1} to receive the bonus (Black, 1992). This is equivalent to a concentration of 3.3 ppm ($\mu g\ g^{-1}$) in the top 90 cm. In Minnesota, postharvest soil NO_3–N concentrations above 4 ppm ($\mu g\ g^{-1}$) were generally associated with excessive fertilizer N rates in 0 to 4 ft (0–1.2 m) samples (Schmitt & Randall, 1994b).

Although we are unaware of published standards for excess NO_3^-, there have been a number of published recommendations for limits for soil P. These are reviewed by Sharpley et al. (1996) and have been developed as part of manure management programs in a number of states. Because manure is typically applied to meet crop N needs, the low N/P ratio of manure results in over-application of manure P. In Michigan and Wisconsin, at soil test values above 75 ppm ($\mu g\ g^{-1}$) Bray P-1, the recommendation is to reduce manure application rates. At an upper limit of 150 ppm ($\mu g\ g^{-1}$) soil P, no further manure or other P sources should be applied until levels decrease. Other states also recommend no further P application above extractable P concentrations ranging from 120 to 200 ppm ($\mu g\ g^{-1}$), depending on the state (Sharpley et al., 1996). These authors suggest that soil test P in the surface 2 in (5 cm) can be a good indicator of dissolved P concentration in runoff, but this information should be integrated with local climatic, topographic, and management information to assess environmental risk.

APPLICATIONS FOR ON-FARM ASSESSMENT

There is increased interest among some crop producers in doing their own soil analysis. This is especially true in the corn belt where the use of the presidedress soil NO_3 test (Magdoff et al., 1984; Blackmer et al., 1989) was implemented in the late 1980's. The concept of on-farm analyses of soils dates back at least 30 years (Ohlrogge, 1962) when a soil and plant analysis kit was developed at Purdue University. The procedures used in the Purdue kit were semi-quantitative. Most kits that are offered today are quantitative and based on two general approaches: (i) use of ion-selective electrodes and (ii) colorimetric methods. These are the same methods used by some commercial laboratories. Various companies produce prepackaged chemicals in amounts suitable for performing the analyses one at a time. Until recently, the colorimetric test results were only qualitative, due to the subjectivity of color comparison charts. With the advent of reflectometers (e.g., Reflectoquant Analysis System from EM Science, Gibbstown, NJ), rapid, quantifiable results are possible in field tests. Reflectometers are available for about $500 and batches of 50 test strips are presently available for pH, NO_3, phosphate, and K at about $55 per batch. Many of the procedures described in this chapter require that the soil samples be dried and ground before extraction of N, P, and K. If sample extracts are prepared in the field and subsamples for moisture content are taken, it may be possible to use the test strips and reflectometer in the field and adjust back to a soil dry weight or volumetric basis later. All procedures must be carefully followed to ensure accurate results. Even with careful and consistent application, in-field test results may be quite variable and not easily reproduced.

Nitrogen

Nitrate

Ion-selective electrodes. Portable meters have been developed that can be used in the field. Some require that samples be filtered, while others can be immersed in the extracting solution-soil mixture directly. The electrode provides precise measurement of NO_3^-–N, but the required equipment costs from $300 to more than $1000.

Colorimetric. This method reduces NO_3^- to NO_2^-. The NO_2^- is then determined using a modification of the Cd reduction method (Barnes & Folkard, 1951). This can be accomplished in a test tube and the intensity of the resulting color estimated using transmitted light. Nitrite sensitive strips also can be used. In this method, the amount of NO_2^- is determined by comparing the color on the strip with reference strips or using a hand held reflectometer (Schaefer, 1986; Jemison & Fox, 1988; Roth et al., 1991).

A procedure to use either or both of these quick-test methods in the field is described by Hartz (1993). Field moist soil cores are collected and mixed. Thirty milliliters of 0.025 M $Al_2(SO_4)_3 \cdot 12\ H_2O$ extracting solution is added to a 50-mL, volumetrically marked centrifuge tube. Moist soil is added to the 40-mL mark and a subsample taken for dry-weight determination. The tube is capped, shaken to disperse aggregates and allowed to settle for 1 to 3 h, until a clear supernatant forms on top. Supernatant NO_3^- concentration can be tested with NO_3-sensitive colorimetric test strips or by a battery-operated NO_3-selective electrode. Across a range of soil textures, moisture contents and NO_3–N concentrations, Hartz obtained good correlations with conventional laboratory analysis of 2 M KCl soil extracts ($r^2 = 0.94$). Jemison and Fox (1988) and Roth et al. (1991) also obtained good correlations of test strip results with standard laboratory analyses of soil N ($r^2 = 0.96$ and 0.98, respectively) using air-dried, ground soil samples. Although correlations were good, exact quantitative agreement was not obtained. Regression equations must be applied to correct the test values.

The calculations to determine kilograms of NO_3–N ha^{-1} or ppm (mg L^{-1}) NO_3–N for the test strips (calibrated in mg NO_3^- L^{-1}) are given below:

$$\text{ppm } (\mu g\ g^{-1})\ NO_3\text{–N in soil} = \text{test strip reading} \times \frac{14\ \text{mg N}}{62\ \text{mg } NO_3^-} \times \frac{\text{volume extracting solution plus soil water (mL)}}{\text{g dry soil}}$$

$$\text{kg NO}_3^- \text{-N ha}^{-1} = \text{test strip reading} \times \frac{14 \text{ mg N}}{62 \text{ mg NO}_3^-}$$

$$\times \frac{\text{volume extracting solution plus soil water (mL)}}{\text{vol soil added (mL)}} \times \frac{\text{depth of sample (cm)}}{10}$$

$$\text{lb NO}_3^- \text{-N ha}^{-1} = (\text{kg NO}_3^- \text{-N ha}^{-1}) \times 0.89$$

Ammonium

The only on-farm method currently available for ammonium determination is use of an ion-selective electrode. This electrode actually measures the amount of ammonia generated by addition of sodium hydroxide to the extracting solution.

Phosphorus

The P procedures for on-farm testing are all based on the development of the blue, phospho-molybdate color and estimating the intensity of the color by comparing with standard color charts or compared with a continuous gradient standard color wheel and transmitted light. Differences in procedures are related to the extracting solution used, either the Bray P-1 extractant for neutral and slightly acid soils (Bray & Kurtz, 1945) or the sodium bicarbonate extractant for calcareous soils (Olsen et al., 1954). Test strips for use with hand-held reflectometers are available from EM Science.

Potassium

An ion-selective electrode is perhaps the most quantitative method for determining K for on-farm testing, but should not be used with an ammonium-based extracting solution. A turbidimetric procedure used in the Purdue soil-testing kit (Ohlrogge, 1962) was to extract K from the soil and then add sodium cobaltinitrite to the solution to form a yellow precipitate. The amount of K present was then estimated by comparing the amount of turbidity in the extract to pictures of precipitates formed in standard samples. This procedure is semiquantitative. Test strips for K are available for use with hand-held reflectometers.

One novel multielement soil-extraction method with potential application to field situations is the hot water percolation method reported by Füleky and Czinkota (1993). This method could be used with a coffee percolator and distilled water, and might be adaptable for use with test strips. A ratio of 3:1 soil/sand was used to obtain an appropriate aliquot of extract (100 mL) within a few minutes. Linear correlations with standard laboratory measures of NO_3–N, NH_4–N, P, and K ranged from $r = 0.63$ to 0.86 and were equally high for some other nutrients (Mg, Fe, and Na). Where good correlations were obtained, quantitative agreement with standard laboratory procedures was reasonably close, but regression equations would need to be applied to correct the test values. More field testing is needed to determine the suitability of these on-farm measures to assess N, P, and K for soil quality considerations.

REFERENCES

Barnes, H., and A.R. Folkard. 1951. The determination of nitrites. Analyst 76:599–603.

Black, C.A. 1992. Soil fertility evaluation and control. Lewis Publ., Boca Raton, FL.

Blackmer, A.M., T.F. Morris, and G.D. Binford. 1992. Predicting N fertilizer needs for corn in humid regions: Advances in Iowa. p. 57–72. *In* B.R. Bock and K.R. Kelley (ed.) Predicting N fertilizer needs for corn in humic regions. Natl. Fert. and Environ. Res. Ctr., TVA, Bull. Y-226. Fert. and Environ. Res. Ctr., Ames, IA.

Blackmer, A.M., T.F. Morris, B.G. Meese, and A.P. Mallarino. 1993. Soil testing to optimize nitrogen management for corn. Iowa State Univ. Ext. PM-1521. Iowa State Univ., Ames.

Blackmer, A.M., D. Pottker, M.E. Cerrato, and J. Webb. 1989. Correlations between soil nitrate concentrations in late spring and corn yields in Iowa. J. Prod. Agric. 2:103–109.

Bray, R.H., and L.T. Kurtz. 1945. Determination of total, organic and available forms of phosphorus in soil. Soil Sci. 59:39–45.

Brown, J.R., and D. Warncke. 1988. Recommended cation tests and measures of cation exchange capacity. p. 15–16. *In* W.C. Dahnke (ed.) Recommended chemical soil test procedures for the North Central region. North Dakota Agric. Exp. Stn. Bull. 499. North Dakota State Univ., Fargo.

Bundy, L.G., and J.J. Meisinger. 1994. Nitrogen availability indices. p. 951–984. *In* R.W. Weaver et al. (ed.) Methods of soil analysis. Part 2. SSSA Book Ser. 5. SSSA, Madison, WI.

Carlson, R.M. 1986. Continuous flow reduction of nitrate to ammonia with granular zinc. Anal. Chem. 58:1590–1591.

Dahnke, W.C., and G.V. Johnson, 1990. Testing soils for available nitrogen. p. 127–139. *In* R.L. Westerman (ed.) Soil testing and plant analysis. 3rd ed. SSSA Book Ser. 3. SSSA, Madison, WI.

Dick, R.P., D.P. Breakwell, and R.F. Turco. 1996a. Soil enzyme activities and biodiversity measurements as integrative microbiological indicators. p. 247–271. *In* J.W. Doran and A.J. Jones (ed.) Methods for assessing soil quality. SSSA Spec. Publ. 49. SSSA, Madison, WI.

Dick, R.P., D.R. Thomas, and J.J. Halvorson. 1996b. Standardized methods, sampling, and sample pretreatment. p. 107–121. *In* J.W. Doran and A.J. Jones (ed.) Methods for assessing soil quality. SSSA Spec. Publ. 49. SSSA, Madison, WI.

Drinkwater, L.E., C.A. Cambardella, J.D. Reeder, and C.W. Rice. 1996. Potentially mineralizable nitrogen as an indicator of biologically active soil nitrogen. p. 217–229. *In* J.W. Doran and A.J. Jones (ed.) Methods for assessing soil quality. SSSA Spec. Publ. 49. SSSA, Madison, WI.

Fixen, P.E., and J.H. Grove. 1990. Testing soils for phosphorus. p. 141–180. *In* R.L. Westerman (ed.) Soil testing and plant analysis. 3rd ed. SSSA Book Ser. 3. SSSA, Madison, WI.

Füleky, G., and I. Czinkota. 1993. Hot water percolation (HWP): A new rapid soil extraction method. Plant Soil 157:131–135.

Gelderman, R.H., and P.E. Fixen. 1988. Recommended nitrate-N tests. p. 10–12. *In* W.C. Dahnke (ed.) Recommended chemical soil test procedures for the North Central Region. North Dakota Agric. Exp. Stn. Bull. 499. North Dakota State Univ., Fargo.

Haby, V.A., M.P. Russelle, and E.O. Skogley. 1990. Testing soils for potassium, calcium and magnesium. p. 181–227. *In* R.L. Westerman (ed.) Soil testing and plant analysis. 3rd ed. SSSA Book Ser. 3. SSSA, Madison, WI.

Hartz, T.K. 1993. A quick test procedure for soil nitrate–nitrogen. Commun. Soil Sci. Plant Anal. 25:511–515.

Jemison, J.M., and R.H. Fox. 1988. A quick-test procedure for soil and plant tissue nitrates using test strips and a hand-held reflectometer. Commun. Soil Sci. Plant Anal. 19:1569–1582.

Keeney, D.R., and D.W. Nelson, 1982. Nitrogen: Inorganic forms. p. 643–698. *In* A.L. Page et al. (ed.) Methods of soil analysis. Part 2. 2nd ed. Agron. Monogr. 9. ASA and SSSA, Madison, WI.

Knudsen, D., and D. Beegle, 1988. Recommended phosphorus tests. p. 12–15. *In* W.C. Dahnke (ed.) Recommended chemical soil test procedures for the North Central Region. North Dakota Agric. Exp. Stn. Bull. 499. North Dakota State Univ., Fargo.

Knudsen, D., G.A. Peterson, and P.F. Pratt. 1982. Lithium, sodium, and potassium. p. 225–246. *In* A.L. Page et al. (ed.) Methods of soil analysis. Part 2. 2nd ed. Agron. Monogr. 9. ASA and SSSA, Madison, WI.

Lachat Instruments. 1993. QuikChem automated ion analyzer methods manual. Lachat Instruments, Milwaukee, WI.

Lindsay, W.L., and W.A. Norvell. 1978. Development of a DTPA soil test for zinc, iron, manganese and copper. Soil Sci. Soc. Am. J. 42:421–428.

Magdoff, F.R. 1991. Understanding the Magdoff pre-sidedress nitrate test for corn. J. Prod. Agric. 4:297–305.

Magdoff, F.R., D. Ross, and J. Amadon. 1984. A soil test for nitrogen availability to corn. Soil Sci. Soc. Am. J. 48:1301–1304.

Mehlich, A. 1984. Mehlich No. 3 soil test extractant: A modification of Mehlich No. 2. Commun. Soil Sci. Plant Anal. 15:1409–1416.

Ohlrogge, A.J. 1962. The Purdue soil and plant tissue tests. Purdue Univ. Exp. Stn. Bull. 635. Purdue University, West Lafayette, IN.

Olsen, S.R., C.V. Cole, F.S. Watanabe, and L.A. Dean. 1954. Estimation of available phosphorus in soils by extraction with sodium bicarbonate. USDA Circ. 939:1–19.

Olsen, S.R., and L.E. Sommers. 1982. Phosphorus. p. 403–430. In A.L. Page et al. (ed.) Methods of soil analysis Part 2. 2nd ed. Agron. Monogr. 9. ASA and SSSA, Madison, WI.

Page, A.L., R.H. Miller, and D.R. Keeney (ed.). 1982. Methods of soil analysis. Part 2. 2nd ed. ASA and SSSA, Madison, WI.

Practical Farmers of Iowa. 1996. 1995 PFI Membership Meeting Shared Visions Conf. and on-Farm Trials Rep. 5–6 Jan. 1996. Iowa State Univ., Ames.

Rehm, G.W., M. Schmitt, and R. Munter. 1994. Fertilizer recommendations for agronomic crops in Minnesota. Minnesota Ext. Serv. BU-6240-E. Univ. of Minnesota, St. Paul.

Roth, G.W., D.G. Beegle, R.H. Fox, J.D. Toth, and W.P. Piekielek. 1991. Development of a quick test kit method to measure soil nitrate. Commun. Soil Sci. Plant Anal. 22:191–200.

Schaefer, N.L. 1986. Evaluation of a hand held reflectometer for rapid quantitative determination of nitrate. Commun. Soil Sci. Plant Anal. 17:937–951.

Schmitt, M.A., and G.W. Randall. 1994a. Developing a soil nitrogen test for improved recommendations for corn. J. Prod. Agric. 7:328–334.

Schmitt, M.A., and G.W. Randall. 1994b. Evaluation of harvest-time plant, stalk and soil nitrogen measurements in nitrogen rate studies of corn. p. 396. In Agronomy abstract. ASA, Madison, WI.

Sharpley, A., T.C. Daniel, J.T. Sims, and D.M. Pote. 1996. Determining environmentally sound soil phosphorus levels. J. Soil Water Conserv. 51:160–166.

Soil and Plant Analysis Council. 1992. Handbook on reference methods for soil analysis. Georgia Univ. Exp. Stn., Athens.

Sikora, L.J., and D.E. Stott. 1996. Soil organic carbon and nitrogen. p. 157–167. In J.W. Doran and A.J. Jones (ed.) Methods for assessing soil quality. SSSA Spec. Publ. 49. SSSA, Madison, WI.

Soltanpour, P.N., and A.P. Schwab. 1977. A new soil test for simultaneous extraction of macro- and micro-nutrients in alkaline soils. Commun. Soil Sci. Plant Anal. 8:195–207.

12 Role of Microbial Biomass Carbon and Nitrogen in Soil Quality[1]

Charles W. Rice

Kansas State University
Manhattan, Kansas

Thomas B. Moorman

USDA-ARS National Soil Tilth Laboratory
Ames, Iowa

Mike Beare

New Zealand Institute for Crop and Food Research
Christchurch, New Zealand

Microbial biomass in soil is the living component of soil organic matter. Many models of organic matter formation include microbial biomass as a precursor to the more stable fractions of organic matter (Parton et al., 1987). Because as much as 95% of the total soil organic matter is nonliving and, therefore, relatively stable or resistant to change, decades may be required to observe a measurable change in soil organic matter. Microbial biomass has a turnover time of <1 yr (Paul, 1984) and therefore, responds rapidly to conditions that eventually alter soil organic matter levels. Thus, the size of the microbial biomass may indicate degradation or aggradation of soil organic matter (Powlson et al., 1987; Sparling, 1992).

As an active component of soil organic matter, soil microbial biomass is involved in nutrient transformations and storage. Nutrients released during turnover of the microbial biomass are often plant available. In native terrestrial ecosystems, where internal cycling of N predominates, microbial biomass is responsible for transforming organic N to plant-available forms. Agricultural systems that rely upon internal sources of N require microbial biomass and its activity to supply N to the crop. In fertilized systems, microbial biomass can be a significant source and sink of N. Carbon contained within the microbial biomass is

---

[1] This work was partially supported by NSF grant BSR-9011662 to Kansas State University. Contribution no. 95-392-B from Kansas Agric. Exp. Stn., Manhattan, KS.

Copyright © 1996 Soil Science Society of America, 677 S. Segoe Rd., Madison, WI 53711, USA. *Methods for Assessing Soil Quality*, SSSA Special Publication 49.

stored energy for microbial processes. Therefore, microbial biomass C may indicate potential microbial activity.

Because of its rapid turnover, microbial biomass is a sensitive indicator of changes in climate (Insam et al., 1989; Insam, 1990), tillage systems (Lynch & Panting, 1980; Carter, 1991), crop rotations (Anderson & Domsch, 1989; Campbell et al., 1991), and pollutant toxicity (Chander & Brookes, 1991a,b, 1993). Microbial biomass content is a function of other soil properties, including pH, texture, and soil water content. Microbial biomass varies with soil texture, probably because of the effect of texture on aggregate formation, thus protecting organic C (Schimel, 1986; Burke, 1989; Gregorich et al., 1991; Zagal, 1993). Microorganisms also play a major role in the formation and maintenance of soil aggregates and structure (Tisdall & Oades, 1982). Microbial biomass C and soil aggregate stability are strongly related (Ross et al., 1982; Haynes & Swift, 1990; Robertson et al., 1991). Because microbial biomass integrates soil physical and chemical properties and responds to anthropogenic activities, it may be considered a suitable biological indicator of soil quality.

The importance of microbial biomass to soil has generated interest from the research community resulting in the development of several analytical procedures. The most widely used procedure is some form of soil fumigation with chloroform. Fumigation results in lysis of the microbial cells. The traditional method involves incubation of fumigated soil for 10 d where the surviving microorganisms convert the organic C and N of lysed cells to CO_2 and to NH_4^+, respectively. This procedure is known as the fumigation-incubation (FI) technique (Jenkinson & Powlson, 1976; Voroney & Paul, 1984). A more recent alternative to the conventional FI procedure is known as the fumigation-extraction (FE) technique (Brookes et al., 1985; Vance et al., 1987a; Amato & Ladd, 1988; Sparling & West, 1988). In this method, the C and N released from cells by fumigation are extracted with a dilute salt solution (0.5 M K_2SO_4) and measured directly. Both methods are now widely used.

Direct counts of microorganisms also can be made by separating the cells from soil, staining with fluorescent dyes, and counting by microscopic observation (Schmidt & Paul, 1982). Microbial biomass C and N can then be estimated by converting cell counts and their estimated C, N, and water content to biomass C and N equivalents. These techniques require moderately priced equipment, skilled personnel, and are time-consuming.

Substrate-induced respiration (SIR) is an alternate technique that relies upon the response of the microbial population to the addition of a readily degraded substrate such as glucose (Anderson & Domsch, 1978; Beare et al., 1990). This method provides an indication of the activity of the microbial biomass and hence the size of the physiologically active biomass. Disadvantages of this technique include increased sensitivity to sample handling, the need to standardize soil water content and temperature, and the inability to measure microbial biomass N.

We recommend the fumigation techniques because of their ease of use, existence of a large database, lower equipment costs, and the ability to measure mineralizable C and N from the unfumigated soil samples. The FI technique permits use of existing resources in many laboratories and also gives an estimate of

C and N mineralization as discussed by Parkin et al. (1996, this publication). Laboratories with resources and/or soil conditions that limit the use of FI (low pH, too low or high soil water content) should consider the FE technique. The reader should be aware of the advantages and disadvantages of other existing techniques and development of new techniques to measure microbial biomass in soils. More detailed reviews of the various techniques for estimating microbial biomass are available in the literature (Jenkinson & Ladd, 1981; Jenkinson, 1988; Hulm et al., 1991; Horwath & Paul, 1994).

RECOMMENDED PROCEDURE

Equipment and Reagents: Fumigation Incubation–Extraction

Materials

ethanol-free chloroform

Note: All work must be done in an adequate fume hood because chloroform has carcinogenic and volatile properties. Commercially available ethanol-free chloroform preserved with heptachlor epoxide has obtained similar results to that of purified chloroform (Voroney et al., 1991).

vacuum pump
vacuum desiccator
125 mL Erlenmeyer flasks
canning jars (1 L capacity)
2 M KCl

Procedure

Soil Storage and Preparation

After sampling, the soil should be kept moist and cool. The soil can be stored overnight at 15°C or at 4°C for up to 10 d. Freezing and complete drying is not recommended because of their potential biocidal effects. The soil sample is normally sieved through a 2- to 6-mm sieve to improve sample uniformity. A 2-mm sieve is preferable due to conformity with the other soil quality analyses. Soil samples that are water-saturated should be partially dried before sieving.

Weigh out duplicate soil samples, usually 25 to 50 g, into 125 mL Erlenmeyer flasks. A third sample should be weighed for gravimetric soil moisture determination (105°C for 24 h). Soil water content can affect the determination of microbial biomass. Soil samples should be adjusted to optimum soil water content to maximize the response after $CHCl_3$ fumigation. Horwath and Paul (1994) recommend 55% of water-holding capacity (WHC). If water-holding capacity is known, then water is added to field-moist soil or removed from (by air-drying at 5 to 15°C) field-moist soil to attain 55% WHC. Alternatively, Doran recommends adjusting the soil water content to 55% water-filled pore space (WFPS). To do this, the volume of field-moist soil is measured and the soil is packed to a known bulk density, usually 1.1 g cm^{-3}. The following equation can be used to calculate the water-filled pore space:

Soil water-filled pore space =

$$\frac{((\% \text{ soil water content}/100, \text{ g g}^{-1}) \times \text{bulk density, g cm}^{-3}))}{(1 - (\text{soil bulk density, g cm}^{-3}/\text{particle density, g cm}^{-3}))}$$

If not directly measured, the density of mineral soil particles is commonly assumed to be 2.65 g cm^{-3}. Water then is added to adjust the WFPS to 55%.

Changes in the size of the microbial biomass may result from rewetting of dry soil, releasing organic C and N available for mineralization. To avoid this pulse of microbial activity, the samples should be pre-incubated for 7 to 10 d at 22 to 25°C after adjustment of soil water. Pre-incubation also may overcome storage effects.

Chloroform–Fumigation

After the pre-incubation period, one set of the duplicate soil samples is placed in a vacuum desiccator. The flasks need to be labeled with a pencil, because the chloroform will dissolve ink. The desiccator is lined with a moist paper towel to prevent drying of the sample during fumigation. A beaker containing 50 mL of ethanol-free chloroform and antibumping granules is placed in the desiccator. The desiccator is evacuated until the chloroform boils vigorously for approximately 30 s; air then is allowed to pass into the desiccator to distribute the chloroform into the soil. This process is repeated three times. After the fourth evacuation, the chloroform is allowed to boil for 2 min., and the desiccator valve is closed. The samples are fumigated for 18 to 24 h for the FI procedure and 24 to 48 h for the FE procedure. For most soils, 24 h is suitable; for FE however, 48 h of fumigation improves the "kill" in some soils and may be the method of choice where the optimum period of fumigation is not known. After the fumigation period, the paper towel and the chloroform are removed, and the desiccator is evacuated eight times for 3 min, letting air pass into the desiccator after each evacuation.

The other set of duplicate samples serves as the control or unfumigated sample. In the FI procedure, the samples are incubated under similar conditions as the fumigated samples except they are not evacuated or treated with $CHCl_3$. In the FE procedure, the controls samples are extracted immediately as described later.

Incubation Procedure

The fumigated and unfumigated samples are placed into airtight containers, a few milliliters of water are added to the bottom of the jar to prevent desiccation, and the containers are incubated for 10 d in the dark at 22°C (room temperature). Canning jars (1 L) make excellent containers for this purpose. A vial containing 1 mL of 2 M NaOH is placed in the jar to trap respired CO_2. To account for CO_2

in air, blanks are CO_2 traps in jars without soil maintained during the incubation period. After 10 d, $BaCl_2$ is added at an equivalent amount of the initial NaOH concentration (1 mL of 2 M $BaCl_2$). The excess NaOH is titrated with 0.1 M HCl to pH 7 or to a phenolphthalein endpoint. The volume of acid required to neutralize a blank minus that required to neutralize the sample is used to calculate the amount of CO_2 respired. One milliliter of 2 M NaOH will trap 12 mg CO_2–C. For soils with high microbial biomass, such as grassland soils or soils amended with organic materials, greater volumes of NaOH may be required (2 to 5 mL). The accumulated CO_2 in the headspace also can be measured by gas chromatography or with an infrared gas analyzer, if available. To determine the amount of microbial biomass N, the soil is extracted with 2 M KCl at a solution to soil ratio of 4:1. The solution is shaken for 30 min and filtered with acid-washed Whatman no. 42 filter paper. The filtrate then is analyzed for NH_4^+ and NO_3^-.

Extraction Procedure

To each of the fumigated and nonfumigated sample flasks, add 100 mL of 0.5 M K_2SO_4, stopper and shake for 30 min. After shaking, the solutions are filtered through preleached or acid-washed Whatman no. 42 filter paper. The filtrates are then analyzed for their concentrations of total C and N by wet chemical procedures. Organic C in the filtrates is measured by a dichromate digestion procedure (Jenkinson & Powlson, 1976; Vance et al., 1987a). Briefly, 8 mL of the filtrate is boiled gently under reflux (30 min) with a dichromate oxidizing reagent (2 mL 66.7 mM $K_2Cr_2O_7$ + 70 mg HgO + 15 mL [2 parts H_2SO_4 (98%) + 1 part H_3PO_4 (88%)]), allowed to cool and diluted with 20 to 25 mL of H_2O. The excess dichromate is determined by back-titration with ferrous ammonium sulphate (33.3 mM) in 0.4 M H_2SO_4, using 25 mM 1,10-phenanthroline-ferrous sulphate-complex solution as an indicator. The organic C concentration is calculated assuming that 1 mL 66.7 mM $K_2Cr_2O_7$ is equivalent to 1200 µg C. To avoid the cost of dichromate disposal, soluble C also may be analyzed on a commercial soluble C analyzer.

Total N in the filtrate can be determined by the method of Cabrera and Beare (1993). Briefly, 5 mL of the filtrate is combined with 5 mL of a persulfate oxidizing reagent (50 g low-N $K_2S_2O_8$ + 30g H_3BO_4 + 100 mL of 3.75 M NaOH made up to 1 L in H_2O) in a 15 mL pyrex glass tube sealed tightly with a teflon-lined screw cap. The tubes are weighed, autoclaved for 30 min at 120°C, allowed to cool to room temperature and reweighed. Any loss in weight (i.e., water) is used to correct the measured N concentrations. The digested samples are then analyzed for their NO_3^-–N concentrations using standard colorimetric procedures (Keeney & Nelson, 1982).

Calculations

Microbial biomass C (mg C kg^{-1}) is calculated from the equation:

$$\text{Microbial biomass C} = (F_C - UF_C)/K_C$$

where, F_C = CO_2 released from the fumigated sample (mg CO_2– C kg^{-1}) in 10 d

UF_C = CO_2 released from the unfumigated sample (mg CO_2 – C kg^{-1}) in 10 d

K_C = fraction of biomass C mineralized to CO_2 or extracted

For FI procedure, $K_c = 0.41$
For FE procedure, $K_c = 0.35$

According to the approach by Shen et al., (1984) microbial biomass N (mg N kg^{-1}) is calculated from the equation:

$$\text{Microbial N} = (F_N - UF_N)/K_N$$

where, F_N = flush of inorganic N ($NH_4^+ + NO_3^-$) released by fumigation (mg N kg^{-1})

UF_N = inorganic N released (mg N kg^{-1})

$K_N = 0.68$

To compare different ecosystems or agricultural management systems, microbial biomass should be expressed on a volumetric basis for a given depth of soil. This is done by:

Microbial biomass (kg C or N ha^{-1} $depth^{-1}$) = mg biomass C or

N kg^{-1}soil × [bulk density (g cm^{-3})] × [sampling depth (cm)] × (0.10)

Cautions

One assumption of the fumigation–incubation technique is that the surviving microbial population will use the organic C released from the killed cells within a 10-d period. Therefore, an active microbial population must be present to use this C. Soil water content is a critical factor. The soil water effect can be avoided by preincubation of the soil after adjusting the water content to 55% of water-filled pore space. Acid soils or subsoils also are known for limited microbial response after fumigation (Vance et al., 1987b). In this situation, an alternative method needs to be considered. Reinoculation of fumigated soil before incubation can be accomplished by adding 0.2 g of unfumigated soil to 50 g of fumigated soil (Vance et al., 1987b; Horwath & Paul, 1994). The fumigation–extraction procedure also has been successfully used in these situations. Any persistent organic toxin or heavy metal also may inhibit the microbial response after fumigation, thus causing an underestimation of microbial biomass.

INTERPRETATION OF RESULTS

Microbial biomass is very dynamic in soil and responds to weather, crop input, and season (McGill et al., 1986; Bristow & Jarvis, 1991; Garcia & Rice, 1994). This variation is accentuated in cropping systems that include tillage, fertilizer applications, and irrigation (McGill et al., 1986). As much as a 40% change in the values can be realized in native ecosystems (Garcia & Rice, 1994) and in agricultural systems (Buchanan & King, 1992; Van-Gestel et al., 1992). Therefore, the microbial biomass value is subject to the time of sampling and the antecedent soil water and temperature conditions, as well as plant dynamics. Single-point in time samples should be taken before tillage, nutrient addition, and planting or postharvest to account for seasonal changes in microbial biomass. Depending upon objectives, sampling should generally be conducted in early spring or late fall in temperate regions. In native or perennial systems, sampling should be done before initiation of plant growth. To determine if a soil is aggrading, degrading, or at equilibrium, it is important to establish a baseline or reference condition. Due to its dynamic nature, the amount of microbial biomass at any one time cannot indicate whether soil organic matter; (i.e., soil quality) is increasing, decreasing, or at equilibrium. Monitoring microbial biomass with time can provide information on the changes in amount and nutrient content of the microbial biomass. Relating microbial biomass to other soil parameters has been successful in assessing changes in soil quality as discussed in the next section. Variation in microbial biomass also can be attributed to variations in soil type or landscape position. Variations in biomass C as large as 35% can be expected at the field scale (Cambardella et al., 1994). Soil sampling from homogenous soil types and the use of composite samples can reduce variation considerably and make the desired comparisons more clear.

RELEVANCE TO OTHER SOIL ATTRIBUTES

Expressing the size of the microbial biomass in relation to other parameters, such as total soil organic C and N, mineralizable C and N or respiration, may provide a measure of soil organic matter dynamics and thus soil quality. Soil microbial biomass comprises 1 to 4% of the total organic C (Anderson & Domsch, 1989; Sparling, 1992) and 2 to 6% of the total organic N (Jenkinson, 1988) in soil. The proportion of organic C as microbial biomass varies with climate, such that within a climatic region, this ratio will have an equilibrium value (Anderson & Domsch, 1989; Insam et al., 1989; Insam, 1990). Sparling (1992) suggested that trends in soil organic matter quality are more apparent with this ratio. Deviations from this ratio indicate long-term degradation or aggradation of soil organic matter (Anderson & Domsch, 1989). As with C, similar ratios of microbial biomass N to organic N can be calculated to alleviate the problems associated with the temporal dynamics of the microbial biomass. Beside differences in climatic regions, vegetation, soil mineralogy, and texture also can influence the ratio microbial biomass to organic C and N. Therefore, interpretation of the ratio should be defined within similar soil types, climatic regions, and cropping systems (Gregorich et al., 1994).

The ratio of microbial C to mineralizable C and microbial biomass N to mineralizable N also may be a useful index for soil quality. Mineralizable C and N may be easily obtained by including a second unfumigated control as described in the microbial biomass procedure except incubated for 20 d (Parkin et al., 1996, this publication). Mineralizable N also can be obtained by the procedures outlined in Drinkwater et al. (1996, this publication). These ratios have been used to estimate the quality of the soil organic matter (Bonde et al., 1988; Rice & Garcia, 1994; Rice et al., 1994).

The rate of respiration (i.e., CO_2 production) from soil can be related to microbial biomass as an indicator of microbial activity. Under stress conditions, many soil microorganisms become dormant; the respiratory ratio (CO_2 production per unit of microbial biomass C, q CO_2) is indicative of general microbial activity (Anderson & Domsch, 1990). The q CO_2 can be calculated easily from the control (unfumigated soil) during the FI procedure or from CO_2 evolution (basal respiration) rate as discussed by Parkin et al. (1996, this publication). The value of q CO_2 is illustrated in the work by Dinwoodie and Juma (1988). They found that one soil had low microbial biomass but a high proportion of microbial C to soil organic C. The respiratory ratio, q CO_2, also was high, possibly indicating that more C was being lost and that greater care was needed to maintain soil organic matter levels.

Microbial biomass also is related to N mineralization (Doran, 1987; Myrold, 1987; Smith & Paul, 1990); however, this relationship is not as strong as the relationship between microbial biomass and soil organic matter. The relationship between microbial biomass and N mineralization can be influenced by N fertilization history and the contribution of nonliving microbial biomass to N mineralization.

APPLICATION

The following examples demonstrate the utility of microbial biomass measurements in soil quality assessment.

Rotations

In semiarid areas, continuous cropping reduces the decline in soil organic matter and microbial biomass compared with a wheat (*Triticum aestivum* L.)–fallow rotation (Campbell et al., 1991; Collins et al., 1992). Campbell et al. (1991) reported that the microbial biomass N was more useful in predicting a change in soil quality than microbial biomass C. In contrast, Jordan and Kremer (1994) reported microbial biomass C to be a better indicator of soil quality. This conflict demonstrates the need for both measures and careful interpretation with other measures for accurate assessment of soil quality. Comparison of corn (*Zea mays* L.) and sorghum [*Sorghum bicolor* (L.) Moench] cropping systems grown continuously and in rotation with soybean [*Glycine max* (L.) Merr.] generally shows an increase in microbial biomass in the rotation (Table 12–1). However one soil, a sandy loam, showed a significant decrease in microbial biomass with rotation. This negative effect may have been due to less residue produced with the soybean crop on a lower organic matter soil.

Tillage

Reduction in tillage intensity will conserve plant residues and may eventually increase soil organic matter. Microbial biomass has been used as an early

Table 12–1. Effect of crop rotation on microbial biomass C and its relation to organic C.

Previous crop	Soil texture	Microbial biomass rotation		% Difference[†]	Organic C rotation		% Difference
		Monoculture	Soybeans		Monoculture	Soybean	
		—— mg C kg^{-1} ——			—— g C kg^{-1} ——		
Sorghum[‡]	SiCL	600	650	+ 8.3	14.8		0
Corn[§]	SiCL`	108	128	+18.5	16.7	15.6	6.6
Corn[§]	SL	115	105	-8.7	8.7	7.7	11.5

[†] Data from Roder et al. (1988).
[‡] Data from Omay et al. (1992).
[§] Percentage of change relative to continuous cropping of corn or sorghum.

indicator of the increase in soil organic matter. A decrease in tillage intensity results in a greater proportion of microbial biomass to organic C (Lynch & Panting, 1980; Carter, 1991; Table 12–2). This may indicate that organic matter is increasing (Carter, 1991), although soil organic C may not be significantly different between tillage systems. Angers et al. (1993) measured 1.2% organic C as microbial biomass in a plowed treatment compared with 3.5 to 5.1% in a minimum-tillage treatment after 11 yr.

Toxins

Microbial biomass has been used to indicate heavy metal toxicity from sludge applications (Chander & Brookes, 1991a, b, 1993; Table 12–3). High levels of Cu or Zn alone (1.4 times the permitted limits) decreased microbial biomass C by 12%, but Zn and Cu combined decreased microbial biomass C by 29 to 53%, thereby suggesting an interactive effect. The percentage of organic C as microbial biomass decreased by more than 50% in the heavy metal-contaminated soils compared with the controls.

Ecosystems

Microbial biomass is related to ecosystem types. Values for microbial biomass C range from 20 kg ha^{-1} in desert grasslands to 1340 kg ha^{-1} in tallgrass

Table 12–2. Effect of tillage on microbial biomass and soil organic C.

Crop	Tillage[†]	Microbial biomass C	Soil organic C	Microbial C / Soil organic C
		mg kg^{-1}	g kg^{-1}	%
Corn[‡]	P	120	19.6	0.61
	NT	237	21.5	1.10
Wheat/Barley[‡]	P	150	22.3	0.67
	NT	299	25.5	1.18
Wheat[§]	P	760	44	1.73
	NT	940	48	1.96

[†] P = plowed, NT = no-tillage.
[‡] Carter, 1991.
[§] Lynch and Panting, 1980.

Table 12–3. Effect of heavy metals on microbial and total organic C (adapted from Chander & Brooks, 1993).

Heavy metal	Application rates	Microbial biomass C	Organic C	Microbial C Soil organic C
	kg ha^{-1}	mg C kg^{-1}	g kg^{-1}	%
Control	0	169	10.4	1.6
Zn	3000	108	15.5	0.7
Cu	3000	82	19.9	0.4
Ni	200	182	12.2	0.1
Zn/Cu	2300/1600	79	18.8	0.4
Zn/Ni	600/100	181	12.2	1.5

prairie (Gallardo & Schlesinger, 1992; Garcia & Rice, 1994; Zak et al., 1994). Forest ecosystems had intermediate values. Microbial biomass C also has been shown to be related positively to aboveground net primary productivity (Myrold et al., 1989; Zak et al., 1994). In the study by Zak et al. (1994), texture did not influence microbial C, but other studies within the shortgrass prairie region have reported a textural effect (Schimel, 1986; Burke, 1989).

CONCLUSION

Soil is a vital natural resource and maintaining its quality is essential to the survival of any society. Soil organic matter is a key indication of soil quality because it impacts other soil properties; however, organic matter levels in soil change slowly, on the order of decades. Microbial biomass is the dynamic, living component of soil organic matter. Therefore, microbial biomass may be an early indicator of the direction of change in soil organic matter levels. Unfortunately, since microbial biomass levels are affected by climatic variables, soil type, and season, direct interpretation of the values need to be done carefully. The ratio of C and N in microbial biomass relative to organic C and N may be a more useful parameter for assessing soil quality. The recommended procedure for measuring microbial biomass C and N uses the chloroform fumigation technique followed by incubation or direct extraction. The incubation technique has the advantage in that mineralizable C and N also can be measured simultaneously, which is another useful indicator of soil organic matter and its quality. In summary, microbial biomass and its relation to other soil properties is a useful measurement for assessing soil quality.

REFERENCES

Amato, M., and J.N. Ladd. 1988. Assay for microbial biomass based on ninhydrin-reactive nitrogen extracts of fumigated soil. Soil Biol. Biochem. 20:107–114.

Anderson, J.P.E., and K.H. Domsch. 1978. A physiological method for the quantitative measurement of microbial biomass in soils. Soil Biol. Biochem. 10:215–221.

Anderson, J.P.E., and K.H. Domsch. 1989. Ratios of microbial biomass carbon to total carbon in arable soils. Soil Biol. Biochem. 21:471–479.

Anderson, J.P.E., and K.H. Domsch. 1990. Application of eco-physiological quotients (qCO_2 and qD) on microbial biomasses from soils of different cropping histories. Soil Biol. Biochem. 22:251–255.

Angers, D.A., A. N'dayegamiye, and D. Cote. 1993. Tillage-induced differences in organic matter of particle-size fractions and microbial biomass. Soil Sci. Soc. Am. J. 57:512–516.

Beare, M.H., C.C. Neely, D.L. Coleman, and W.L. Hargrove. 1990. A substrate-induced respiration (SIR) method for measurement of fungal and bacterial biomass on plant residues. Soil Biol. Biochem. 22:585–594.

Bonde, T.A., J. Schnurer, and T. Rosswall. 1988. Microbial biomass as a fraction of potentially mineralizable nitrogen in soils from long-term field experiments. Soil Biol. Biochem. 20:447–452.

Bristow, A.W., and S.C. Jarvis. 1991. Effects of grazing and nitrogen fertilizer on the soil microbial biomass under permanent pasture. J. Sci. Food Agric. 54:9–21.

Brookes, P.C., A. Landman, G. Pruden, and D.S. Jenkinson. 1985. Chloroform fumigation and the release of soil nitrogen: A rapid direct extraction method to measure microbial biomass nitrogen in soil. Soil Biol. Biochem. 17:837–842.

Buchanan, M., and L.D. King. 1992. Seasonal fluctuations in soil microbial biomass carbon, phosphorus, and activity in no-till and reduced-chemical-input maize agroecosystems. Biol. Fert. Soils 13:211–217.

Burke, I.C. 1989. Control of nitrogen mineralization in a sagebrush steppe landscape. Ecology 70:1115–1126.

Cabrera, M.L., and M.H. Beare. 1993. Alkaline persulfate oxidation for determining total nitrogen in microbial biomass extracts. Soil Sci. Soc. Am. J. 57:1007–1012.

Cambardella, C.A., T.B. Moorman, J.M. Novak, T.B. Parkin, D.L. Karlen, R.F. Turco, and A.E. Konopka. 1994. Field-scale variability of soil properties in central Iowa soils. Soil Sci. Soc. Am. J. 58:1501–1511.

Campbell, C.A., V.O. Biederbeck, R.P. Zentner, and G.P. Lafond. 1991. Effect of crop rotations and cultural practices on soil organic matter, microbial biomass and respiration in a thin Black Chernozem. Can. J. Soil Sci. 71:363–376.

Carter, M.R. 1991. The influence of tillage on the proportion of organic carbon and nitrogen in the microbial biomass of medium-textured soils in a humid climate. Biol. Fertil. Soils 11:135–139.

Chander, K., and P.C. Brookes. 1991a. Microbial biomass dynamics during the decomposition of glucose and maize in metal-contaminated and non-contaminated soils. Soil Biol. Biochem. 23:917–925.

Chander, K., and P.C. Brookes. 1991b. Effects of heavy metals from past application of sewage sludge on microbial biomass and organic matter accumulation in a sandy loam and a silty loam U.K. soil. Soil Biol. Biochem. 23:927–932.

Chander, K., and P.C. Brookes. 1993. Residual effects of zinc, copper, and nickel in sewage sludge on microbial biomass in a sandy loam. Soil Biol. Biochem. 25:1231–1239.

Collins, H.P., P.E. Rasmussen, and C.L. Douglas, Jr. 1992. Crop rotation and residue management effects on soil carbon and microbial dynamics. Soil Sci. Soc. Am. J. 56:783–788.

Dinwoodie, G.D., and N.G. Juma. 1988. Allocation and microbial utilization of C in two soils cropped to barley. Can. J. Soil Sci. 68:495–505.

Doran, J.W. 1987. Microbial biomass and mineralizable nitrogen distribution in no-tilled and plowed soils. Biol. Fertil. Soils 5:68–75.

Drinkwater, L.E., C.A. Cambardella, J.D. Reeder, and C.W. Rice. 1996. Potentially mineralizable N as an indicator of active soil nitrogen. p. 217–229. In J.W. Doran and A.J. Jones (ed.) Methods for assessment of soil quality. SSSA Spec. Publ. 49. SSSA, Madison, WI.

Gallardo, A., and W.H. Schlesinger. 1992. Carbon and nitrogen limitations of soil microbial biomass in desert ecosystems. Biogeochemistry. 18:1–17.

Garcia, F.O., and C.W. Rice. 1994. Microbial biomass dynamics in tallgrass prairie. Soil Sci. Soc. Am. J. 58:816–823.

Gregorich, E.G., M.R. Carter, D.A. Angers, C.M. Monreal, and B.H. Ellert. 1994. Towards a minimum data set to assess soil organic matter quality in agricultural soils. Can. J. Soil Sci. 74:367–385.

Gregorich, E.G., R.P. Voroney, and R.G. Kachanoski. 1991. Turnover of carbon through the microbial biomass in soils with different textures. Soil Biol. Biochem. 23:799–805.

Haynes, R.J., and R.S. Swift. 1990. Stability of soil aggregates in relation to organic constituents and soil water content. J. Soil Sci. 41:73–83.

Horwath, W.R., and E.A. Paul. 1994. Microbial biomass. p. 753–773. In R.W. Weaver et al. (ed.) Methods of soil analysis. Part 2. SSSA Book Ser. 5. SSSA, Madison, WI.

Hulm, S.C., D.L. Castle, K. Cook, A. Self, and M. Wood. 1991. Evaluation of soil microbial biomass methodology. Toxicol. Environ. Chem. 30:183–192.

Insam, H. 1990. Are the soil microbial biomass and basal respiration governed by the climatic regime? Soil Biol. Biochem. 22:525–532.

Insam, H., D. Parkinson, and K.H. Domsch. 1989. Influence of macroclimate on soil microbial biomass. Soil Biol. Biochem. 21:211–221.

Jenkinson, D.S. 1988. Determination of microbial biomass carbon and nitrogen in soil. p. 368–386. *In* J.R. Wilson (ed.) Advances in nitrogen cycling in agricultural ecosystems. CAB Int., Wallingford, England.

Jenkinson, D.S, and J.N. Ladd. 1981. Microbial biomass in soil: Measurement and turnover. p. 415–471. *In* E.A. Paul and J.N. Ladd (ed.) Soil biochemistry. Vol. 5. Marcel Dekker, New York.

Jenkinson, D.S., and D.S. Powlson. 1976. The effects of biocidal treatments on metabolism in soil V. A method for measuring microbial biomass. Soil Biol. Biochem. 8:209–213.

Jordan, D., and R.J. Kremer. 1994. Potential use of soil microbial activity as an indicator of soil quality. p. 243–249. *In* C.E. Pankhurst et al. (ed.) Soil biota: Management in sustainable farming systems. CSIRO, Australia.

Keeney, D.R., and D.W. Nelson. 1982. Nitrogen: Inorganic forms. p. 643–698. *In* A.L. Page et al. (ed.) Methods of soil analysis. Part 2. 2nd ed. Agron. Monogr. 9. ASA and SSSA, Madison, WI.

Lynch, J.M., and L.M. Panting. 1980. Cultivation and the soil biomass. Soil Biol. Biochem. 12:29–33.

McGill, W.B., K.R. Cannon, J.A. Robertson, and F.D. Cook. 1986. Dynamics of soil microbial biomass and water soluble organic C in Breton L after 50 years of cropping to two rotations. Can. J. Soil Sci. 66:1–19.

Myrold, D.D. 1987. Relationship between microbial biomass nitrogen and a nitrogen availability index. Soil Sci. Soc. Am. J. 51:1047–1049.

Myrold, D.D., P.A. Matson, and D.L. Peterson. 1989. Relationships between soil microbial properties and aboveground stand characteristics of conifer forests in Oregon. Biogeochemistry 8:265–281.

Omay, A.B., C.W. Rice, L. Maddux, and B. Gordon. 1992. Changes in soil microbial and chemical properties under long-term crop rotation. p. 264. *In* Agronomy abstracts. ASA, Madison, WI.

Parkin, T.B., J.W. Doran, and E. Franco-Vizcaino. 1996. Field and laboratory tests of soil respiration. p. 231–245. *In* J.W. Doran and A.J. Jones (ed.) Methods for assessment of soil quality. SSSA Spec. Publ. 49. SSSA, Madison, WI.

Parton, W.J., D.S. Schimel, C.V. Cole, and D.S. Ojima. 1987. Analysis of factors controlling soil organic matter levels in Great Plains grasslands. Soil Sci. Soc. Am. J. 51:1173–1179.

Paul, E.A. 1984. Dynamics of organic matter in soils. Plant Soil 76:275–285.

Powlson, D.S., P.C. Brookes, and B.T. Christensen. 1987. Measurement of soil microbial biomass provides an early indication of changes in total soil organic matter due to straw incorporation. Soil Biol. Biochem. 19:159–164.

Rice, C.W., and F.O. Garcia. 1994. Biologically active pools of soil C and N in tallgrass prairie. p. 201–208. *In* J. Doran et al. (ed.) Defining soil quality for a sustainable environment. SSSA Spec. Publ. 35. SSSA, Madison, WI.

Rice, C.W., F.O. Garcia, C.O. Hampton, and C.E. Owensby. 1994. Soil microbial response in tallgrass prairie to elevated CO_2. Plant Soil 165:67–74.

Robertson, E.B., S. Sarig, and M.K. Firestone. 1991. Cover crop management of polysaccharide-mediated aggregation in an orchard soil. Soil Sci. Soc. Am. J. 55:734–739.

Roder, W., S.C. Mason, M.O. Clegg, J.W. Doran, and K.R. Kniep. 1988. Plant and microbial responses to sorghum-soybean cropping systems and fertility management. Soil Sci. Soc. Am. J. 52:1137–1142.

Ross, D.J., K.R. Tate, and A. Cairns. 1982. Biochemical changes in a yellow-brown loam and a central gley soil converted from pasture to maize in the Wailcato area. New Zeal. J. Agric. Res. 25:35–42.

Schimel, D.S. 1986. Carbon and nitrogen turnover in adjacent grassland and cropland ecosystems. Biogeochemistry 2:345–357.

Schmidt, E.L., and E.A. Paul. 1982. Microscopic methods for soil microorganisms. p. 803–814. *In* A.L. Page et al. (ed.) Methods of soil analysis. Part 2. 2nd ed. Agron. Monogr. 9. ASA and SSSA, Madison, WI.

Shen, S.M., G. Pruden, and D.S. Jenkinson. 1984. Mineralization and immobilization of nitrogen in fumigated soil and the measurement of microbial biomass nitrogen. Soil Biol. Biochem. 16:437–444.

Smith, J.L., and E.A. Paul. 1990. The significance of soil biomass estimations. p. 357–396. *In* J.-M. Bollag and G. Stotsky (ed.) Soil biochemistry. Vol. 6. Marcel Dekker, New York.

Sparling, G.P. 1992. Ratio of microbial biomass to soil organic carbon as a sensitive indicator of changes in soil organic matter. Aust. J. Soil Res. 30:195–207.

Sparling, G.P., and A.W. West. 1988. A direct extraction method to estimate soil microbial C: Calibration *in situ* using microbial respiration and ^{14}C labelled cells. Soil Biol. Biochem. 20:337–343.

Tisdall, J.M., and J.M. Oades. 1982. Organic matter and water stable aggregates in soils. J. Soil Sci. 33:141–163.

Van-Gestel, M., J.N. Ladd, and M. Amato. 1992. Microbial biomass responses to seasonal changes and imposed drying regimes at increasing depths of undisturbed topsoil profiles. Soil Biol. Biochem. 24:103–111.

Vance, E.D., P.C. Brookes, and D.S. Jenkinson. 1987a. An extraction method for measuring soil microbial biomass C. Soil Biol. Biochem. 19:703–707.

Vance, E.D., P.C. Brookes, and D.S. Jenkinson. 1987b. Microbial biomass measurements in forest soils: Determination of K_c values and tests of hypotheses to explain the failure of the chloroform-incubation method in acid soils. Soil Biol. Biochem. 19:689–696.

Voroney, R.P., and E.A. Paul. 1984. Determination of K_c and K_n *in situ* for calibration of the chloroform incubation method. Soil Biol. Biochem. 16:9–14.

Voroney, R.P., J.P. Winter, and E.G. Gregorich. 1991. Microbe/plant soil interactions. p. 77–99. *In* D.C. Coleman and B. Fry (ed.) Carbon isotope techniques. Academic Press, New York.

Zagal, E. 1993. Measurement of microbial biomass in rewetted air-dried soil by fumigation–incubation and fumigation–extraction techniques. Soil Biol. Biochem. 25:553–559.

Zak, D.R., D. Tilman, R.R. Parmenter, C.W. Rice, F.M. Fisher, J. Vose, D. Milchunas, and C.W. Martin. 1994. Plant production and the biomass of soil microorganisms in late-successional ecosystems: A continental-scale study. Ecology 75:2333–2347.

13 Potentially Mineralizable Nitrogen as an Indicator of Biologically Active Soil Nitrogen

Laurie E. Drinkwater
Rodale Institute
Kutztown, Pennsylvania

Cynthia A. Cambardella
National Soil Tilth Laboratory
Ames, Iowa

Jean D. Reeder
Crops Research Laboratory
Fort Collins, Colorado

Charles W. Rice
Kansas State University
Manhattan, Kansas

Almost all of the N in surface soils is present in the form of organic compounds that cannot be used directly by plants and also are not susceptible to loss through leaching. The amount of N converted from organic to mineral forms (mineralization) on an annual basis varies, depending on the past management history, annual climatic variation and inherent soil properties (Sprent, 1987; Paul & Clark, 1989). This capacity of the soil to supply plant-available N is an important indicator of soil quality and many chemical and biological methods have been developed in an effort to provide a simple, reliable indicator of potentially mineralizable N (Keeney, 1982; Bundy & Meisinger, 1994). In this chapter, we discuss the use of N mineralization potential as an indicator of soil quality and the advantages and disadvantages of the various methods available. We then recommend and describe two biologically-based laboratory methods of determining N mineralization potential.

USE OF N MINERALIZATION POTENTIAL AS AN INDICATOR OF SOIL QUALITY

Nitrogen mineralization potential is useful in conjunction with total N or soil C as an indicator of soil organic matter quality. Soil organic matter (SOM) is

---

Copyright © 1996 Soil Science Society of America, 677 S. Segoe Rd., Madison, WI 53711, USA.
Methods for Assessing Soil Quality, SSSA Special Publication 49.

composed of a continuum of readily decomposable and resistant components. The accessibility of these various SOM fractions to mineralization varies, ranging from extremely labile microbial biomass, labile nonliving organic residues, to chemically recalcitrant and physically protected pools (Paul, 1984; Parton et al., 1987). We are using the term *active soil N* to refer to the biologically dynamic, labile organic N fractions that can be mineralized in the course of one growing season (Duxbury et al., 1991). While some soil properties, such as cation-exchange capacity are dependent on the more stable, humic organic matter fractions, many desirable soil properties are linked to biological activity. Properties such as aggregate stability and N availability are often highly correlated with indicators of biological activity such as microbial biomass, but are frequently poorly correlated with total organic C and N (Tisdall & Oades, 1982; Kay, 1990; Roberson et al., 1991). Soils that have been managed differently can have similar levels of total soil N but very different N mineralization potentials, indicating differences in SOM quality. A recent study comparing soils receiving inputs of organic residues in place of mineral fertilizers concluded that even after 10 yr of management with organic residues, qualitative rather than quantitative differences in SOM were more prominent (Wander et al., 1994). Thus, indicators that are sensitive to SOM quality and that reflect the size of the labile SOM pools (i.e., microbial biomass, N mineralization potential and soil respiration) are needed in order to fully assess overall soil quality.

COMPARISON OF METHODS FOR ESTIMATING POTENTIALLY MINERALIZABLE NITROGEN

Numerous chemical and biological methods have been developed for determination of N mineralization potential. Chemical methods for estimating potentially mineralizable N range from fairly mild treatments such as extraction of N with concentrated salt solutions, to drastic treatments that use acids to partially hydrolyze organic N compounds. Although these chemically-based techniques are usually more rapid and convenient than biological methods, they do not simulate the activities of soil microorganisms nor do they selectively release the fraction of soil organic N that is made available for plant growth by microbial activity (Bremner, 1965). Thus chemical methods are entirely empirical and limited by the degree to which they correlate with reliable biological measurements of soil N availability, e.g., N uptake, crop yield or mineralizable N (Stanford, 1982). Several reviews are available that discuss the pros and cons of specific chemical methods for potentially mineralizable N (Keeney, 1982; Bundy & Meisinger, 1994; Rice & Havlin, 1994).

Biological methods for estimating potentially mineralizable N involve measuring the amount of mineral N released in the soil by microbial activity during incubation. These procedures are usually more time consuming and tedious than chemical methods, but generally are considered to provide a more realistic assessment of the potential ability of soils to provide N for plant growth, since they rely on the normal biological mineralization processes (Keeney, 1982); however, laboratory procedures cannot simulate field conditions. First, N mineralization in the field is strongly influenced by climate- and management-induced changes in soil temperature and soil moisture, and by the availability of organic N substrates (Duxbury & Nkambule, 1994). Secondly, actively growing roots also effect mineralization and can either increase or decrease the mineralization rate of SOM (Helal & Sauerbeck, 1986; Liljeroth et al., 1990; 1994). The effect roots have on mineralization of native SOM depends on crop species and soil environmental conditions such as N availability (Liljeroth et al., 1994). Thus, while laboratory incubations offer better estimates of mineralizable N than chemical procedures, they are sometimes poorly correlated with N mineralization dynamics in the field; however, insensitivity to environmental conditions is a positive attribute for a soil quality indicator, particularly if a wide array of soils are compared across climatic zones and crop rotations.

A range of incubation procedures have been published involving short-term (7–28 d) or long-term (30–200 d) incubations under either aerobic or anaerobic (waterlogged) conditions (see Keeney, 1982 or Bundy & Meisinger, 1994 for extensive reviews of these published procedures). All of these incubation procedures measure *net* mineralization potential that is the mineralized N remaining after microbial immobilization. Recently, Duxbury et al. (1991) developed an incubation method based on pulse labeling with ^{15}N followed by aerobic incubation to estimate the size of the biologically active N pool. This method relies on microbial immobilization–mineralization to distribute ^{15}N to all biologically active pools. Therefore this method reflects microbial activity as well as biologically active N in the nonliving organic pools. In situ incubation methods have been developed to measure net N mineralization under field conditions (Hart & Firestone, 1988). These in situ procedures can provide excellent information on the temporal dynamics of N mineralization.

Aerobic Incubation

Standford and Smith (1972) proposed a 30-wk aerobic incubation method to estimate soil N mineralization potential. The soil is incubated under optimum conditions and inorganic N is removed at various times during the incubation by leaching with a dilute Ca salt solution. Cumulative mineralized N after 30 wk of incubation is used to calculate potentially mineralizable N (N_o) by assuming the mineralization of labile organic N follows first order kinetics. It has been suggested that long-term aerobic incubations (12–30 wk) may be better indicators of total mineralization potential than short-term incubations (14–28 d) because active nonbiomass N and stabilized, labile organic N pools are mineralized in addition to the most active microbial biomass pool (Stanford & Smith, 1972); however, while the contribution of microbial biomass N to total mineralizable N during these long-term incubations is sometimes minimal (Juma & Paul, 1984) others have found that a major portion of the N mineralized is derived from the microbial biomass (Bonde et al., 1988; Harris, 1993). Bonde et al. (1988) reported that the proportion of mineralized N derived from the microbial biomass during long-term incubations depends on management history and the initial size of the microbial biomass N pool. Harris (1993) found that up to 60% of the N mineralized after 200 d originated from the microbial biomass in soils with a history of leguminous residue inputs. Paustian and Bonde (1987) and Duxbury and Nkambule (1994) review the pros and cons of long-term aerobic incubations in detail.

Short-term aerobic incubations quantify a portion of the total potentially mineralizable N pool and reflect the relative N- supplying capacities of soil under specific environmental conditions (Stanford & Smith, 1972; Stanford et al., 1974). Net mineralizable N from short-term incubations has been related to ryegrass (*Lolium perenne* L.) uptake in the greenhouse (Keeney & Bremner, 1966, 1967), ryegrass and barley (*Hordeum vulgare* L.) N uptake in the greenhouse and in the field (Jenkinson, 1968), oat (*Avena sativa* L.) N uptake in the greenhouse (Stanford & Legg, 1968), and maize (*Zea mays* L.) yield response to N in the field (Walmsley & Forde, 1976). Typically, as with long-term incubations, correlation with field data is often less satisfactory than correlation with greenhouse data. Short-term production of net mineralizable N during aerobic laboratory incubation also has been used successfully by ecologists to quantify biologically active soil organic N pools (Woods et al., 1982; Schimel et al., 1985; Schimel, 1986; Wood et al., 1990; Davidson et al., 1991; Cambardella & Elliott, 1994; Cambardella & Kanwar, 1995).

Anaerobic Incubation

Aerobic incubation procedures provide optimal temperature, moisture, and aeration conditions for the microbial population responsible for mineralization of soil organic N under most field conditions. The major disadvantage of aerobic procedures is the difficulty in maintaining optimal soil water content during incubation. A second disadvantage is the need to measure both NH_4^+ and NO_3^- concentrations following incubation. An anaerobic incubation procedure was first proposed by Waring and Bremner (1964) as a simpler, more rapid alternative to aerobic incubations. Conducting the incubation under water-logged conditions eliminates the need for establishing and maintaining a standard soil water content during the incubation. A further advantage is that since nitrification is prevented, all N mineralized during the incubation will be in the form of NH_4^+. Since mineralization is occurring under anaerobic conditions, immobilization may be reduced compared with aerobic conditions because of the reduced energetic efficiency (C is not completely oxidized to CO_2). This difference in energetic efficiency does not fundamentally change the mineralization of N.

As with other laboratory incubations, many questions remain about the source of the mineralized N produced during anaerobic incubations. Some have suggested that the anaerobic incubation method may serve as a substitute for microbial biomass determinations since it apparently involves killing and mineralizing the obligate aerobes, a process somewhat analogous to the fumigation–incubation method of microbial biomass determinations (Doran et al., 1987; Myrold, 1987). Myrold (1987) labeled the resident microbial biomass in several forest soils with ^{15}N and then determined net mineralized N using anaerobic incubation and microbial biomass using chloroform fumigation–incubation. He found a highly significant correlation between both the amount and the proportion of ^{15}N released by the two methods and concluded that the anaerobic

Table 13–1. Relationship between microbial biomass and anaerobic N-mineralization potential under differing fertility input regimes. Pearson correlations and P values are given.

Fertility management	n	Nmin† vs. MBN‡	MBN vs. MBC§	Source
Inorganic N 1992	36	NS¶	$0.68, P = 0.0001$	Doran, unpublished#
1993	33	NS	$0.57, P = 0.0003$	
Cover crop residues, varying C/N ratios	31	$0.91, P = 0.0001$	$0.95, P = 0.0001$	Sarrantonio, unpublished#
Inorganic N	42	$-0.35, P = 0.02$	NS	Gunapala & Scow,††
Cover crop residues, manure	52	$0.52, P = 0.0001$	$0.52, P = 0.0001$	(in press)

† Anaerobic net N-mineralization potential.
‡ Microbial biomass N.
§ Microbial biomass C.
¶ NS = $P > 0.05$.
# Fumigation–incubation method.
†† Fumigation–extraction method.

incubation measures N released primarily from the microbial biomass. This kind of detailed isotope experiment has not yet been done in other soil types or ecosystems such as grasslands or agricultural systems. Soil microbial dynamics in agricultural systems differ from those in natural ecosystems because management interventions disrupt C turnover equilibrium resulting in boom/bust cycles of microbial activity (Drinkwater et al., 1995b; Wyland et al., 1995; Gunapala & Scow, 1997). A number of studies have compared anaerobic net N mineralization and fumigation–incubation microbial biomass N and C determinations in agricultural soils with inconsistent results. Sometimes the correlation is very strong, consistent with the hypothesis that both techniques are measuring N mineralized from the microbial biomass N pool, however in many cases there appears to be no relationship between the two (Table 13–1). Thus, while anaerobic N mineralization potential is probably a good indicator of the potential for soil to deliver N, it does not necessarily reflect microbial biomass N levels and cannot be considered a dependable substitute for the fumigation–incubation method. A probable explanation for these discrepancies is that other properties that are not being measured, such as the composition of the microbial community and its metabolic status (i.e., dormant vs. active, C-limited vs. N-limited) or composition of the organic substrate being mineralized are differentially affecting these measurements.

Sample Handling and Cautions

Sample handling presents the greatest challenge in terms of standardization of these incubation procedures. Quantification of mineralized N using incubation methods can be affected by: (i) soil sampling procedures; (ii) soil drying; (iii) grinding or sieving the soil and the mesh size used; and (iv) length of storage time and temperature of storage. Samples collected shortly after management practices that stimulate microbial activity, i.e., tillage or residue additions, will have greater N mineralization potentials (Drinkwater et al., 1995a,b; Wyland et al., 1995). Likewise, seasonal effects on microbial biomass will effect N mineralization potentials (Patra et al., 1990). Thus, the timing of sample collection is an important consideration (Dick et al., 1996, this publication).

Sieving soil is somewhat analogous to tillage and increases N mineralization potential, particularly in soils from no-till systems, by exposing protected labile organic matter (Ross et al., 1985). Sieving soils prior to incubation has the advantage of increasing homogeneity by mixing composite samples and ensuring removal of large pieces of undecomposed organic debris such as roots. If the main goal is to use N mineralization potential as an indicator of SOM quality, then sieving soils to improve standardization is probably desirable; however, if the goal is to estimate the potential for N release in the field, then incubations with sieved soils from untilled systems will usually result in overestimations of N mineralization (Duxbury & Nkambule, 1994).

There is an on-going debate about the pros and cons of conducting biological incubations on field-moist vs. air-dried soils. Air-drying has been shown to increase the amount of mineral N produced compared with field-moist soil and this increase is positively correlated with the length of time the air-dry sample is stored prior to incubation (Keeney & Bremner, 1966; Stanford & Legg, 1968); however, in some cases, estimates of available N obtained from aerobic laboratory incubations of air-dry soil have been found to be better correlated with crop response than aerobic incubations of field-moist soil (Keeney & Bremner, 1966). Perhaps the greatest problem with using air-dried soils in these incubations is that the impact of air drying and rewetting on SOM accessibility and the soil biota will vary, depending on soil texture and climate and possibly management history.

Dry–wet cycles can influence N mineralization through aggregate disruption and the release of soluble C from the microbial biomass (Kemper & Rosenau, 1986; Kieft et al., 1987). One option is to minimize moisture variability by collecting samples at field capacity, either following irrigation or rainfall. This may not be feasible for some rainfed systems especially in semiarid and arid areas. In these situations, since surface soils become air-dry in the field at some point during the growing season, it seems appropriate to recommend air-drying prior to incubation for routine laboratory analysis in order to standardize the incubations. Soil samples (field-moist or air-dry) should be stored at 4°C prior to incubation (Chaudhry & Cornfield, 1971). Since mineralization potential is strongly controlled by soil moisture at the time of sampling, we encourage researchers to adopt the pretreatment that is most appropriate to the dominant climate–soil–cropping scenario of a given region.

Interpretation

The interpretation of net N mineralization potential must be done in conjunction with other soil measurements such as organic N and C and mineral N pools. In general, short-term incubations are useful for estimating labile soil N but may not reflect differences in the active soil N from the larger, more stable pools (Duxbury & Nkambule, 1994; Wander et al. 1994). Greater mineralization potentials have been found in a variety of cropping systems under organic management (Doran et al., 1987; Drinkwater et al., 1995a; Gunapala & Scow, 1997) reflecting the increased role of decomposers in determining N availability in systems based on organic soil amendments. The ratio of N mineralized to total organic N can serve as a sensitive indicator of SOM quality differences. The proportion of total soil N mineralized in short-term, anaerobic incubations was more than two-fold greater in organically compared with conventionally-managed soils (1.2 vs. 0.5%) indicating significant qualitative differences in the SOM (Drinkwater et al., 1995a).

Taken together, microbial biomass N, N mineralization potentials and mineral N pools can serve as indicators on the status of N dynamics in the soil. Large mineralization potentials in conjunction with high concentrations of mineral N, especially during times of reduced crop uptake, could indicate susceptibility to N losses through leaching. Organically managed soils are sometimes characterized by higher levels of microbial activity and potentially mineralizable N, in conjunction with lower instantaneous mineral N pools compared with soils receiving conventional mineral fertilizers (Drinkwater et al., 1995a). The combination of low instantaneous mineral N pools and enhanced microbial activity in the organic soils are indicative of a more tightly coupled N cycle (Sprent, 1987; Jackson et al., 1989; Jenkinson & Parry, 1989), with higher turnover rates of mineral N pools than in conventional soils.

RECOMMENDED AEROBIC INCUBATION PROCEDURE

During the aerobic incubation, results are affected by soil water content, temperature, and pH, primarily through the effects on microbial activity. Optimal soil conditions for aerobic incubation are given as soil water content at 60% water-filled pore space (WFPS, Linn & Doran, 1984), soil temperature between 30 and 35°C, and soil pH between 6.6 and 8.0 (Paul & Clark, 1989). Adjustment of soil water content requires preliminary analysis to determine the amount of water required for incubation. Keeney and Bremner (1967) suggest mixing 10 g of air-dried soil with 30 g of 30- to 60-mesh acid-washed quartz sand and then moistening the mixture with 6 mL of water. The resulting water content is between 40 and 60% WFPS. We have found it almost impossible to get the soil–sand uniformly mixed so H_2O contents vary among the small microcosms of heterogeneous soil–sand mixtures during incubation (Cambardella & Reeder, unpublished data).

Schimel et al. (1985) suggested an alternative approach that eliminates some of the problems associated with repeated leachings that are used during long-term incubations. Subsamples are extracted with 2.0 M KCl at 10, 20, and 60 d. Cumulative mineralized N (NH_4^+ + NO_2^- + NO_3^-) is calculated after subtraction of initial soil inorganic N. Leaving samples sealed for long periods may result in low atmospheric O_2 levels and high CO_2 that restrict mineralization. Modifications of both the Keeney and Bremner (1967) method and the method of Schimel et al. (1985) by Cambardella (1994) have produced the following method for aerobic determination of mineralizable N in soil.

Soil Sample Preparation

1. Field-moist soil is passed through 2- to 8-mm sieve. Large pieces of organic material and rocks are removed. The mesh size that is practical for sieving field-moist soil will depend on soil type and water content. Always use the same sieve size and report results based on the sieve size used, e.g., mineralizable N (<2 mm).
2. Soil water content is determined gravimetrically while the soil is stored at 4°C until incubation.
3. Soil baseline mineral N [(NO_2^- + NO_3^-) + NH_4^+] is determined using a colorimetric method (Keeney & Nelson, 1982) after a 30 min extraction in 2 M KCl (1:5 soil/solution ratio).

Laboratory Incubation

4. 10-g subsamples of sieved soil are weighed into each of four scintillation vials for incubations.
5. Soil is tamped down, where possible, to a uniform bulk density of 1.0 g cm^{-3}. Soil is adjusted to 60% WFPS with deionized or distilled water. The volume of water needed to obtain 60% WFPS is calculated using bulk density and initial water content (see Lowery et a., 1996, this publication).
6. The vials are placed in a quart Mason jar (about 950 mL) to which a small amount of water has been added at the bottom to maintain 100% relative humidity inside the sealed Mason jar. (Four empty jars containing everything but soil are run as blanks if mineralizable C is being quantified along with mineralizable N). The mason jars with the four vials are incubated at 30°C.
7. One vial is removed from each Mason jar at time zero plus 1 wk and extracted with 2 M KCl (1:5 soil/solution ratio). Again, mineral N is measured using a colorimetric method (Keeney & Nelson, 1982). Baseline mineral N is subtracted to give net cumulative mineral N.
8. One vial is removed at time zero plus 2 wk, plus 3 wk, and plus 4 wk and mineral N assessed as in Step 7.
9. Mineralizable N is reported as net cumulative mineral N after 4 wk incubation.
10. *Mineralizable C also can be assessed by using NaOH base traps in the Mason jars to trap CO_2–C. Mineralizable C is quantified by back-titration using HCl after addition of $BaCl_2$ to precipitate the CO_3–C. See Rice et al., 1996, this publication on soil biomass C determinations.

* Samples are removed each week for 4 wk in order to ascertain that the slope of the net cumulative mineral N curve vs. time (in weeks) remains constant over the 4 wk incubation. Most often, if the slope is different at all, the slope for the first 1 or 2 wk of the incubation will be different than for the last 2 wk. If this occurs, mineralizable N is reported as net cumulative mineral N at 4 wk minus net cumulative mineral N at 1 or 2 wk. In practice, however, the slope is generally the same across the 4 wk. If possible duplicate or triplicate samples are recommended.

RECOMMENDED ANAEROBIC INCUBATION PROCEDURE

The recommended anaerobic method for N mineralization potential is based on the method of Waring and Bremner (1964). The anaerobic incubation method determines net potentially mineralizable N with a 7 d anaerobic incubation at a standard temperature, usually between 25 and 37°C. After the incubation, NH_4^+ is extracted with 2 M KCl and analysis of NH_4^+ concentration with a Lachat Flow Injection system. The amounts of NO_3^- and NH_4^+ originally in the soil are determined by a 2 M KCl extraction on separate soil samples prior to incubation. Potentially mineralizable N is calculated by subtracting the initial amount of NH_4^+ in the soil from the amount of NH_4^+ released during the incubation. It is essential that soil samples remain completely anaerobic during incubation to eliminate possible nitrification–denitrification reactions at the soil-water interface that would lead to low results (Keeney, 1982). For this reason, in the recommended procedure described below, incubation containers are purged with N_2 and sealed with rubber stoppers and electrical tape prior to incubation. We recommend sieving soils and starting the incubations–extractions out in the field if possible, or immediately upon return to the laboratory. Samples should be transported to the laboratory on ice. This approach is most appropriate when accurate determinations of mineral N pools in conjunction with N mineralization potentials are a priority.

We have recommended two sizes of soil samples and incubation containers. The smaller sample version is appropriate for most soils, particularly if a large number of incubations will be performed. The larger sample size (80 g field moist soil in a 500 mL plastic centrifuge bottle) is recommended for soils that have recently received fresh residues.

The main advantage of the centrifuge tube method is that it avoids the need to transfer samples from one container to another and then to filter funnels. If samples are to be filtered rather than centrifuged, an alternate method using 60 mL syringes as incubation–extraction vessels accomplishes the same thing and might be more convenient (Lober & Reeder, 1993).

Reagents

2.0 M KCl: 149 g L^{-1}
2.67 M KCl: 199 g L^{-1}

Materials

Plastic, disposable 50 mL polypropylene centrifuge tubes (25/tray) work well as incubator containers for most applications and usually last for two field seasons before they begin to leak. The incubation tubes are sealed with no. 6 rubber stoppers and electrical tape. For larger samples, 500 mL plastic centrifuge bottles can be used. For sieving the moist soils, we recommend the aluminum seed sieves, 2 to 6 mm (depending on soil type), from Seedsburo (New Jersey). Traditional mesh soil sieves clog easily when soils are moist if the clay content is high. A centrifuge, or a filtering system is needed to remove soil from the extract. Extracts can be stored in 20-mL scintillation vials with polypropylene caps (caps without liners are needed). A cylinder of N gas is needed to purge the incubations.

Procedure

Prior to collecting soil samples:

1. Prepare solutions.
2. Rinse all labware twice with deionized or distilled H_2O.
3. Prepare two sets of sample containers: one set for initial (T_o) extraction and one for anaerobic incubation. Add 2 M KCl to T_o containers and deionized H_2O to containers for incubations. For 50 mL centrifuge tubes: Add 40 mL 2 M KCl to T_o tubes and 10 mL deionized H_2O to incubations using repipettes or graduated cylinders. Get a tare weight for each tube (tube + liquid + cap).

    For 500 mL bottles: Add 400 mL 2 M KCl to T_o bottles and 100 mL deionized H_2O to incubations. These don't need to be preweighed because the soil is weighed into them.

The day of sampling:

4. Field moist soil should be sieved as soon after sampling as possible and measured into extraction–incubation containers immediately after sieving. Sieving should remove large organic debris, roots, and stones.

    For 50 mL centrifuge tubes: Add about 8 g of moist soil (equal to a volume change of 5–6 mL to each tube. Weigh again with caps on as before to determine how much soil was added.

    For 500 mL bottles: Add exactly 80 g of moist soil.
5. Place extractions (KCl) on the shaker for 1 h.
6. Remove KCl extractions from shaker and stand upright so the soil can begin to settle. Swirl so that the solution cleans off any soil that has accumulated in the top of the centrifuge tube. Make sure the inside of the tops are clean before placing the tubes in the centrifuge.

    For 50 mL centrifuge tubes: Centrifuge for 10 min at 400 to 500 × g. The amount of time and gravitational force needed to clear the extracts by centrifugation may vary with soil type so adjust as needed. Collect supernatant from below the surface (avoid floating organic debris) using a pipetman and put into labeled scintillation vials. These extracts also can be filtered.

    For 500 mL bottles: Set up filter paper and funnels with tall 100 mL beakers or wide-mouthed 250 mL flasks.
7. Store all extracts in the freezer until NO_3^- and NH_4^+ can be analyzed.
8. While the KCl extractions are shaking, the incubations can be purged with N_2 and sealed with rubber stoppers and electrical tape. A Pasteur pipette is connected to the tank with flexible tubing. During purging,

the tip of the Pasteur pipette should be just above the water level, and should not be submerged. Purge for 45 to 60 s, withdrawing the pipette tip as you put the stopper in place (or cover the bottle opening with the cap). It is important to completely seal the tubes. Usually two pieces of tape are needed, one over the top of the stopper and the other is wrapped around the tube–stopper seal. For the large samples, bottle caps should be secured tightly. Place sealed incubation vessels in 37°C incubator for exactly 7 d.
9. Determine soil water content gravimetrically by weighing sieved, field moist soil subsamples and placing them in the drying oven.

After one week:

1. Remove samples from incubator and add 30 mL of 2.67 M KCl to each tube.
2. Replace caps tightly and loosen the soil in the bottom of the tube or bottle by shaking or thumping tubes on the counter. Go to Steps 5 to 7 above to extract NH_4^+. Incubations are analyzed for NH_4^+ only.

REFERENCES

Bonde, T.A., J. Schnurer, and T. Rosswall. 1988. Microbial biomass as a fraction of potentially mineralizable nitrogen in soils from long-term field experiments. Soil Biol. Biochem. 20:447–452.

Bremner, J. M. 1965. Nitrogen availability indices. p. 1324–1341. *In* A. Black et al. (ed.) Methods of soil analysis. Part 2. 2nd ed. Agron. Monogr. 9. ASA and SSSA, Madison, WI.

Bundy, L.G., and J.J. Meisinger. 1994. Nitrogen availability indices. p. 951–984. *In* R.W. Weaver et al. (ed.) Methods of soil analysis. Part 2. SSSA Book Ser. 5. SSSA, Madison, WI.

Cambardella, C.A. 1994. Nutrient use efficiency: An ecological approach. p. 229–232. *In* Proc. of the Integrated Crop Management Conf., Ames, IA. 30 Nov.– 1 Dec. 1994. Iowa State Univ., Ames.

Cambardella, C.A., and E.T. Elliott. 1994. Carbon and nitrogen dynamics of soil organic matter fractions from cultivated grassland soils. Soil Sci. Soc. Am. J. 58:123–130.

Cambardella, C.A., and R.S. Kanwar. 1995. N mineralization potential in soil and its relationship to soluble organic N in tile drain water. p. 297. *In* Agronomy abstracts. ASA, Madison, WI.

Chaudhry, I.A., and A.H. Cornfield. 1971. Low-temperature storage for preventing changes in mineralizable nitrogen and sulphur during storage of air-dried soils. Geoderma 5:165–168.

Davidson, E.A., S.C. Hart, C.A. Shanks, and M.K. Firestone. 1991. Measuring gross nitrogen mineralization, immmobilization, and nitrification by ^{15}N isotopic pool dilution in intact soil cores. J. Soil Sci. 42:335–349.

Dick, R.P., D.R. Thomas, and J.J. Halvorson. 1996. Standardized methods, sampling, and sample pretreatment. p. 107–121. *In* J.W. Doran and A.J. Jones (ed.) Methods for assessing soil quality. SSSA Spec. Publ. 49. SSSA, Madison, WI.

Doran, J.W., D.G. Fraser, M.N. Culik, and W.C. Liebhardt. 1987. Influence of alternative and conventional agricultural management on soil microbial processes and nitrogen availability. Am. J. Alt. Agric. 2(3):99–106.

Drinkwater, L.E., D.K. Letourneau, F. Workneh, A.H.C. van Bruggen, and C. Shennan. 1995a. Fundamental differences in organic and conventional agroecosystems in California. Ecol. Appl. 5:1098–1112.

Drinkwater, L.E., C. Rieder, and W. Herdman. 1995b. Effect of compost quality on nitrogen dynamics. p. 241. *In* Agronomy abstracts. ASA, Madison, WI.

Duxbury, J.M., J.G. Lauren, and J.R. Fruci. 1991. Measurement of the biologically active soil N fraction by an ^{15}N technique. Agric. Ecosys. Environ. 34:121–129.

Duxbury, J.M., and S.V. Nkambule. 1994. Assessment and significance of biologically active soil organic nitrogen. p. 125–146. *In* J.W. Doran et al. (ed.) Defining soil quality for a sustainable environment. SSSA Spec. Publ. Number 35. SSSA and ASA, Madison, WI.

Gunapala, N., and K. Scow. 1997. Dynamics of soil microbial biomass and activity in conventional and organic farming systems. Soil Biol. Biochem. (In press.)
Harris, G.H. 1993. Nitrogen cycling in animal-, legume-, and fertilizer-based cropping systems. Ph.D. diss. Michigan State Univ., East. Lansing.
Hart, S.C., and M.K. Firestone. 1988. Evaluation of three *in situ* soil nitrogen availability assays. Can. J. For. Resour. 19:185–191.
Helal, H.M., and D.R. Sauerbeck. 1986. Effect of plant roots on carbon metabolism of soil microbial biomass. Z. Pflanzenernaehr Bodenkd. 149:181–188.
Jackson, L.E., J.P. Schimel, and M.K. Firestone. 1989. Short-term partitioning of ammonium and nitrate between plants and microbes in an annual grassland. Soil Biol. Biochem. 21:409–415.
Jenkinson, D.S. 1968. Chemical tests for potentially available nitrogen in soil. J. Sci. Food Agric. 19:160–168.
Jenkinson, D.S., and L.C. Parry. 1989. The nitrogen cycle in the Broadbalk Wheat Experiment: A model for the turnover of nitrogen through the soil microbial mass. Soil Biol. Biochem. 21(4): 535–541.
Juma, N.G., and E.A. Paul. 1984. Mineralizable soil nitrogen: Amounts and extractability ratios. Soil Sci. Soc. Am. J. 48:76–80.
Kay, B.D. 1990. Rates of change of soil structure under different cropping systems. Adv. Soil Sci. 12:1–52.
Keeney, D.R. 1982. Nitrogen availability indices. p. 711–733. *In* A.L. Page et al. (ed.) Methods of soil analysis. Part 2. 2nd ed. Agron. Monogr. 9. ASA and SSSA, Madison. WI.
Keeney, D.R., and J.M. Bremner. 1966. Comparison and evaluation of laboratory methods of obtaining an index of soil nitrogen availability. Agron. J. 58:498–503.
Keeney, D.R., and J.M Bremner. 1967. Determination of isotopic-ratio analysis of different forms of nitrogen in soil: VI. Mineralizable nitrogen. Soil Sci. Soc. Am. Proc. 31:34–39.
Keeney, D.R., and D.W. Nelson. 1982. Nitrogen: Inorganic forms. p. 643–698. *In* A.L. Page et al. (ed.) Methods of soil analysis. Part 2. 2nd ed. Agron. Monogr. 9. ASA and SSSA, Madison. WI.
Kemper, W.D., and R.C. Rosenau. 1986. Aggregate stability and size distribution. p. 425–442 *In* A. Klute (ed.) Methods of soil analysis. Part 1. 2nd ed. Agron. Monogr. 9. ASA and SSSA, Madison, WI.
Kieft, T.L., E. Soroker, and M.K. Firestone. 1987. Microbial biomass response to a rapid increase in water potential when dry soil is wetted. Soil Biol. Biochem. 19:119–126.
Liljeroth, E., P. Kuikman, and J.A. Van Veen. 1994. Carbon translocation to the rhizosphere of maize and wheat and influence on the turnover of native soil organic matter at different soil nitrogen levels. Plant Soil. 161:233–240.
Liljeroth E., J.A. Van Veen, and H.J. Miller. 1990. Assimilate translocation to the rhizosphere of two wheat cultivars and subsequent utilization by rhizosphere microorganisms at two soil nitrogen levels. Soil Biol. Biochem. 22:1015–1021.
Linn, D.M., and J.W. Doran. 1984. Effect of water-filled pore space on carbon dioxide and nitrous oxide production in tilled and nontilled soils. Soil Sci. Soc. Am. J. 48:1267–1272.
Lober, R.W., and J.D. Reeder. 1993. Modified waterlogged incubation method for assessing nitrogen mineralization in soils and soil aggregates. Soil Sci. Soc. Am. J. 57:400–403.
Lowery, B. M.A. Arshad, R. Lal, and W.J. Hickey. 1996. Soil water parameters and soil quality. p. 143–155. *In* J.W. Doran and A.J. Jones (ed.) Methods for assessing soil quality. SSSA Spec. Publ. 49. SSSA, Madison, WI.
Myrold, D.D. 1987. Relationship between microbial biomass nitrogen and a nitrogen availability index. Soil Sci. Soc. Am. J. 51: 1047–1049.
Parton, W.J., D.S. Schimel, C.V. Cole, and D.S. Ojima. 1987. Analysis of factors controlling soil organic matter levels in Great Plains grasslands. Soil Sci. Soc. Am. J. 51:1173–1179.
Patra, D.D., P.C. Brookes, K. Coleman, and D.S. Jenkinson. 1990. Seasonal changes of soil microbial biomass in an arable and a grassland soil which have been under uniform management for many years. Soil Biol. Biochem. 22(6):739–742.
Paul, E.A. 1984. Dynamics of organic matter in soils. Plant Soil. 76:275–285.
Paul, E.A., and F.E. Clark. 1989. Soil microbiology and biochemistry. Academic Press, New York.
Paustian, K., and T.A. Bonde. 1987. Interpreting incubation data on nitrogen mineralization from soil organic matter. INTECOL Bull. 15:101–112.
Rice, C.W., and J.L. Havlin. 1994. Integrating mineralizable nitrogen indices into fertilizer nitrogen recommendations. p. 1–13. *In* J.L. H. Havlin and J.S. Jacobsen (ed.) Soil testing: Prospects for improving nutrient recommendations. SSSA Spec. Publ. 40. SSSA, Madison, WI.

Rice, C.W., T. Moorman, and M. Beare. 1996. Role of microbial biomass carbon and nitrogen in soil quality. p. 203–215. *In* J.W. Doran and A.J. Jones (ed.) Methods for assessing soil quality. SSSA Spec. Publ. 49. SSSA, Madison, WI.

Roberson, E.B., S. Sarig, and M.K. Firestone. 1991. Cover crop management of polysaccharide-mediated aggregation in an orchard soil. Soil Sci. Soc. Am. J. 55: 734–739.

Ross, D.J., T.W. Speir, K.R. Tate, and V.A. Orchard. 1985. Effects of sieving on estimations of microbial biomass, and carbon and nitrogen mineralization, in soil under pasture. Aust. J. Soil Res. 23:319–324.

Schimel, D.S. 1986. Carbon and nitrogen turnover in adjacent grassland and cropland ecosystems. Biogeochemistry 2:345–357.

Schimel, D.S., D.C. Coleman, and K.A. Horton. 1985. Soil organic matter dynamics in paired rangeland and cropland toposequences in North Dakota. Geoderma 36:201–214.

Sprent, N. 1987. The ecology of the nitrogen cycle. Cambridge Press, Cambridge.

Stanford, G., 1982. Assessment of soil nitrogen availability. p. 651–688. *In* F.J. Stevenson (ed.) Nitrogen in agricultural soils. Agron. Monogr. 22. ASA, Madison, WI.

Stanford, G., J.N. Carter, and S.J. Smith. 1974. Estimates of potentially mineralizable soil nitrogen based on short-term incubations. Soil Sci. Soc. Am. Proc. 38:99–102.

Stanford, G., and J.O. Legg. 1968. Correlation of soil N availability indexes with N uptake by plants. Soil Sci. 105:320–326.

Stanford, G., and S.J. Smith. 1972. Nitrogen mineralization potentials of soils. Soil Sci. Soc. Am. Proc. 36:465–472.

Tisdall, J.M., and J.M. Oades. 1982. Organic matter and water-stable aggregates in soils. J. Soil Sci. 33:141–163.

Walmsey, D., and S.C.M. Forde. 1976. Further studies on the evaluation and calibration of soil analysis methods for N, P, and K in the Eastern Caribbean. Trop. Agric. (Trinidad) 53:281–291.

Wander, M.M., S.J. Traina, B.R. Stinner, and S.E. Peters. 1994. The effects of organic and conventional management on biologically-active soil organic matter pools. Soil Sci. Soc. Am. J. 58:1130–1139.

Waring, S.A., and J.M. Bremner. 1964. Ammonium production in soil under waterlogged conditions as an index of nitrogen availability. Nature (London) 201:951–952.

Wyland, L.J., L.E. Jackson, and K.F. Schulbach. 1995. Soil-plant nitrogen dynamics following incorporation of a mature rye cover crop in a lettuce production system. J. Agric. Sci. 124:17–25.

Wood, C.W., D.G. Westfall, G.A. Peterson, and I.C. Burke. 1990. Impacts of cropping intensity on carbon and nitrogen mineralization under no-till dryland agroecosystems. Agron. J. 82:1115–1120.

Woods, L.E., C.V. Cole, E.T. Elliott, R.V. Anderson, and D.C. Coleman. 1982. Nitrogen transformations in soil as affected by bacterial-microfaunal interactions. Soil Biol. Biochem. 14:93–98.

14 Field and Laboratory Tests of Soil Respiration

Timothy B. Parkin
USDA-ARS, National Soil Tilth Laboratory
Ames, Iowa

John W. Doran
USDA-ARS, University of Nebraska
Lincoln, Nebraska

E. Franco-Vizcaíno
Michigan State University
East Lansing, Michigan

Soil supports a vast amount and diversity of biological activities, therefore, it is generally thought that some measure of soil biological activity would be a valuable indicator of the health or quality of the soil. There have been many techniques used to measure the activity of soil biota. Some of these techniques reflect the activities of specific organisms or groups of organisms whereas other techniques provide an assessment of overall biological activity. Microorganisms play a key role in soil ecology. By decomposing dead plant and animal material, soil microorganisms recycle essential nutrients. Because of these characteristics, an active microbial population is an attribute that is often cited as a key component of good soil quality (Howard, 1947; Waksman, 1927; Turco et al., 1994; Kennedy & Papendick, 1995).

Respiration is a process that reflects biological activity. A distinction must be made between microbial respiration and soil respiration, as often times these two terms are used synonymously. In this chapter, we define soil respiration as the production of CO_2 or consumption of O_2 as a result of the metabolic processes of living organisms in soil. Here microbial respiration is defined as the production of CO_2 or the uptake of O_2 as a result of the metabolism of microorganisms such as bacteria, fungi, algae, and protozoa. Whereas the term microbial respiration refers to the metabolic activity of microorganisms, soil respiration has a more general implication, and indicates the biological activity of the entire soil biota including microorganisms, macroorganisms (such as earthworms, nematodes, and insects), and plant roots. This distinction is important because a vari-

---

[1]Names of products are included for the benefit of the reader and do not imply endorsement or preferential treatment by USDA.

Copyright © 1996 Soil Science Society of America, 677 S. Segoe Rd., Madison, WI 53711, USA.
Methods for Assessing Soil Quality, SSSA Special Publication 49.

ety of different methods for measuring respiration have been developed, and some are more appropriate for determining microbial respiration, while others better represent total soil respiration.

This chapter describes methods for assessment of soil and microbial respiration, and to provide a framework for interpreting these measurements relative to soil quality. It must be pointed out that we will not discuss in detail the limitations and implications of all the available methodology for determining biological and microbial activity in soil. For additional information related to soil respiration measurements, the reader is referred to past review and methods articles (Anderson, 1982; Zibilske, 1994). Rather, we propose two basic methods, both based on quantification of the rate of CO_2 production. The first method, used to assess total biological activity, is based on determination of CO_2 flux rates using chambers placed over the soil surface. The second procedure is used to assess microbial activity, and is based on CO_2 production from sieved, mixed soil in the laboratory under a fixed temperature and moisture regime.

METHODS DESCRIPTION

Soil Respiration: Field Carbon Dioxide Flux Using Soil Covers

Principles of Measurement

Many techniques have been proposed and applied to the quantification of gas flux from soil, but the simplest is the closed chamber method. This technique involves covering the soil with a canister or chamber and measuring gas flux rate by determining changes in concentration of the gas in the headspace of the chamber over time. Typically, soil respiration is determined by monitoring the rate of accumulation of CO_2 in the chamber (as opposed to monitoring decrease in O_2 concentration). The CO_2 accumulating in the chamber is not necessarily only the result of soil microorganisms; CO_2 also is produced by other soil fauna and plant roots.

There are several possible ways to determine CO_2 concentrations. Discrete gas samples can be removed from the chamber using a syringe, and the CO_2 content of the gas samples can be measured using a gas chromatograph or infrared gas analyzer. Carbon dioxide can be trapped during the incubation period in an alkali solution, and the amount of CO_2 determined by titration or by weighing soda lime traps (Anderson & Ingram, 1993). Real-time determinations of CO_2 can be performed by recirculating the chamber headspace gas through a portable infrared gas analyzer. The critical aspect of soil respiration measurements is that the amount of CO_2 emanating from the soil is quantified over a known time period.

There are many tradeoffs that must be considered in deciding upon the appropriate methodology for measuring CO_2. Among the factors that must be considered are cost, ease of application, and biases associated with CO_2 flux determinations. Tradeoffs exist between measurement bias that may exist in the closed chamber technique and the simplicity of this technique. It has often been observed that CO_2 accumulation in closed chambers is not linear, but rather the rate decreases with time. This effect has been attributed to alteration of the CO_2 gradient within the soil profile, because of CO_2 accumulation within the chamber headspace (Anthony et al., 1995), or CO_2 depletion in the headspace if soda lime is used (Nay et al., 1994). Measurement bias also can result from alterations in temperature and pressure within the chamber, as a result of chamber placement on the soil surface.

Tradeoffs also exist in relation to method of CO_2 measurement. Gas chromatographs and infrared gas analyzers yield accurate, reproducible results; however, the cost of the instrumentation to perform these analyses is high. Alternatively, CO_2 can be determined using Draeger gas detection tubes which cost approximately $3.50 each. We have found the accuracy of these gas detection tubes to be comparable to that obtained with a gas chromatograph (Table 14–1).

Depending upon the CO_2 measurement technology available, a determination must be made regarding number of CO_2 measurements required to quantify the CO_2 flux rate. If instrumentation such as a gas chromatograph or an infrared gas analyzer is available, it is advantageous to make several CO_2 determinations over the course of the incubation. With soda lime trapping or gas analysis tube determinations, usually, only a single time point analysis is performed.

Due to chamber effects, the duration of the period when the soil is covered should be as short as practically possible. Again a tradeoff exists between the length of time required to allow enough accumulation of CO_2 to enable accurate assessment vs. biases that may occur due to prolonged soil coverage. The limitations associated with closed chamber methods do not preclude its use if the method is applied judiciously. It is by far the simplest, and least expensive soil respiration methodology currently available. The following discussion provides a list of materials and procedures required to perform soil respiration measurements using a simple closed chamber coupled with a single time point CO_2 measurement using a Draeger gas analysis tube. This approach is part of the on-farm soil quality methods described in Sarrantonio et al., (1996, this publication).

Table 14–1. Draeger tube vs. gas chromatographic determinations of CO_2 from respiration chambers.

Chamber	Gas chromatograph[†]	Draeger tube
	% CO_2	
1	0.13	0.15
2	0.17	0.15
3	0.16	0.20
4	0.21	0.25
5	0.18	0.25
6	0.17	0.25
7	0.17	0.25
8	0.14	0.10
Mean[‡]	0.17	0.20
Std. dev.	0.025	0.06

[†] Gas chromatograph determinations performed on 8 mL of soil chamber headspace gas collected in evacuated vials using a thermal conductivity detector.
[‡] Means are not significantly different ($P > 0.05$) as determined by a Mann Whitney test.

Method 1: Soil Respiration

Materials Required

1. Infiltration ring. Aluminum irrigation pipe, 14.9 cm i.d. (~6 in), cut to 12.7 cm (5 in) length, edge on one end beveled, outside of ring marked at 7.6 cm (3 in) from the bottom of the ring. (See description of soil infiltration measurement).
2. Chamber lid. Can bottom from Number 10 food can (15.3 cm diam.) cut with 2.54 cm (1 in) lip, containing three holes fit with rubber stoppers. Stoppers are red rubber for serum or vaccine bottles and are 1.9 cm long by 0.3 cm thick, tapered from 1.63 cm diam. at the top to 1.43 cm diam. at the bottom.
3. Soil thermometer. Any metal thermometer with a range from 0 to 100°C
4. Draeger gas detection tubes. 0.1% CO_2 sampling tubes.
5. A 140 cc Syringe (plastic).
6. Latex tubing. Two pieces approximately 10 cm (4 in) each, Tubing has inner diameter of 0.64 cm (1/4 in) and a wall thickness of 0.48 cm (3/16 in).
7. Hypodermic needles. 18 to 22 gauge, 2.5 to 3.8 cm (1 in to 1.5 in) long.
8. Watch or timer.

Procedure

1. Install infiltration ring (beveled edge down) to the line marked at 7.6 cm (3 in) into soil using wood block and hand sledge (Fig. 14–1A).
2. Place chamber lid on infiltration ring (Fig. 14–1B) and record time or start timer.
3. After ½ h, attach the latex tubing-Draeger tube-Syringe assembly to the soil chamber by inserting hypodermic needle through one of the chamber lid's rubber stoppers (Fig. 14–1C). As a precaution another syringe needle is placed into one chamber stopper that is at least 4 in away from the rubber stopper used for sampling to prevent formation of a vacuum during the sampling procedure.
4. Slowly draw 100 mL of headspace sample (do this over about a 15 s time period) through an opened 0.1% CO_2 Draeger gas detection tube using the 140 mL syringe for suction (Fig. 14–1D). Note, the Draeger detection tube is a sealed glass tube, and each end must be broken off before gas can be sampled. The Draeger tube is opened by using a 0.16 cm (1/16 in) diameter hole drilled in the syringe plunger handle to break off each end of the tube.
5. After 100 mL of chamber headspace gas has been drawn through the Draeger tube, read CO_2 as percentage by volume on the $N = 1$ scale (100 mL) of the Draeger tube as indicated by the furthest advance of a violet color change down the tube. If the advancing color line is not parallel with the gradation lines, estimate it's average position.

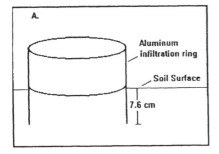
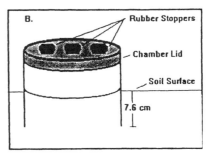
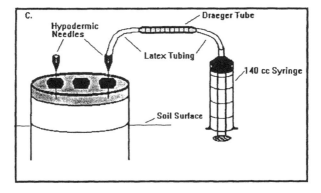
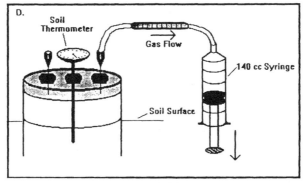

Fig. 14–1. Schematic diagram of soil respiration methodology.

6. Install soil thermometer through the central stopper in the lid of the chamber to a depth of 5 cm (2 in) in the soil (Fig. 14–1D). Record soil temperature and percentage of CO_2 at time of sampling.

Calculations

The CO_2 reading from the Draeger tube must be converted from units of volumetric percentage to grams of CO_2–C per square meter of soil. The critical factor that influences this calculation is the chamber headspace volume, which is a function of how high the chamber extends above the soil surface. According to the procedure above, the chamber headspace height should be approximately 5 cm (2 in), however, if the soil surface is uneven the average soil can height above the surface must be determined by measuring the soil chamber height at several places within the ring. Once the average height of the chamber above the surface is determined (in centimeters) it is inserted into the following formula.

$$g\ CO_2\text{--}C\ m^{-2}\ d^{-1} = (h/5.1) \times PF \times [(ST + 273)/273]$$

$$\times [\%CO_2 - 0.035] \times 13.0 \qquad [1]$$

where h is the average headspace height in cm, PF equals inches Hg barometric pressure/29.9, and ST is the soil temperature at 5 cm (°C). Note the barometric pressure factor, PF, can be ignored if the elevation is <2000 ft. The constant of 13.0 used in this equation assumes a chamber height of 5.1 cm and a measurement time of 0.5 h. Derivation of this constant is as follows:

$$(\%CO_2 - 0.035)/100 \times (12\ g\ C/22\ 400\ cm^3) \times (889\ cm^3/\text{chamber})$$

$$\times [(10^4\ cm^2/m^2)/(174.8\ cm^2/\text{chamber})] \times (48\ \text{one-half h d}^{-1})$$

$$= (\%CO_2 - 0.035) \times 13 = g\ CO_2\text{--}C\ m^{-2}\ d^{-1}.$$

Respiration in units of kg CO_2–C ha d^{-1} can be obtained by multiplying g CO_2–C m^{-2} d^{-1} by 10.

Methodological Variations

In many situations it may be desirable to compare CO_2 flux values from different sites or at the same site at different times, however, differences in soil temperature and soil water content, may introduce variability into such comparisons. In order to facilitate site or time comparisons it is desirable to normalize CO_2 flux values to some defined standard conditions of soil temperature and water content. Soil temperature corrections can be performed using the general rule that biological activity increases by a factor of 2 with each 10°C increase in temperature. If normalization of CO_2 flux rates is desired we suggest use of a standard temperature of 25°C. The following formula indicates how this temperature correction is made.

$$\text{Standardized } CO_2 \text{ Flux Rate} = R \times 2^{[(25 - T)/10]} \qquad [2]$$

where R is the measured CO_2–C flux rate, and T is the measured temperature in °C at the time of sampling. This standardization formula is only recommended when the measured soil temperature is between 15 and 35°C. Between 0 and 15°C Eq. [3] is used.

$$\text{Standardized } CO_2 \text{ Flux Rate} = R \times 4^{[(25 - T)/10]} \qquad [3]$$

Soil water content is another factor that influences CO_2 flux rate. Standardization of field soil respiration determinations is based on laboratory obser-

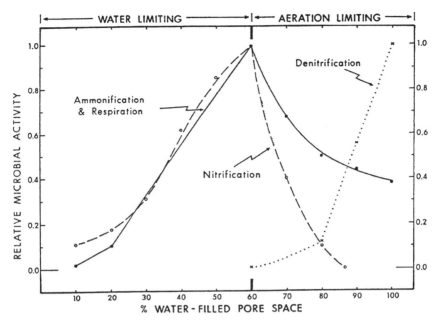

Fig. 14–2. Relationship between aerobic and anaerobic microbial processes and the percentage of water-filled pore space (after Linn & Doran, 1984).

vations for a wide range of soils that indicate maximum aerobic microbial respiration occurs when 60% of the soil pores are filled with water (Doran et al., 1990; Linn & Doran, 1984). As illustrated in Fig. 14–2, aerobic microbial activity increases linearly with water-filled pore space (WFPS) up to about 60% WFPS, and then decreases at higher water contents, apparently due to limited aeration. The base level of 0.4 relative activity for respiration (CO_2 production) and ammonification at saturated conditions is apparently due to a reduced rate of ammonification and CO_2 production under anaerobic conditions. Respiration rates can be adjusted to equivalent values at 60% WFPS through use of the following equations, where R_{60} is the adjusted respiration rate normalized to 60% WFPS.

For WFPS between 30 and 60%;

$$R_{60} = \text{measured respiration rate} \times (60/\text{measured \%WFPS}) \quad [4]$$

For WFPS between 60 and 80% WFPS;

$$R_{60} = \text{measured respiration rate}/[(80 - \%\text{WFPS}) \times 0.03] + 0.4 \quad [5]$$

These equations are empirical derivations from the respiration response observed in Fig. 14–2. Respiration rates measured in the field when soil water status exceeds 80% WFPS are not dependable since diffusion of CO_2 into the chamber may be restricted by wet conditions. We caution that although microbial respiration shows a strong relationship to %WFPS in the laboratory, the relationship between %WFPS and soil respiration has not been extensively evaluated in the field.

Other methodological variations are possible, especially related to chamber size and CO_2 detection techniques. Larger chambers should, theoretically, reduce the spatial variability associated with field CO_2 flux measurements. Also, portable infrared gas analyzers enable real time CO_2 determinations that allow for short term (i.e., 2 min or less) flux determinations that should reduce biases associated with changes in soil conditions resulting from chamber placement.

Method 2: Microbial Respiration

In many cases field respiration measurements do not directly indicate microbial respiration, due to the presence of plant roots, and other soil organisms; however, microbial respiration can be measured if these other sources of CO_2 are removed through the sieving and mixing process. It is proposed that microbial respiration be assessed in laboratory incubations on sieved soil. Laboratory measurement should be conducted on sieved soil packed in a beaker at bulk density of 1.0 g cm^{-3}, (or to the natural reconsolidation density of the soil being tested), incubated at 60% WFPS, and at a temperature of 25°C. Microbial respiration is the cumulative CO_2–C produced during a 0- to 10- or a 0- to 20-d incubation period. Details on implementation of this method are presented in the description of the microbial biomass measurement (Rice et al., 1996, this publication).

INTERPRETING SOIL RESPIRATION MEASUREMENTS

The most difficult task in the development of a soil quality index is ascribing an interpretation to those soil attributes identified as important or valuable soil quality indicators. Interpretation of soil indicator measurements relative to soil quality is entirely dependent upon the precise definition or perception of soil quality. In the development of a definition of soil quality, a critical question that must be answered is "Why does the soil have value?". Only after the value of soil has been defined, can soil attributes that contribute to this ascribed value be identified.

Two general approaches have been proposed for interpretation of soil quality indicators. One of these approaches promotes the use of the soil characteristics of natural or undisturbed soils as the benchmark by which soil quality can be judged. With this approach it is assumed that the natural or undisturbed system is *best*. A second approach has been to interpret soil quality indicators in relation to soil function. Doran and Parkin (1994) proposed a general framework for assessment of soil function based on measured soil attributes. In this chapter, we adopt the second approach, and attempt an interpretation of soil respiration based on soil function.

In general terms, soil respiration represents the activity of the soil biotic component (Fig. 14–3), including microbial activity (bacteria, actinomycetes, fungi, algae, and protozoa), invertebrate activity (e.g., nematodes, gastropods, earthworms, and insects), and plant root activity. This biological activity is a

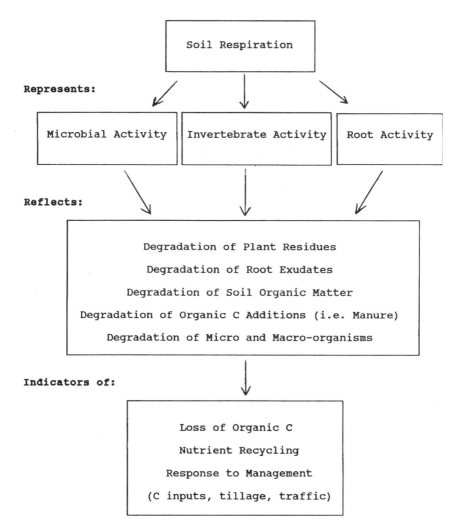

Fig. 14–3. Diagrammatic interpretation of respiration as a soil quality indicator.

direct reflection of the degradation of organic C compounds in soil. These organic C compounds may reside in a variety of different forms or pools including plant residues and root excretions, soil organic matter, organic C amendments such as manure, and the residues of dead micro- and macro-organisms. Interpretation of soil respiration measurements relative to soil quality must identify how soil respiration relates to soil function. Figure 14–3 indicates that organic C decomposition reflected by soil respiration is indicative of two important processes in soil: (i) loss of soil C and (ii) the turnover (release and stabilization) of nutrients. In addition, soil respiration may represent a sensitive indicator of the response of the soil biotic component to management such as plant residue or animal manure addition, tillage, and traffic.

Table 14–2. Beneficial effects of microorganism on soil.

Recycle nutrients (release nutrients to plants)
Promote soil structure
Degrade toxic compounds
Build stable soil organic matter
Degrade crop residues
Degrade animal material
Control the activities–populations
  of harmful microorganisms
    (through competition–inhibition–predation)
Fix atmospheric N
Consume greenhouse gasses
Provide a pool of diverse genetic material
Provide a pool of readily available nutrients
Promote plant nutrient and water uptake

Microorganisms are key components of soil, and it has been proposed that soil respiration may be a potentially valuable indicator of microbial activity in soil. Because microorganisms perform many beneficial functions in soil (Table 14–2), it is generally regarded that high microbial activity is a positive indicator of soil quality; however, this perspective may be too simplistic. In addition to the positive functions of microorganisms in soil, microorganisms have detrimental impacts (Table 14–3). The situation is complicated by the fact that assessment of whether or not a given activity is positive or negative, depends upon when and where in the soil profile the activity occurs. For example, high pesticide degradation activity in soil may represent a positive function of soil in terms of serving as an environmental buffer; however, if high pesticide degradation activity is expressed too soon after the pesticide is applied, loss of pesticide efficacy may result. Indeed, soils in which this phenomenon has been observed to occur have often been referred to as problem soils. Thus, for one to accept the notion that the higher the soil respiration, the higher the soil quality, one must assume that soil respiration measurements only indicate the positive attributes microorganisms impart to soil and not the negative ones.

A method or framework for interpreting soil respiration measurements relative to specific functions carried out by soil microorganisms is needed. Unfortunately, due to the physical and chemical complexity of the soil environment, the diversity of soil microbial populations and other soil fauna, the variety of microbial processes that occur in soil, and the complex nature of soil organic C, development of a universal framework that relates a general measure of soil biotic

Table 14–3. Detrimental effects of microorganisms on soil.

Release nutrients at the wrong time
Degrade soil organic matter
Degrade crop residues
Degrade pesticides (loss of efficacy)
Plant pathogens
Human–animal pathogens
Produce greenhouse gasses
Control the activities–populations
  of beneficial microorganisms (through
    competition–inhibition–predation)

activity, such as soil respiration, to all the functions carried out by soil microorganisms appears doubtful.

Whereas identification of specific microbial function based on soil respiration may be an unrealistic expectation, soil respiration measurements do provide a direct assessment of a process of critical importance to soil function: loss of organic C. Organic C has a positive impact on soil in many ways. Organic C has been shown to be positively related with soil structure, water penetration, water retention, root development, and nutrient storage (Brady, 1984). Since a positive relationship between soil organic matter and a variety of soil functions (enhanced productivity, enhanced structure, and enhanced water entry and retention) can be conceptualized, and since soil respiration represents a loss of organic C, the overall relationship between soil respiration and soil quality may be a negative one.

A strict negative relationship between soil respiration and soil quality is too simplistic. In terms of crop production, the positive relationship between crop productivity and soil organic matter is due in part to the fact that plant nutrients are released as a result of organic matter degradation. Also, in terms of the structural benefits of organic matter (OM) only certain types of OM are beneficial (Roberson et al., 1995; Arshad & Schnitzer, 1987). This implies that microbial processing of organic residues in soil must occur before beneficial effects on structure are attained. Thus, soil respiration may indicate two opposing aspects of the relationship between soil organic matter and soil quality. In the long term, loss of soil organic matter can be viewed as a negative result of soil respiration, however, in the short term, respiration represents the release of plant available nutrients. From this perspective assignment of either a *more is better* relationship or a *more is worse* relationship between soil respiration and the plant productivity component of soil quality is not appropriate. Clearly, some optimum soil respiration must be defined that balances the long-term detrimental aspects of soil C loss through respiration, and the soil nutrient turnover that respiration represents.

To develop a soil quality interpretation of soil respiration relative to loss of organic C, and release of nutrients, each of these soil quality components must be assessed. Ideally, it would be advantageous to use a single measurement for assessment of both of these processes, however, currently there is no quantitative relationship between field CO_2 fluxes and nutrient release, and conversely laboratory measurements of CO_2 production may not adequately represent total C loss. Thus, at this stage it is advised that these two components, soil C loss, and soil nutrient release, be determined separately, by different methodology. Specifically, we propose: (i) the use of soil respiration measurements as determined by CO_2 flux in the field as an indicator of the soil organic matter storage potential of the soil, and (ii) the use of laboratory measurements of microbial respiration as an indicator of nutrient release, specifically, the N supplying potential of soil.

Respiration as a Predictor of Soil Organic Matter Storage

As previously mentioned, from a C storage perspective, soil respiration can be viewed as a negative attribute. Simply looking at absolute respiration rates may not be adequate to characterize a given soil system. High respiration rates exhibited by soils receiving high organic C inputs may provide a biased picture

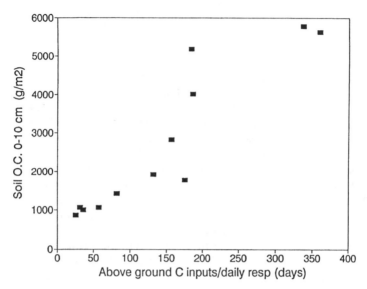

Fig. 14–4. Relationship between the ratio of aboveground plant inputs/soil respiration and soil organic matter (Parkin & Colvin, 1994, unpublished data).

of organic C storage. For example, in systems receiving high C inputs such as animal manures or green manure cover crops, the magnitude of the respiration response observed may be dominated by organic C amendments to soil. High respiration responses in such systems may indicate release of plant required nutrients but it indicates little in terms of whether or not the manure is being accumulated as soil organic matter. What is needed is some way to normalize or adjust soil respiration rates to account for differences in C inputs to the system.

In the assessment of a given soil's role in C storage it is not total C loss that is important, rather it is C loss relative to C inputs. From this C balance perspective if losses of organic C exceed C inputs, soil organic matter is being depleted. Thus, representation of soil respiration C losses in relation to C inputs may provide an indicator of the rate and direction of change of the soil organic matter pool. In practice it is difficult to quantify C inputs from plant production, primarily because assessment of below ground production is a tedious operation. An approximation may be obtained from above ground C inputs. As a rough estimate, the mass of C in the roots can be assumed to be approximately 62% of the aboveground plant residue C for wheat and 58% of the aboveground plant residue C for soybeans and corn (Buyanovsky & Wagner, 1986). Ideally, the amount of aboveground plant residue C should be measured directly, however, if C analyses are not available it can be assumed that the plant residue C is approximately 37% for wheat (*Triticum aestivum* L.), and 41% for corn (*Zea mays* L.), and soybeans [*Glycine max* (L.) Merr.; Buyanovsky & Wagner, 1986].

In a recent assessment of soil respiration in an agricultural field, a strong relationship was observed between the ratio of above ground C inputs to respiration and soil organic C content (Fig. 14–4).This relationship implies that the longer it takes for the aboveground C inputs to decompose, the greater the chances that more organic C will be retained and hence the greater the probability of increased soil organic C. A similar relationship between soil respiration, crop production, and soil organic matter can be observed in data from studies conducted in Minnesota (Reicosky & Lindstrom, 1993; Reicosky et al., 1995). These investigators observed a burst of CO_2 from soils immediately following tillage, and a positive relationship was observed between the crop residue/CO_2 flux ratio and decreasing tillage intensity suggesting the soil organic C content of the plots with less tillage should increase soil organic matter.

Currently, precise values for a soil quality rating based on soil respiration cannot be set without additional data, however, we recommend that for comparative ranking purposes, the number of days to degrade the aboveground inputs may be a useful comparative tool. This information, if coupled with climate data (temperature and rainfall) may provide an indicator of changing organic C content of soils. This concept has been recently advanced in a soil quality rating system developed by the Natural Resources Conservation Service (NRCS, formerly the SCS) where regional residue inputs required to maintain soil organic matter at current levels are estimated (Argabright et al., 1991).

Microbial Respiration Relation to Nutrient Turnover

Microorganisms play a key function in nature through the release of nutrients from decomposition of dead plant and animal material. Control of the release of nutrients, and especially N is an important aspect of soil quality, from plant productivity and water quality perspectives. There have been several laboratory studies that show strong correlations between microbial respiration and net N mineralization. In 4-wk incubations of soil amended with a variety of animal manures, nearly a 1:1 response between CO_2 production and the amount of N released was observed by Castellans and Pratt, (1981). Similarly, Gilmour et al. (1985) found linear relationships between CO_2 production and N released in soil amended with organic materials. Slopes of the regression equations were nearly 1 for N-rich materials such as sewage sludge, alfalfa (*Medicago sativa* L.) and clover (*Trifolium* sp.), but for bermudagrass [*Cynodon dactylon* (L.) Pers] and rye (*Secale cereale* L.) amendments regression slopes were 0.43 and 0.53, respectively. These data were used to develop a model employing the C and N content of the organic substrate, and the respiration rate to predict N mineralization from added substrate. A strong relationship between CO_2 production and N-mineralization also exists for nonamended soils. Data from Smith et al.(1986) was used to compute ratios of cumulative CO_2–C produced net N mineralized during a 34-d incubation of 20.6 and 17.3 for a Palouse silt loam and a Walla Walla silt loam, respectively. Similar values have been observed for soils of central and south western Iowa (Table 14–4).

In terms of a soil quality index, there is little advantage gained in using laboratory derived microbial respiration data to predict N mineralization, when in fact net N mineralization can be measured directly in such incubations. The real value of these studies is the indication that a possible relationship between soil respiration and N mineralization in the field may be possible.

Table 14-4. Ratios of cumulative CO_2-C respired to net N mineralized For several Iowa soils (0 to 15 cm). Values were obtained from 30-d laboratory incubations of sieved soil packed to a bulk density of 1.0, at 60% water-filled pore space (WFPS), and an incubation temperature of 22°C.

Soil	Management	Cumulative CO_2/N-mineralized
Clarion loam	Restored Prairie	15.5
Clarion loam	Cultivate corn/soybean	17.3
Clearfield slc	CRP converted to cultivated corn/soybean†	16.9
Clearfield scl	CRP	13.8
Canisteo scl	Cultivated corn/soybean	16.5

† CRP, Conservation Reserve Program.

SUMMARY

In the process of developing of soil quality indicators it is critical to set targets or establish criteria that allow for interpretation of indicator measurements. Because respiration is an indicator of organic matter decomposition in soil, it reflects two general processes: (i) loss of C from the soil system, and (ii) recycling of nutrients. Either of these processes can be viewed as detrimental or beneficial depending upon the intended use of the soil, the magnitude of the respiration activity, and the temporal and spatial distributions exhibited by these processes. In this chapter we briefly outlined two proposed interpretations for field and laboratory respiration measurements. Exact targets or values that allow precise interpretation of soil respiration need to be established. This may have to be done on a site by site basis, to specifically account for the intended use of the soil, soil management, and climatic factors.

REFERENCES

Anderson, J.M., and J.S.I. Ingram. 1993. Tropical soil biology and fertility: A handbook of methods. p. 41–43. *In* Soil CO_2 evolution. CAB Int., Wallingford, Oxon, England.

Anderson, J.P.E. 1982. Soil respiration. p. 831–871. *In* A.L. Page et al. (ed.) Methods of soil analysis. Part 2. 2nd ed. Agron. Monogr. 9. ASA and SSSA, Madison, WI.

Anthony, W.H., G.L. Hutchinson, and G.P. Livingston. 1995. Chamber measurement of soil-atmosphere gas exchange: Linear vs. diffusion-based flux models. Soil Sci. Soc. Am. J. 59:1308–1310.

Argabright, S., D. Breitbach, D. Shoup, D. Lightle, L. Oyer, C.L. Girdner, and L. Samson. 1991. Soil quality ratings for cropland management systems in the Midwest states. Midwest Soil Condition Rating Committee Tech. Rep. USDA-NRCS, Lincoln, NE.

Arshad, M.A., and M. Schnitzer. 1987. Characteristics of the organic matter in a slightly and in a severely crusted soil. Z. Pflanzenernahr. Bodenk. 50:412–416.

Brady, N.C. 1984. The nature and properties of soils. 9th ed. Macmillan Publ. Co., New York.

Buyanovsky, G.A., and G.H. Wagner. 1986. Post-harvest residue input to cropland. Plant Soil 93:57–65.

Castellans, J.Z., and P.F. Pratt. 1981. Nitrogen mineralization of manure nitrogen-correlation with laboratory indexes. Soil Sci. Soc. Am. J. 45:354–357.

Doran, J.W, L.N. Mielke, and J.F. Power. 1990. Microbial activity as regulated by soil water-filled pore space. p. 94–100. *In* Trans. of the 14th Int. Congress of Soil Sci., Kyoto, Japan. 12–18 Aug. 1990. ISSS, Wagenigen, the Netherlands.

Doran, J.W, and T.B. Parkin. 1994. Defining and assessing soil quality. p. 1–22. *In* J.W. Doran et al. (ed.) Defining soil quality for a sustainable environment. SSSA Spec. Publ. 35. SSSA, Madison, WI.

Gilmour, J.T., M.D. Clark, and G.C. Sigua. 1985. Estimating net nitrogen mineralization from carbon dioxide evolution. Soil Sci. Soc. Am. J. 49:1398–1402.

Howard, A. 1947. The soil and health. Devin-Adair, New York.

Kennedy, A.C., and R.I. Papendick. 1995. Microbial characteristics of soil quality. J. Soil Water Conserv. 50:243–248.

Linn, D.M., and J.W. Doran. 1984. Effect of water-filled pore space on carbon dioxide and nitrous oxide production in tilled and nontilled soils. Soil Sci. Soc. Am. J. 48:1267–1272.

Nay, S.M, K.G. Mattson, and B.T. Bormann. 1994. Biases of chamber methods for measuring soil CO_2 efflux demonstrated with a laboratory apparatus. Ecology 78:2460–2463.

Reicosky, D.C., W.D. Kemper, G.W. Langdale, C.L. Douglas, Jr., and P.E. Rasmussen. 1995. Soil organic matter changes resulting from tillage and biomass production. J. Soil Water Conserv. 50:253–261.

Reicosky, D.C., and M.J. Lindstrom. 1993. Fall tillage method: Effect on short-term carbon dioxide flux from soil. Agron. J. 85:1237–1243.

Rice, C.W., T. Moorman, and M. Beare. 1996. Role of microbial biomass carbon and nitrogen in soil quality. p. 203–215. *In* J.W. Doran and A.J. Jones (ed.) Methods for assessing soil quality. SSSA Spec. Publ. 49. SSSA, Madison, WI.

Roberson, E.B., S. Sarig, C. Shennan, and M.K. Firestone. 1995. Nutritional management of microbial polysaccharide production and aggregation in an agricultural soil. Soil Sci. Soc. Am. J. 59:1587–1594.

Sarrantonio, M., J.W. Doran, M.A. Liebig, and J.J. Halvorson. 1996. On-farm assessment of soil quality and health. p. 83–105. *In* J.W. Doran and A.J. Jones (ed.) Methods for assessing soil quality. SSSA Spec. Publ. 49. SSSA, Madison, WI.

Smith, J.L., B.L. McNeal, H.H. Cheng, and G.S. Campbell. 1986. Calculation of microbial maintenance rates and net nitrogen mineralization in soil at steady-state. Soil Sci. Soc. Am. J. 50:332–338.

Turco, R.F., A.C. Kennedy, and M.D. Jawson. 1994. Microbial indicators of soil quality. p. 73–90. *In* J.W. Doran et al. (ed.) Defining soil quality for a sustainable environment. SSSA Spec. Publ. 35. SSSA, Madison, WI.

Waksman, S.A. 1927. Microbiological analysis of soil as an aid to soil characterization and classification. J. Am. Soc. Agron. 19:297–311.

Zibilske, L.M. 1994. Carbon mineralization. p. 835–864. *In* R.W. Weaver et al. (ed.) Methods of soil analysis. Part 2. SSSA Book Ser. 5. SSSA, Madison, WI.

15 Soil Enzyme Activities and Biodiversity Measurements as Integrative Microbiological Indicators

Richard P. Dick

Oregon State University
Corvallis, Oregon

Donald P. Breakwell

Snow College
Ephraim, Utah

Ronald F. Turco

Purdue University
West Lafayette, Indiana

The quality of soil can impact land use capability, agronomic sustainability, environmental buffering, and farm productivity. Human and animal health is closely linked to soil productivity and environmental quality. The biological component is central to innumerable processes and functions that are carried out in soils such as decomposition of organic residues, nutrient cycling, synthesis of soil humic substances, aggregation, degradation of xenobiotics, N fixation, and many other processes. Soil biology is an assemblage of diverse, interrelated groups of organisms representing different trophic levels. Although the complexity and importance of biological diversity in soils is not well understood, it is reasonable that stable and functionally diverse microbial communities in soils are an important consideration in assessing soil quality. Furthermore, soil microorganisms can be sensitive biological markers and be used to assess disturbed or contaminated soils.

In order to assess biological properties of soil, it is important to first determine whether the goal is to measure activity, size (mass), diversity, or biochemically mediated processes of soil biology. Although there are relatively straight forward methods for measuring microbial activity and size (see Allen & Killorn, 1996, this publication; Parkin et al., 1996, this publication), these provide no information about composition, species diversity, or function of the microbial population. In this chapter we offer two approaches for assessing soil biological

---

Copyright © 1996 Soil Science Society of America, 677 S. Segoe Rd., Madison, WI 53711, USA. *Methods for Assessing Soil Quality*, SSSA Special Publication 49.

properties: process level biological indexes and functional community diversity. Soil enzyme assays are process level indicators and are presented as a means of determining the potential of a soil to degrade or transform substrates (see review by Dick, 1994). This can be useful to indicate how well a soil can carry out important steps in nutrient cycling, nitrification, oxidation, and other processes. Furthermore, they can be an integrative index of past soil biological activity as influenced by soil management and many enzyme assays are operationally simple to run; however, enzyme assays provide little direct information on community diversity.

The traditional method to evaluate microbial form and function or diversity is dilution plating from soils, but this approach suffers in that it is impossible to recover a cross section of the entire soil microbial population on one selective microbial growth medium. Therefore, the use of plate counts tells a lot about growth of the microorganism on that particular medium but little about the community structure. New approaches to measure soil microbial diversity are being developed but considerably more research is needed before these can have practical applications in soil quality assessment. Consequently, we will provide background information on the emerging biodiversity methods of phospholipid ester-linked fatty acid pattern (PLFA) and DNA characterization; and only present the specific methodology for the Biolog C use method because it is a relatively straight forward and standardized method. This latter approach characterizes the functional microbial diversity by the ability of a soil to hydrolyze a diverse set of C substrates.

SAMPLING AND SAMPLE HANDLING

Sampling for soil biological properties requires some special considerations because of the dynamic nature of the biological component of the soil. A major factor is the time of year to sample. Ideally it would be best to sample several times during the year because biological activity naturally varies on an annual basis and from year to year because of climatic influences. If sampling can only be done at a single time, it is best to avoid sampling just after a tillage or residue incorporation operation. For example, sampling just after residue incorporation may show elevated levels of enzyme activity, but this may only be a transitory effect that does not reflect the long-term trajectory of the soil system. Therefore, it is best to sample at a time of year when there has been a period of constant climatic conditions (soil moisture and temperature) and no recent soil disturbance. For agricultural systems, this might be at harvest time prior to any post-harvest tillage operations. For sampling over a period of several years, it is important to sample at the same time each year to reduce effect of in-season variability.

Spatial variability also must be considered and it is important to have a representative sample. Biological properties vary naturally as a function of both within and among soil types. Therefore it is important to sample within the same soil type and obtain a representative sample by standard sampling protocols (see Dick et al., 1996, this publication). After obtaining a representative sample, it is best to sieve the soil through a 2-mm sieve and then either air dry or store field moist samples. In both cases, unless the analysis can be run immediately, which is preferable on field moist samples if the goal is to obtain results similar to field conditions, it is best to store samples at 4°C until the assay can be done.

Whether the soil can be air dried or if field moist samples should be taken depends on the measurement of interest. If the interest is to obtain results that are similar to field conditions, then it is best to keep samples field moist and immediately cool to 4°C (see Dick et al., 1996, this publication) and for some biological measurements it is essential that a soil not be air dried. For each procedure in this chapter, the proper pretreatment is described. The effect of air drying on enzyme activity varies with enzyme type and can cause an increase in activity, but for most enzymes, the activity is reduced 40 to 60%. Screening of a wide array of enzymes has shown that although activity may decrease with air drying, the relative change in activity among samples from different soil management systems within the same soil type remains the same (Dick, 1996, unpublished data). Activity for some enzymes is much more stable in air-dried soils than field moist samples and remains unchanged (or small decreases) for weeks or months after sampling. Air drying greatly facilitates sample handling and allows for timely analysis. Combining this with the relative simplicity of many enzyme assays makes it possible to run a large number of samples on a routine basis.

ENZYME ASSAYS

The conceptual rationale for soil enzyme activity as a soil quality indicator is that enzyme activities: (i) are often closely related to important soil quality parameters such as organic matter, soil physical properties and microbial activity or biomass (Dick, 1994); (ii) can begin to change much sooner (1 to 2 yr) than other properties (e.g., soil organic C) thus providing an early indication of the trajectory of soil quality with changes in soil management; (iii) can be an integrative soil biological index of past soil management; and (iv) involve procedures that are relatively simple compared to other important soil quality properties (e.g., physical and some biological measurements) and therefore have potential to be done routinely by soil testing or analytical environmental laboratories.

In biological systems, nearly all reactions are catalyzed by enzymes that are proteins. Enzymes have enormous catalytic power and greatly accelerate reactions by forming a complex with a substrate that decreases the energy required for a reaction to proceed. Because enzymes are difficult to extract from soils and usually loose their integrity, enzymes in soils are characterized by measuring their activity under a strict set of conditions (e.g., temperature, pH buffer, and substrate concentration). Therefore, an enzyme assay measures potential activity and not in situ activity.

Soil enzymes are thought to be largely of microbial origin and obviously are associated with viable cells; however, many enzymes can remain catalytic in cell debris, in soil solution or complexed with clay or organic colloids. Although the ecological role of cell debris or extracellular complexed enzymes has yet to be determined, Burns (1982) hypothesized that humic-enzyme complexes may benefit some organisms by hydrolyzing substrates that are too large or insoluble for microbial uptake. This latter aspect of soil enzymes provides a conceptual model for the role of enzyme assays in measuring soil quality. Soils that have been managed to promote soil quality (e.g., minimum tillage, organic amendments, crop rotations, and others) should have higher biological activity that would be reflected in greater enzyme production and probably greater potential to stabilize and protect enzymes complexed in the soil matrix (i.e., through greater production of organic colloids and aggregation there is increased complexation and protection of enzymes).

This has been reflected in many studies where enzyme activities have been sensitive to soil management comparisons (see review by Dick, 1994). A few studies have shown that enzyme activities can be temporally sensitive within the first 2 yr of changes in soil management, long before there are measurable differences in soil organic matter (Dick, 1994); but not all enzymes are capable of distinguishing soil management effects and activities vary as function of soil type. We present methods that have been shown to be sensitive discriminators of soil management effects; however, even these at times have shown inconsistent results, a likely reflection of the complexity soil biological systems. Therefore, it is important to identify the specific goal for assessing soil quality and it is best to make several chemical, physical and other biological tests in combination with enzyme assays. For more extensive and in-depth information regarding background and methodology of soil enzymes, the reader is referred to Burns (1978), Dick (1994), and Tabatabai (1994).

SHORT-TERM SOIL BIOLOGICAL ACTIVITY VERSUS SOIL QUALITY TRAJECTORY

Choosing an Enzyme Assay

A few enzymes have potential as indicators of viable soil microbial biomass or activity. Most notable is dehydrogenase, which has a long history as a biological indicator and often is closely related to average activity of microbial populations (Skujins, 1973) because it only exists as an integral part of viable microorganisms. It has been widely used and often shown a relationship to biological properties, so we have included the procedure in this chapter.

Enzymes that correlate closely with microbial activity, however, may be less suited to predict long-term changes or trajectory in soil quality because they would reflect recent management or seasonal (climatic) effects that may be transitory. Therefore enzymes that remain catalytic after air drying and closely correlate with organic matter content may be better indicators of permanent changes in soil quality because such enzymes are probably complexed and protected in the soil humic- or clay-complexes. The interpretation of this characteristic from a soil quality perspective is that soil management that promotes stabilization of enzymes also promotes stabilization of organic matter and associated structural properties (e.g., aggregation and porosity). All the enzymes we have chosen correlate strongly with organic matter and clay content, except for dehydrogenase and FDA hydrolysis (for FDA hydrolysis no such study has yet been conducted).

Although oxidative enzymes would appear to be attractive for soil quality measurement because of their potential role in formation of organic matter, the assays generally only provide a qualitative measurement. This is because these assays cannot always be run under pH buffered conditions or there may be inorganic Mn and Fe compounds that also can catalyze the decomposition of the substrate, thus confounding the interpretation of results. The type of assay to run depends on the goal. If the goal is to indicate the viable biological component of soil, then assays such as dehydrogenase are appropriate. The other enzymes that have been included in this chapter can be assayed on air-dried soils and show less seasonal variability and should provide a better indication of the long-term trajectory of soil quality as affected by soil management.

In general hydrolytic enzymes are good choices as soil quality indices because organic residue-decomposing organisms are probably the major contributors to soil enzyme activity (Speir, 1977; Speir & Ross, 1976). A number of assays have been developed that perform key reactions in the C, N, P, and S cycles. The most widely studied are those associated with enzymes that produce nutrient mineralization end products that are important in plant nutrition (e.g., urease, phosphatases, and sulfatases); however, there is evidence that some of these can be confounded by long-term applications of fertilizers or liming. Phosphatase activity can be depressed by phosphate fertilizers and also is affected by pH, which is independent of other soil quality factors such as organic matter content (Dick, 1994). Also, there is limited evidence that N fertilizers can have a similar effect on certain enzymes involved in the N cycle (e.g., urease and amidase; Dick et al., 1988b; McCarty et al., 1992). Therefore from a soil quality perspective, enzymes involved in the C cycle are probably better choices and we have provided the assay for ß-glucosidase, which is an important enzyme in releasing C for energy (Tabatabai, 1994). A relatively new assay that we have included is the hydrolysis of fluorescein diacetate (3′, 6′-diacetylfluorescein; FDA) (Schnürer & Rosswall, 1982). It has potential to broadly represent soil enzyme activity and accumulated biological effects because FDA is hydrolyzed by a number of different enzymes, such as proteases, lipases, and esterases (Guilbaut & Kramer, 1964; Rotman & Papermaster, 1966) and its hydrolysis has been found among a wide array of the primary decomposers, bacteria and fungi (Lundgren, 1981; Medzon & Brady, 1969; Söderström, 1977).

Interpretation

Soil enzyme activity is operationally defined. If the conditions of the assay (e.g., temperature, buffer pH, buffer type, or ionic strength) are altered, results also will change. Therefore, to make meaningful comparisons among studies or over different time periods, the exact same protocol should be followed for each enzyme assay. The procedures outlined in this chapter are standard procedures that are established in the literature. An inherent characteristic of most soil properties is that they vary as a function of soil type and for the soil enzymes studied so far, there is no exception to this rule. In general, enzyme activity increases with increasing organic matter and clay content and thus sandy soils tend to be lower in enzyme activity.

With each assay method, a range and mean of activity values are given from reports in the literature. These results were obtained from studies where the same protocols were used as those outlined in this chapter. It should be noted that this is a very limited and incomplete data set drawn from diverse environments. Systematic studies across soil types, environments, and soil management systems are needed to fully determine the potential of soil enzyme activity to characterize soil quality and develop calibration data to interpret enzyme activities. Therefore, enzyme activities should be interpreted with caution and be measured along with other soil properties to assess soil quality.

Nearly all of the literature on soil enzyme activity is reported on a soil mass basis, however it may be of interest to convert the results to a volumetric or a unit area-soil depth basis to provide an ecological perspective. Since previous studies do not report bulk densities it is not possible to present the results from the literature on a unit area-soil depth basis; however, we have included the following calculations to make such a conversion:

1. For those assays that measure p-nitrophenol (PNP) as the product, the results can be converted to a nutrient equivalent basis by the following calculation:

$$\left(\frac{\mu g \text{ PNP}}{g \text{ soil} \times \text{unit time}}\right) \times \left(\frac{\mu g \text{ nutrient}/\mu mol}{139.11 \ \mu g \text{ PNP}/\mu mol}\right)$$

$$= \mu g \text{ nutrient g}^{-1} \text{ soil unit}^{-1} \text{ time}$$

Example: 80 mg PNP kg^{-1} soil h^{-1} for arylsulfatase activity

$$\left(\frac{80 \ \mu g \text{ PNP}}{g \text{ soil} \times h}\right) \times \left(\frac{32.06 \ \mu g \text{ S}/\mu mol}{139.11 \ \mu g \text{ PNP}/\mu mol}\right) = 18 \ \mu g \text{ SO}_4\text{-S g}^{-1} \text{ soil h}^{-1}$$

2. To convert results to a volume basis multiply activity units of product mass/mass soil/unit time by the bulk density (g cm^{-3}).

Example: 18 μg SO$_4$-S g^{-1} soil h^{-1} arylsulfatase activity of a soil with a bulk density of 1.2 g cm^{-3}.

18 μg SO$_4$-S g^{-1} soil h^{-1} × 1.2 g cm^{-3} = 21.6 μg SO$_4$-S cm^{-3} h^{-1}

3. The following calculation can be used to convert enzyme activities to a per area basis (ha):

(μg product g^{-1} soil unit^{-1} time) × 10^{-9} kg μg^{-1} (g cm^{-3}) × (10^8 cm^2 × cm soil depth ha^{-1}) = kg product unit^{-1} time ha^{-1}

Example: 18 μg SO$_4$-S g^{-1} soil h^{-1} arylsulfatase activity with a bulk density of 1.2 g cm^{-3} to a depth of 20 cm.

$(18 \ \mu g \ SO_4-S \ g^{-1} \ soil \ h^{-1}) \times 10^{-9} \ kg \ \mu g^{-1} \ (1.2 \ g \ cm^{-3}) \times (10^8 \ cm^2 \times 20 \ cm$ soil depth $ha^{-1}) = 43.2 \ kg \ SO_4-S \ ha^{-1} \ h^{-1}$ to a depth of 20 cm.

In the case of dehydrogenase activity or for other oxidative enzyme assays, the activity rates can be converted to a C oxidation basis by knowing the amount of electrons transferred for each molecule of product produced. For example in the case of dehydrogenase two electrons are transferred for each molecule of triphenyl formazan (TPF) produced. Thus converting the mass of TPF produced to a molar basis and converting this to the number of electrons transferred divided by the four electrons transferred per atom of C oxidized and putting on an area-depth basis provides an indication of the potential of a soil to oxidize organic matter inputs. For example, the average TPF reported below for dehydrogenase is 337 mg TPF g^{-1} soil 24 h^{-1}, which for a soil with a bulk density of 1.2 would be the equivalent of 1.61 g C oxidized m^{-2} 24 h^{-1} at a depth of 20 cm, which over a 6-mo period would oxidize 293 g C m^{-2}. This is in the same order of magnitude of the amount of litter and root C provided to soil in a grassland (420 g C m^{-2} yr^{-1}; Schlesinger, 1977). Converting enzyme activities to an area basis often results in rates that far exceed in situ rates. This is partially related to the fact that assays are run under optimal conditions of pH and temperature and most importantly at substrate concentrations that completely saturate the enzyme. This of course does not happen under natural conditions.

ENZYME ASSAY PROCEDURES

These procedures are adapted from the methods as described in the original paper for each method and from methods outlined by Tabatabai (1994; except FDA method).

Dehydrogenase

Measurement of the activity by dehydrogenase systems is attractive because these systems are an integral part of microorganisms and their apparent role in oxidation of organic matter (Casida et al., 1964); however, its activity does not consistently correlate well with other biological properties such as O_2 uptake, CO_2 evolution or microbial biomass (Skujins, 1978; Frankenberger & Dick, 1983; Ross, 1973; Howard, 1972). Dehydrogenase activity may be confounded by extracellular phenol oxidases, which are known to exist in soils (Howard, 1972) or by inorganic compounds in soils (Bremner & Tabatabai, 1973) that can catalyze the dehydrogenase reaction. Also, Chander and Brookes (1991) found there was an abiological reaction between triphenylformazan (the end product measured in dehydrogenase activity) and Cu, which could result in underestimating of dehydrogenase activity in soils that have been contaminated with various soil pollutants. Nonetheless other studies have found dehydrogenase to be correlated with microbial biomass and other measures of biological activity (Stevenson, 1959; Skujins, 1973; Ladd & Paul, 1973). Although dehydrogenase has been shown to be sensitive to soil management effects (Martens et al., 1992; Dick et al., 1988a; Doran, 1980), it is less suited to project permanent changes in soil quality because it cannot accumulate in a complexed form in soils and is best used as an indication of the viable microbial population.

Sample Pretreatment

Pass soil through 2-mm sieve. Keep field moist and **do not air dry**. If samples cannot be run immediately after sampling, cool at 4°C.

Equipment

1. Test tubes, 16 by 150 mm.
2. Spectrophotometer or colorimeter.

Reagents

1. Methanol, analytical reagent grade.
2. Calcium carbonate ($CaCO_3$), reagent grade.
3. 2,3,5-Triphenyltetrazolium chloride (TTC), 3%:, dissolve 3 g of TTC (Calbiochem, Los Angeles) in 80 mL of water and adjust the volume to 100 mL with water.
4. Triphenyl formazan (TPF) standard solution: In a 100 mL volumetric flask dissolve 100 mg of TPF (Calbiochem, Los Angeles) in about 80 mL of methanol and adjust the volume to 100 mL with methanol. Mix thoroughly.

Procedure

1. Thoroughly mix 0.2 g of $CaCO_3$ and 20 g of air-dried soil (<2 mm), and place 6 g of this mixture in each of three test tubes.
2. To each tube add 1 mL of 3% aqueous solution of TTC and 2.5 mL of distilled water. There should be a small amount of free liquid at the surface of the soil after mixing.
3. Mix the contents of each tube with a glass rod, and stopper the tube and incubate it at 37°C. After 24 h, add 10 mL of methanol, stopper the tube, and shake it for 1 min.
4. Unstopper the tube, and filter the suspension through a glass funnel plugged with absorbent cotton, into a 100-mL volumetric flask. Wash the tube with methanol and quantitatively transfer the soil to the funnel, then add additional methanol (in 10-mL portions) to the funnel until the reddish color has disappeared from the cotton plug. Dilute the filtrate to a 100-mL volume with methanol. Measure the intensity of the reddish color by using a spectrophotometer at a wavelength of 485 nm and a 1-cm cuvette with methanol as a blank.
5. Calculate the amount of TPF produced by reference to a calibration graph prepared from TPF standards. To prepare this graph, dilute 10 mL

of TPF standard solution to 100 mL with methanol (100 µg of TPF mL^{-1}). Pipette 5-, 10-, 15-, or 20-mL aliquotes of this solution into 100-mL volumetric flasks (500, 1000, 1500, and 2000 µg of TPF 100 mL^{-1}, respectively), make up the volumes with methanol, and mix thoroughly. Measure the intensity of the red color of TPF as described for the samples. Plot the absorbance readings against the amount of TPF in the 100-mL standard solutions.

Comments

This assay tends to have a high variability, which is why a minimum of three assays per sample is recommended. This may mean more extensive field sampling is required to obtain representative results. The TTC and TPF are photosensitive and therefore should be stored in the dark but this reaction is slow enough under laboratory conditions that color changes are not detected during the assay. Toluene, which is often used in other enzyme assays, should not be used for dehydrogenase activity because it severely inhibits the activity of this enzyme in soils (Frankenberger & Johanson, 1986).

Activity Values Reported in the Literature

Field moist; 9 to 1760, mean 337 mg TPF kg^{-1} soil 24 h^{-1}, $n = 30$ (USA, Australia, and Europe).

ß-Glucosidase

Glucosidases and galactosidases are widely distributed in nature and are important in the C cycle (Eivazi & Tabatabai, 1988). These enzymes also have been detected in soils (Skujins, 1967, 1976) and ß-glucosidase (EC 3.2.1.21) has been detected in fungi (Jermyn, 1958) and plants (Veibel, 1950). ß-glucosidase is more dominant in soils than α- and ß-galactosidases (Tabatabai, 1994). The hydrolysis products of ß-glucosidases are believed to be important energy sources for microorganisms in soils (Tabatabai, 1994). Bandick et al. (1994) have found this assay to be sensitive in discriminating soil management effects in several soil types and in periods as short as 2 yr. The method is based on colorimetric determination of the *p*-nitrophenol released by ß-glucosidase when soil is incubated with buffered (pH 6.0) *p*-nitrophenyl-ß-D-glucoside. The *p*-nitrophenol released is extracted by filtration and determined colorimetrically.

Pretreatment

Sieve to pass 2-mm sieve and air dry or leave field moist and store at 4°C.

Equipment

1. Incubation flasks (50 mL Erlenmeyer flasks with no. 2 stoppers).
2. Incubator or water bath that has temperature control.
3. Spectrophotometer.

Reagents

1. Toluene, Fisher certified reagent (Fisher Scientific Co., Chicago).
2. Modified universal buffer (MUB) stock solution: Dissolve 12.1 g tris(hydroxymethyl) aminomethane (THAM), 11.6 g of maleic acid, 14.0 g of citric acid, and 6.3 g of boric acid (H_3BO_3) in 488 mL of 1 M sodium hydroxide (NaOH) and dilute the solution to 1 L with water. Store in a refrigerator.
3. Modified universal buffer, pH 6.0: Place 200 mL of MUB stock solution in a 500-mL beaker containing a magnetic stirring bar, and place the beaker on a magnetic stirrer. Titrate the solution to pH 6.0 with 0.1 M hydrochloric acid (HCl), transfer to a 1 L volumetric flask, and adjust the volume to 1 L with water.
4. p-Nitrophenyl-ß-D-glucosidase (PNG), 0.05 M: Dissolve 0.654 g of PNG (Sigma Chemical Co., St. Louis, MO) in about 40 mL of MUB pH 6.0 and dilute to 50 mL with MUB of the same pH. Store the solution in a refrigerator.
5. Calcium chloride ($CaCl_2$) 0.5 M: Dissolve 73.5 g of $CaCl_2 \cdot 2\ H_2O$ in about 700 mL of water, and dilute the volume to 1 L with water.
6. Sodium hydroxide (NaOH), 0.5 M: Dissolve 20 g of NaOH in about 700 mL of water, and dilute the volume to 1 L with water.
7. Tris(hydroxymethyl)aminomethane (THAM), 0.1 M, pH about 10: Dissolve 12.2 g of THAM in about 800 mL of water, and adjust the volume to 1 L with water.
8. THAM extractant solution, 0.1 M, pH 12: Dissolve 12.2 g of THAM in about 800 mL of water, adjust pH to 12 by titration with 0.5 M NaOH, and dilute to 1 L with water.
9. Standard p-nitrophenol solution: Dissolve 1.0 g of p-nitrophenol in about 700 mL of water, and dilute the solution to 1 L with water. Store the solution in refrigerator.

Steps

1. Place 1 g of soil (<2 mm) in a 50-mL Erlenmeyer flask, add 0.25 mL of toluene, 4 mL of MUB pH 6.0, 1 mL of PNG solution, and swirl the flask for a few seconds to mix the contents. Stopper the flask, and place it in an incubator at 37°C.
2. After 1 h, remove the stopper, add 1 mL of 0.5 M $CaCl_2$ and 4 mL of 0.1 M THAM buffer pH 12, swirl the flask for a few seconds, and filter the soil suspension through a Whatman no. 2v folded filter paper.
3. Read yellow color development intensity of the filtrate with a spectrophotometer set at 410 nm.
4. Calculate the p-nitrophenol content of the filtrate by reference to a calibration graph plotted from the results obtained with standards containing 0, 10, 20, 30, 40, and 50 µg of p-nitrophenol. To prepare this graph, dilute 1 mL of the standard p-nitrophenol solution to 100 mL in a volumetric flask and mix the solution thoroughly. Then pipette 0-, 1-, 2-, 3-,

4-, or 5-mL aliquots of this diluted standard solution into a 50 mL Erlenmeyer flask, adjust the volume to 5 mL by addition of water, and proceed as described for *p*-nitrophenol analysis of the incubated soil sample (i.e., add 1 mL of 0.5 M CaCl$_2$ and 4 mL of 0.5 M of NaOH mix and filter the resultant suspension). If the color intensity of the filtrate exceeds that of the 50 μg *p*-nitrophenol standard, an aliquot of the filtrate should be diluted with 0.1 M THAM pH~10 solution until the colorimeter reading falls within the limits of the calibration graph.
5. To perform controls, follow the procedure described for the assay of ß-glucosidase activity, but make the addition of 1 mL of PNG solution after the additions of 0.5 M CaCl$_2$ and 4 mL of 0.1 M THAM buffer pH 12 and immediately before filtration of the soil suspension.

Comment

The PNG solution is stable for several days if stored at 4°C. It is necessary to add CaCl$_2$ to prevent dispersion of clay and extraction of soil organic matter during treatment with THAM buffer pH 12 used for extraction of the *p*-nitrophenol released.

Activity Values Reported in the Literature

Field moist; 38 to 720, mean 148 mg nitrophenol kg^{-1} soil h^{-1}, $n = 8$ (USA).
Air dried; 48 to 169, mean 112 mg *p*-nitrophenol kg^{-1} soil h^{-1}, $n = 7$ (USA).

Urease

Urease (urea amidohydrolase, EC 3.5.1.5) is the enzyme that catalyzes the hydrolysis of urea to CO$_2$ and NH$_3$ (Tabatabai & Bremner, 1972). Urease is widely distributed in nature and it has been detected in microorganisms, plants, and animals. The method proposed by Tabatabai and Bremner (1972) is based on the amount of NH$_4$ released during the assay. This method involves use of toluene, and it has been shown to be applicable to soils that fix NH$_4$ and is described here. Although urease activity has the potential to be confounded by applications of fertilizers containing NH$_4$ (Dick et al., 1988b; McCarty et al., 1992), we have included it because it has been shown to be a strong discriminator between effects on soils amended with plant residues, N fertilizer or animal manures (Bandick et al., 1994) and its activity has been widely reported in the literature.

Pretreatment

As described in ß-glucosidase section.

Equipment

1. Volumetric flasks, 50 mL.
2. Incubator or temperature-controlled water bath.
3. Steam distillation apparatus as described by Keeney and Nelson (1982).

Reagents

1. Toluene, Fisher certified reagent (Fisher Scientific Co., Chicago).
2. Tris(hydroxymethyl)aminomethane (THAM) buffer (0.05 M), pH 9.0: Dissolve 6.1 g of reagent grade THAM in about 700 mL of water, titrate to pH 9.0 with ~0.2 M H_2SO_4, and dilute to 1 L with water.
3. Urea solution 0.2 M: Dissolve 1.2 g of urea (Fisher certified reagent, Fisher Scientific Co., Chicago) in about 80 mL of THAM buffer, and dilute the solution to 100 mL with THAM buffer. Store the solution in a refrigerator.
4. Potassium chloride (2.5 M)-silver sulfate (100 ppm; KCl-Ag_2SO_4) solution: Dissolve 100 mg of reagent grade Ag_2SO_4 in about 700 mL of water, dissolve 188 g of reagent grade KCl in this solution, and dilute to 1 L with water.
5. Reagents for determination of NH_4-N (MgO, H_3BO_3-indicator solution, 0.0025 M H_2SO_4): Prepare as described by Keeney and Nelson (1982).

Steps

1. Place 5 g of soil (<2 mm) in a 50-mL volumetric flask, add 0.2 mL of toluene and 9 mL of THAM buffer, swirl the flask for a few seconds to mix the contents, add 1 mL of 0.2 M urea solution, and swirl the flask again for a few seconds. Stopper the flask, place it in an incubator at 37°C.
2. After 2 h, remove the stopper, add approximately 35 mL of KCl-Ag_2SO_4 solution, swirl the flask for a few seconds, and allow the flask to stand until the contents have cooled to room temperature (about 5 min). Make the contents to 50 mL by addition of KCl-Ag_2SO_4 solution, stopper the flask, and invert it several times to mix the contents.
3. To determine NH_4-N in the resulting soil suspension, pipette a 20-mL aliquot of the suspension into a 100-mL distillation flask, and determine the NH_4-N released by steam distillation of this aliquot with 0.2 g of MgO for 4 min as described by Keeney and Nelson (1982).
4. Controls should be performed in each series of analyses to allow for NH_4-N not derived from urea through urease activity. To perform controls, follow the procedure described for assay of urease but make the addition of 1 mL of 0.2 M urea solution after the addition of 35 mL of KCl-Ag_2SO_4 solution.

Comment

The KCl-Ag_2SO_4 solution must be prepared by the addition of KCl solid to Ag_2SO_4 solution because Ag_2SO_4 will not dissolve in KCl solution. The soil suspension must be thoroughly mixed immediately before sampling for NH_4 analysis. Samples should be analyzed within 2 h after termination of incubation and addition of KCl-Ag_2SO_4 solution or stored at 4°C if longer periods are required before determination of NH_4.

Activity Values Reported in the Literature

Field moist; range of 22 to 422, mean 202 mg NH_4–N kg^{-1} soil 2 h^{-1}, $n = 22$ (New Zealand, USA).

Air dried; range of 22 to 305, mean 89 mg NH_4–N kg^{-1} soil 2 h^{-1}, $n = 7$ (USA).

Amidase

Amidase (acylamide amidohydrolase, EC 3.5.1.4) is an important enzyme in the N cycle that is involved in N mineralization releasing NH_4 from linear amides by acting on C–N bonds other than peptide bonds (Frankenberger & Tabatabai, 1980). It is widely distributed in soils, plants (Frankenberger & Tabatabai, 1982), yeast (Joshi & Handler, 1962), and fungi (Hynes, 1970, 1975).

Pretreatment

As described in ß-glucosidase section.

Equipment

As described in urease section.

Reagents

1. Toluene, Fisher certified reagent (Fisher Scientific Co., Chicago).
2. Tris(hydroxymethyl)aminomethane (THAM) buffer 0.1 M, pH 8.5: Dissolve 12.2 g of THAM (Fisher certified reagent, Fisher Scientific Co., Chicago) in about 700 mL of water, titrate to pH 9.0 with ~0.2 M H_2SO_4, and dilute to 1 L with water.
3. Formamide solution (0.50 M): Add 2.0 mL of formamide (Aldrich certified) into a 100-mL volumetric flask. Make up volume by adding THAM buffer, and mix the contents. Store the solution at 4°C.
4. Potassium chloride (2.5 M)-silver sulfate (100 ppm) solution as described in urease section.
5. Reagents for determination of NH_4–N (MgO, H_3BO_3-indicator solution, 0.0025 M H_2SO_4): Prepare as described by Keeney and Nelson (1982).

Steps

1. Follow the procedure described in urease section but use 1 mL of 0.2 M formamide solution instead of 1 mL of 0.5 M urea solution.
2. Controls should be performed in each series of analyses to allow for NH_4–N not derived from formamide hydrolysis. To perform controls, follow the urease procedure but use 1 mL of 0.5 M formamide solution instead of 0.2 M urea.

Comment

In this procedure, the KCl–Ag_2SO_4 is used to terminate the enzyme activity over the KCl–$UO_2(C_2H_3O_2)_2 \cdot 2\ H_2O$ [as originally proposed by Frankenberger & Tabatabai (1980)]. This change eliminates the use of a low radioactive compound, $KCl-UO_2(C_2H_3O_2)_2 \cdot 2\ H_2O$ that requires greater disposal precautions; however, it is important that KCl is added to Ag_2SO_4 solution because Ag_2SO_4 as a solid will not dissolve in KCl solution. Furthermore, with $KCl-Ag_2SO_4$, NH_4 should be determined immediately after the incubation assay because it does not completely inactivate this enzyme. If this is not possible, $KCl-UO_2(C_2H_3O_2)_2 \cdot 2\ H_2O$ should be used as described by Frankenberger and Tabatabai (1980).

Activity Values Reported in the Literature

Field moist; range of 22 to 422, mean 175 mg NH_4-N kg^{-1} soil 2 h^{-1}, $n = 16$ (USA).

Air-dried; range of 114 to 406, mean 273 mg NH_4-N kg^{-1} soil 2 h^{-1}, $n = 6$ (USA).

Phosphatase

Phosphatases are important in the P cycle because they provide P for plant uptake by releasing PO_4. Phosphomonoesterases are classified as acid (orthophosphoric monoester phosphohydrolase, EC 3.1.3.2) or alkaline (orthophosphoric monoester phosphohydrolase, EC 3.1.3.1) according to the optimal activity at acid and alkaline pH, respectively (Tabatabai & Bremner, 1969; Eivazi & Tabatabai, 1977). Phosphatase activity is strongly influenced by soil pH (Eivazi & Tabatabai, 1977; Juma & Tabatabai, 1977, 1978). Repeated additions of PO_4 fertilizer suppresses phosphatase activity (Dick, 1994). None-the-less, we have included phosphatase activity because it is important in the P cycle and acid phosphatase has been widely studied and it provides a potential index for a soil to mineralize organic P. The phosphatase activity procedures are based on a similar principle to ß-glucosidase in that hydrolysis of *p*-nitrophenyl phosphate releases *p*-nitrophenol, which is extracted quantitatively from soil and measured colormetrically.

Pretreatment

As described in ß-glucosidase section.

Equipment

1. 50-mL Erlenmeyer flasks fitted with no. 2 stoppers.
2. Incubator or temperature-controlled water bath.
3. Colorimeter or spectrophotometer.

Reagents

1. Toluene, MUB pH 6.5 and 11 (use stock MUB and titrate with 0.1 *M* HCl or 0.1 *M* NaOH, respectively), calcium chloride ($CaCl_2$; 0.5 *M*), and standard *p*-nitrophenol solution as described in ß-glucosidase section.
2. *p*-Nitrophenyl phosphate solution (PNP), 0.05 *M*: Dissolve 0.840 g of disodium *p*-nitrophenyl phosphate tetrahydrate (Sigma 104, Sigma

Chemical Co., St. Louis, MO) in about 40 mL of MUB pH 6.5 (for acid phosphatase activity), or pH 11 (alkaline phosphatase activity), and take to 50 mL with the same buffer. Store the solution at 4°C.

Steps

1. Place 1 g of soil (<2 mm) in a 50-mL Erlenmeyer flask, add 0.2 mL of toluene, 4 mL of MUB (pH 6.5 for assay of acid phosphatase of pH 11 for assay of alkaline phosphatase), 1 mL of PNP solution made in the same buffer, and swirl the flask for a few seconds to mix the contents. Stopper the flask, and place it in an incubator at 37°C.
2. After 1 h, remove the stopper, add 1 mL of 0.5 M $CaCl_2$ and 4 mL of 0.5 M NaOH, swirl the flask for a few seconds, and filter the soil suspension through a Whatman no. 2v folded filter paper.
3. Measure the intensity of the yellow color of the filtrate with a spectrophotometer with wavelength adjusted to 410 nm.
4. Calculate the p-nitrophenol content of the filtrate by reference to a calibration graph that plots standards containing 0, 10, 20, 30, 40, and 50 µg of p-nitrophenol. To prepare this graph, dilute 1 mL of the standard p-nitrophenol solution to 100 mL in a volumetric flask and mix the solution thoroughly. Then pipette 0-, 1-, 2-, 3-, 4-, or 5-mL aliquots of this diluted standard solution into a 50-mL Erlenmeyer flask, adjust the volume to 5 mL by addition of water, and proceed as described for p-nitrophenol analysis of the incubated soil sample (i.e., add 1 mL of 0.5 M $CaCl_2$ and 4 mL of 0.5 M of NaOH mix and filter the resultant suspension). If the color intensity of the filtrate exceeds that of 50 µg of the p-nitrophenol standard, an aliquot of the filtrate should be diluted with water until the colorimeter reading falls within the limits of the calibration graph.
5. To perform controls, follow the procedure described for assay of ß-glucosidase activity, but make the addition of 1 mL of PNP solution after the additions of 0.5 M $CaCl_2$ and 4 mL of 0.5 M NaOH, immediately before filtration of the soil suspension.

Acid Phosphatase Activity Values Reported in the Literature

Field moist; 23 to 2100, mean 617 mg p-nitrophenol kg^{-1} soil h^{-1}, $n = 46$ (USA, New Zealand, Europe).

Air dried; 80 to 1112, mean 284 mg p-nitrophenol kg^{-1} soil h^{-1}, $n = 11$ (USA, Australia, Denmark).

Alkaline Phosphatase Activity Values Reported in the Literature

Field moist; 11 to 1000, mean 122 mg p-nitrophenol kg^{-1} soil h^{-1}, $n = 17$ (USA).

Air dried; 18 to 381, mean 144 p-nitrophenol kg^{-1} soil h^{-1}, $n = 11$ (USA, Italy).

Comment

Substrate solution is stable for several days and the standard *p*-nitrophenol solution is stable for a few weeks if stored at 4°C. The $CaCl_2$ is added to prevent dispersion of clay and extraction of organic matter during NaOH treatment, thus removing interferences in color analysis.

Arylsulfatase

Sulfatases release sulfate (SO_4), the plant available form of sulfur, from various organic sulfate esters (Tabatabai & Bremner, 1970). Arylsulfatase (arylsulfatase sulfohydrolase, EC 3.1.6.1) is the most widely studied soil sulfatase and is thought to play an important role in the hydrolysis of ester sulfate (as determined by reduction with hydriodic acid), which comprises 40 to 70% of the total S in many soils (Tabatabai, 1994). Bandick et al. (1994) showed that arylsulfatase was highly effective at discriminating soil management effects. Since microbial ester sulfates are found only in fungi and not in bacteria, elevated sulfatase activity may be related to the stimulation of sulfatases by increased levels of ester sulfate produced by fungi.

This procedure uses *p*-nitrophenyl sulfate as the substrate that upon hydrolysis releases sulfate and *p*-nitrophenol, which can be measured colormetrically.

Pretreatment

As described in ß-glucosidase section.

Equipment

1. 50-mL Erlenmeyer flasks fitted with no. 2 stoppers.
2. Incubator or temperature-controlled water bath.
3. Colorimeter or spectrophotometer.

Reagents

1. Toluene, Fisher certified reagent (Fisher Scientific Co., Chicago).
2. Acetate buffer, 0.5 *M*, pH 5.8: Dissolve 68 g of sodium acetate trihydrate in about 700 mL of water, add 1.70 mL of glacial acetic acid (99%), and dilute the volume to 1 L with water.
3. *p*-Nitrophenyl sulfate solution (PNS), 0.05 *M*: Dissolve 0.614 g of potassium *p*-nitrophenyl sulfate (Sigma Chemical Co., St. Louis, MO) in about 40 mL of acetate buffer, and dilute the solution to 50 mL with buffer. Store this solution at 4°C.
4. Calcium chloride ($CaCl_2$, 0.5 *M*), sodium hydroxide (NaOH, 0.5 *M*), and standard *p*-nitrophenol solution: Prepare as described for ß-glucosidase activity.

Steps

1. To a 50-mL Erlenmeyer flask, add 1 g of soil (<2 mm), 0.25 mL of toluene, 4 mL of acetate buffer, and 1 mL of PNS solution. Then swirl

the flask for a few seconds to mix the contents. Stopper the flask, and place it in an incubator at 37°C.
2. After 1 h, remove the stopper, add 1 mL of 0.5 M CaCl$_2$ and 4 mL of 0.5 M NaOH, swirl the flask for a few seconds, and filter the soil suspension through a Whatman no. 2v folded filter paper.
3. Measure the yellow color intensity of the filtrate with a spectrophotometer set at the wavelength of 410 nm.
4. Calculate the p-nitrophenol concentration by reference to a calibration graph prepared from standard p-nitrophenol as described in ß-glucosidase section. If the color intensity of the filtrate exceeds the limit of the calibration graph, dilute an aliquot of the filtrate with water until the colorimeter reading falls within the limits of the graph.
5. Controls should be performed with each soil analyzed to allow for color not derived from p-nitrophenol released by arylsulfatase activity. To perform controls, follow the procedure described for assay of arylsulfatase activity, but make the addition of 1 mL of PNS solution after the addition of 1 mL of 0.5 M CaCl$_2$ and 4 mL of 0.5 M NaOH, immediately before filtration of the soil suspension.

Comment

Same as phosphatase section.

Activity Values Reported in the Literature

Field moist; 7 to 340, mean 80 mg p-nitrophenol kg^{-1} soil h^{-1}, $n = 30$ (USA, New Zealand, Canada).

Air dried; 2 to 361, mean 85 mg p-nitrophenol kg^{-1} soil h^{-1}, $n = 16$ (USA, Italy).

FDA Hydrolysis

The hydrolysis of fluorescein diacetate (3',6'-diacetylfluorescein) (FDA) has potential to broadly represent soil enzyme activity (Schnürer & Rosswall, 1982) and accumulated biological effects because FDA is hydrolyzed by a number of different enzymes, such as proteases, lipases, and esterases (Guilbaut & Kramer, 1964; Rotman & Papermaster, 1966) and its hydrolysis has been found among a wide array of the primary decomposers, bacteria and fungi (Lundgren, 1981; Medzon & Brady, 1969; Söderström, 1977).

The method is based on hydrolysis of FDA by microbial cells releasing fluorescein dye, which causes active cells to fluoresce a brilliant yellow–green in ultraviolet light, which can be quantified by fluorometry or spectrophotometry.

Sample Pretreatment

As described in ß-glucosidase section.

Equipment

1. 125 mL Erlenmeyer flasks.
2. Shaker incubator.

3. Spectrophotometer.
4. 50 mL centrifuge tubes.

Reagents

1. Fluorescein diacetate (FDA; 3',6'-diacetylfluorescein; Sigma F-7378, Sigma Chemical Co., St. Louis, MO) stock solution, 4.8 mM: Dissolve 2 g FDA in 1 L acetone and store at -20°C.
2. Sodium phosphate buffer (SPB), 60 mM, pH 7.6: Dissolve 16.1 g dibasic heptahydrate sodium phosphate ($Na_2HPO_4 \cdot 7\ H_2O$; DBP) in 1 L water (60 mM) and 8.3 g monobasic monohydrate sodium phosphate (MBP; $NaH_2PO_4 \cdot H_2O$) in 1 L water (60 mM). Place 200 mL MBP in a large 1 L beaker and add DBP while stirring until pH reaches 7.6.
3. Standard fluorescein solution, 2.6 mM: Dissolve 0.5 g fluorescein, sodium salt ($C_{20}H_{10}O_5Na_2$; Sigma F6377) in 500 mL water.
4. Acetone.

Procedure

1. Place 1 g soil in 125 mL Erlenmeyer flask, add 20 mL SPB, and shake for 15 min. Add 100 µL substrate solution and shake for 2 h at 25°C.
2. Add 20 mL acetone, swirl suspension, and transfer to 50 mL centrifuge tube.
3. Centrifuge at 6000 rpm for 5 min, and filter through Whatman No. 4 filter paper.
4. Read absorbance at 490 nm.
5. Controls should be performed in each series of analyses to allow for absorbance not derived from FDA hydrolysis. To perform controls, follow the procedure described for FDA hydrolysis but make the addition of 100 mL FDA solution after the addition of 20 mL acetone.
6. Calculate the fluorescein content of the filtrate by reference to a calibration graph plotted from the results obtained with standards containing 0, 100, 200, 300, 400, and 500 mg of fluorescein. To prepare this graph, dilute 5 mL of the standard fluorescein solution with acetone to 50 mL in a volumetric flask and mix the solution thoroughly. Then pipette 0-, 1-, 2-, 3-, or 5-mL aliquots of this diluted standard solution into 50 mL Erlenmeyer flasks, add 20 mL buffer, and adjust volume to 40 mL with acetone.

Comments

It is important that the buffer be at pH 7.6, otherwise you will not obtain color development. Although some FDA methods use a 50:50 acetone/water solution, we recommend undiluted acetone to ensure termination of the reaction and to fully eliminate colloids from the supernatant.

It should be recognized that this method has not been optimized for soils. Research is in progress to determine the optimal method (D. Stott and R.P. Dick).

Activity Values Reported in the Literature

Field moist; 25 to 125, mean 61 mg fluorescein kg^{-1} soil h^{-1}, $n = 5$ (USA).
Air dried; 16 to 75, mean 42 mg fluorescein kg^{-1} soil h^{-1}, $n = 9$ (Italy, USA).

MICROBIAL COMMUNITY STRUCTURE

Zak et al. (1994) point out that most considerations of biodiversity have concentrated on the macroorganisms and that little attention has been paid to the microorganisms. This preoccupation with the macro scale reflects a type of bias based on size where a greater importance is given to visible organisms. As a consequence of this bias, the interrelationship between the macro and micro scale is dismissed and the microorganisms are unintentionally shown to be self sustaining. In fact, as indicated by Zak et al. (1994) and others (Lee, 1991), these kingdoms are interrelated by a nearly continuous interaction. Systems that consider only macroorganisms as indicators of biodiversity are testing only a portion of the system as microorganisms and their activities drive most ecosystems (Perry et al. 1989; Turco et al., 1994). The types and functioning of the microorganisms present in a system must be assessed for a clear estimation of ecosystem stability.

As a result of their size and distribution in soils, prokaryotic microorganisms are difficult or impossible to assess directly. Bacterial heterogeneity is expressed in diverse biological abilities, not in significant morphological differences. Fungi and to some degree algal diversity can be estimated by morphological differences; however, recent estimations of genetic structure of fungal populations have shown a significant level of previously unresolved diversity (Hawksworth & Mound, 1991). Although bacterial diversity is not expressed in cellular morphological differences, colony morphology has been used because of its simplicity. This approach assumes that microorganisms recovered on solid nutrient medium and expressed by the formation of visible colonies, are representative of those present and active in a given soil. There is evidence that many soil organisms do not form colonies on solid agar but are important in the functioning of the soil system (Pillai et al., 1991). For the most part, indexing of soil microbial populations based on colony morphology defines the diversity of the microorganism that can grow on a given medium to form a visible colony. Therefore, to better characterize diversity with this approach, numerous media formulations are needed, which allows a range of organisms to be challenged and observed. Yet, even with a wide array of media, some soil microorganisms may not grow on any of the media or express important phenotypic characteristics. Therefore, genetic and/or biochemical approaches that involve multiple tests of functions and estimations of key biochemical differences may provide the best estimation of microbial diversity in a soil system.

Phospholipid Analysis

Phospholipids are found in the bacterial cell membrane. It has been recognized for many years that differences in the pattern of the phospholipids in the cell membrane are indicative of a resident population. This approach has been used to identify microorganisms responsible for infections and other human and animal microbial pathogens; however, this approach has generally relied on the isolation and enrichment of the organisms of interest. In contrast, the phospholipid method as suggested by White and Frerman (1967) and White et al., (1979) has been used to study the size of the microbial biomass resident in soils and the subsurface to depths of 200 cm in the profile (Federle, 1988). The approach uses organic solvents to extract lipids from the environmental sample and does not require an enrichment step. Therefore, this method is viewed as a direct indicator the structure of the microbial populations. Factors that affect the population should cause changes in the phospholipid pattern and content. This method is advantageous because it can give estimates of both biomass size and community composition. The extracted phospholipids can be characterized for type and amount. Phospholipid turnover in environmental samples is fairly rapid as materials in dead cells will be readily consumed by the living biomass. Hence, the total phospholipid content gives an estimation of the living biomass size (Findlay et al., 1989, 1990) and can give some indication of the community structure (Zelles et al., 1992).

The analysis of phospholipids is time-consuming and requires extensive capital investment in order to both capture and evaluate the data. A number of commercial firms presently are available to conduct the procedures.

DNA Analysis

In much the same manner as PLFA can be used to estimate biochemical diversity within a population, direct extraction and characterization of the soil system DNA also can be used. Like PLFA, the content of DNA in a given unit of soil is extremely small, has a rapid rate of turnover, and should be indicative of the active biomass; however, the low levels of DNA in soil makes quantification and description difficult. Recent developments including polymerase chain reactions (PCR) and CsCL-bisbizimide gel gradients may provide the tools to understand the population structure expressed within the DNA. Because the extraction and recovery of DNA from soil is difficult, requires both extensive laboratory capabilities and further research before it can have practical applications, we are not presenting this method and refer the reader to Hinton et al. (1995) for details.

Functional Microbial Diversity

A functional diversity index for a soil community can be obtained by determining its potential to use a series of C substrates (Zak et al., 1994). The utilization pattern is developed into a score or functional diversity index which is the basis of the Biolog C substrate utilization system. The Biolog system was originally developed to characterize isolates of single strains of bacteria originating from medical sources and has been extended to testing of mixed populations of bacteria in soil and water systems (Garland & Mills, 1994; Zak et al., 1994). The Biolog system tests the ability of a microorganism (or a community of microorganisms) to use 95 different C substrates in microtiter plate wells. The number of

C substrates can be expanded to 128 if both the gram negative and gram positive plates are used. The substrates are mixed with tetrazolium dye and distributed in the wells of a microtiter plate. Use of a substrate is determined by the concurrent reduction of tetrazolium dye (changing from clear to a purple color) and substrate oxidization. This allows a binary decision concerning the use a given substrate to be made.

Equipment

1. 160 mL wide-mouth dilution bottle 1 per sample, containing 95 mL (after autoclaving) of diluent solution.
2. 160 mL milk dilution bottles, 8 per sample, containing 90 mL (after autoclaving) of diluent solution.
3. Sterile glass or plastic disposable 10 mL graduated, serological pipettes and a pipette bulb. The initial transfer is facilitated with the use of a wide-bore pipette.
4. Mechanical or electronic balance.
5. Mechanical pipette to transfer diluted materials to Biolog plates.
6. Automated microplate spectrophotometer (optional).

Reagents

1. Sterile-distilled water. 1000 mL of distilled water is placed into a 2-L container. The container is loosely capped and steam autoclaved at 1.05 kg cm^{-3}, 121°C for 20 min. The containers should be cooled before removing from the autoclave. If Erlenmeyer flasks are used, cotton plugs wrapped in cheese cloth serve well as the cap. Plugs should be covered with aluminum foil before autoclaving.
2. Phosphate-buffered saline solution: Sodium chloride 8.5 g, 0.3 g anhydrous potassium dihydrogen phosphate (KH_2PO_4), 0.6 g anhydrous, sodium hydrogen phosphate (Na_2HPO_4). Adjust pH to 6.8 using either 0.1 M sodium hydroxide (NaOH) or hydrochloric acid (HCl). Autoclave as above. Gelatine agar (0.2%) also can be used as diluent. Agar (0.2 g) is added to 1000 mL of distilled water and the mixture autoclaved as above.
3. The Biolog system is extremely simple to use. The test plates are purchased (Biolog Inc., Hayward, CA) and stored cold (4°C) until used. The plates come in a number of forms but most ecological assessments have been made using either the GN gram negative or gram positive. Plates containing only the tetrazolium dye also are available. These can be used to construct a specialty assay where the researcher adds their own C substrates or other test media.

Procedure

1. A soil or a series of soils to be tested is selected. Handling the soil should follow the same rules as indicated for enzyme assays.
2. If soils from different locations are to be tested, an initial evaluation of the size of the standing biomass in the soil should be made. That can be

accomplished by using a direct count procedure. This information can be used to resolve a major problem encountered when testing soils from different locations. If the size of the standing biomass is different, the rate at which the dye is reduced will appear to be different. That is, in any given incubation, the larger biomass may give a positive result sooner than a smaller biomass. Therefore, these differences should be normalized so that the same size biomass is added to all of the wells (Zak et al., 1994). This is accomplished using a dilution scheme that will provide approximately the same number of cells to each well, regardless of the source of soil.

3. The soils are diluted to achieve a desired loading rate in the wells of the microtiter plate. Zak et al. (1994) recommend that soils be diluted in 0.2% water agar and mixed with a small blender at high speeds (1 min) to aid dispersion. Therefore, once some indication of the size of the standing biomass is made, the soils can be normalized by dilution into the water agar.

4. Once the cells are diluted they should be pre-incubated for 24 h before application to the plates. This allows microbial utilization of any dissolved organic C derived from the soil. This eliminates a source of reducing power and possible interference in the detection of lower level responses.

5. After 24 h, the dilution bottles are shaken and 100 µL of the proper dilution is applied to each well of the plate. (The exact volume depends on the subsequent handling of the plates. Overfilling the wells will make it difficult to transport or move the plates without splashing liquid between the wells).

6. A single tip Pipetman can be used to inoculate the plates. To speed the inoculation process we suggest that a multitip mechanical pipette be used. They are available from most laboratory supply vendors. Be sure to purchase a mechanical pipette that is adjustable over the full range of volumes that you will be making. The multitip mechanical pipette will allow up to six wells to be inoculated at the same time. Because the multitip mechanical pipette will not fit into most dilution bottles, pour the diluted materials into a sterile petri dish and use this as a source of inoculum for the plates.

7. The inoculated plates are incubated at 25°C and examined on a 12 h time step. As indicated by Zak et al. (1994), the exact incubation time is a function of the ecosystem that is being tested which must be determined experimentally. It is possible, using an automated plate reader, to develop a rate estimation for utilization of each substrate; however, we have generally scored the wells as positive or negative following a 72-h incubation and not used an automated plate reader. The best way to score the plates is using a score sheet that simulates the C patterns in the plates. Each well is scored, marked with a pen and the score is then written on the tally sheet. The plates should be scored by the same person each time as this eliminates any color bias that could exist between individuals. The plates also should be read at the same incubation time and it is best

to incubate them in a controlled temperature environment. Alternatively one can use an automated microplate spectrophotometer. An arbitrary absorbance such as 0.4 can be used to define positive wells and avoid false positives.

Comments

This procedure can generate extensive quantities of data. If both the gram positive and gram negative plates are used, 128 different nutrients are evaluated. These data need to be consolidated and scored, because simply reporting a usage pattern is fairly uninformative and is discouraged. The goal should be to construct a fingerprint or numerical index that describes the ecological contribution of the microbial biomass to a soil. Treatment effects, impacts of management, chemicals or other perturbations can be assessed by evaluation of the changes in the index number. A number of approaches to generate numerical assessments are available. Zak et al. (1994) coded their data as to carbohydrate usage and then used four measures of similarity assessment. These included cluster and principle components analysis. We have used both cluster and principle components analysis to establish trends in pattern data and found them to be workable systems. Zak et al. (1994) and Garland and Mills (1994) showed that either approach would reveal the patterns present in the Biolog data. It is suggested that a minimum data analysis include some sort of grouping analysis.

REFERENCES

Allan, D. and R. Killom. 1996. Assessing soil nitrogen, phosphorus, and potassium for crop nutrition and environmental risk. p. 187–201. *In* J.W. Doran and A.J. Jones (ed.) Methods for assessing soil quality. SSSA Spec. Publ. 49. SSSA, Madison, WI.

Bandick, A., M. Miller, and R.P. Dick . 1994. Soil enzyme stability as an indicator of soil quality. p. 292. *In* Agronomy abstracts. ASA, Madison, WI.

Bremner, J.M., and M.A. Tabatabai. 1973. Effect of some inorganic substances on TTC assay of dehydrogenase activity in soils. Soil Biol. Biochem. 5:385–386.

Burns, R.G. 1978. Soil enzymes. Academic Press, London.

Burns, R.G. 1982. Enzyme activity in soil: Location and a possible role in microbial activity. Soil Biol. Biochem. 14:423–427.

Casida, L.E., Jr., D.A. Klein, and T. Santoro. 1964. Soil dehydrogenase activity. Soil Sci. 98:371–376.

Chander, K., and P.S. Brookes. 1991. Is the dehydrogenase assay invalid as a method to estimate microbial activity in Cu-contaminated and non-contaminated soils? Soil Biol. Biochem. 23:901–915.

Dick, R.P. 1994. Soil enzyme activities as indicators of soil quality. p. 107-124. *In* Doran et al. (ed.) Defining soil quality for a sustainable environment. SSSA Spec. Publ. 35. SSSA and ASA, Madison, WI.

Dick, R.P., D.D. Myrold, and E.A. Kerle. 1988a. Microbial biomass and soil enzyme activities in compacted and rehabilitated skid trail soils. Soil Sci. Soc. Am. J. 52:512–516.

Dick, R.P., P.E. Rasmussen, and E.A. Kerle. 1988b. Influence of long-term residue management on soil enzyme activity in relation to soil chemical properties of a wheat–fallow system. Biol. Fert. Soils 6:159–164.

Dick, R.P., D.R. Thomas, and J.J. Halvorson. 1996. Standardized methods, sampling , and pretreatment. p. 107–121. *In* J.W. Doran and A.J. Jones (ed.) Methods for assessing soil quality. SSSA Spec. Publ. 49. SSSA, Madison, WI.

Doran, J.W. 1980. Soil microbial and biochemical changes associated with reduced tillage. Soil Sci. Soc. Am. J. 44:765–771.

Eivazi, F., and M.A. Tabatabai. 1977. Phosphatases in soils. Soil Biol. Biochem. 9:167–172.

Eivazi, F., and M.A. Tabatabai. 1988. Glucosidases and galactosidases in soils. Soil Biol. Biochem. 20:601–606.

Federle, T.W. 1988. Mineralization of monosubstituted aromatic compounds in unsaturated and saturated subsurface soils. Can. J. Microbiol. 34:1037–1042.

Findlay, R.H., G. King, and L. Watling 1989. Efficacy of phospholipid analysis in determining microbial biomass in sediments. Appl. Environ. Microbiol. 55:2888–2893.

Findlay, R.H., M.B. Trexler, J.B. Guckert, and D.C. White. 1990. Laboratory study of disturbance in marine sediments: Response of microbial community. Marine Ecol. Prog. Ser. 62:121–133.

Frankenberger, W.T., Jr., and W.A. Dick. 1983. Relationships between enzyme activities and microbial growth and activity indices in soil. Soil Sci. Soc. Am. J. 47:945–951.

Frankenberger, W.T., Jr., and J.B. Johanson. 1986. Use of plasmolytic agents and antiseptics in soil enzyme assays. Soil Biol. Biochem. 18:209–213.

Frankenberger, W.T., Jr., and M.A. Tabatabai. 1980. Amidase activity in soils: I. Methods of assay. Soil Sci. Soc. Am. J. 44:282–287.

Frankenberger, W.T., Jr., and M.A. Tabatabai. 1982. Amidase and urease activities in plants. Plant Soil 64:153–166.

Garland J.L., and A.L. Mills. 1994. A community-level physiological approach for studying microbial communities. p. 77–83. *In* K. Ritz et al. (ed.) Beyond the biomass. British Soc. of Soil Sci. Wiley-Sayce Publication.

Guilbaut, G.G., and D.N. Kramer. 1964. Fluorometric determination of lipase, acylase, alpha- and gama-chymotrypsin and inhibitors of these enzymes. Anal. Chem. 36:409–412.

Hawksworth, D.L., and Mound, L. A. 1991. Biodiversity database: The crucial significance of collections. p. 17–31. *In* D.L. Hawksworth (ed.) The biodiversity of microorganism and invertebrates: Its role in sustainable agriculture. CAB Int., England.

Hinton, S.M., V. Minak-Bernero, and L. G. Keim. 1995. Innovative molecular tools for understanding bioremediation: 16s RNA sequences and panmers. p. 103–117. *In* H.D. Skipper and R.F. Turco (ed.) Bioremediation: Science and applications. SSSA Spec. Publ. 43. SSSA and ASA, Madison, WI.

Howard, P.J.A. 1972. Problems in the estimation of biological activity in soil. Oikos 23:23–240.

Hynes, M.J. 1970. Induction and repression of amidase enzymes in *Aspergillus nidulans*. J. Bacteriol. 103:482–487.

Hynes, M.J. 1975. Amide utilization in *Aspergillus nidulans*: Evidence for a third amidase enzyme. J. Gen. Microbiol. 91:99–109.

Jermyn, M.A. 1958. Fungal cellulases. Aust. J. Biol. Sci. 11:114–126.

Joshi, J.G., and P. Handler. 1962. Purification and properties of nicotinamidase from *Torula cremoris*. J. Biol. Chem. 237:929–935.

Juma, N.G., and M.A. Tabatabai. 1977. Effects of trace elements on phosphatase activity in soils. Soil Sci. Soc. Am. J. 41:343–346.

Juma, N.G., and M.A. Tabatabai. 1978. Distribution of phosphomonoesterases in soils. Soil Sci. 126:101–108.

Keeney, D.R., and D.W. Nelson. 1982. Nitrogen-inorganic forms. p. 643–698. *In* A.L. Page et al. (ed.) Methods of soil analysis. Part 2. 2nd ed. Agron. Monogr. 9. ASA and SSSA, Madison, WI.

Ladd, J.N., and E.A. Paul. 1973. Changes in enzymic activity and distribution of acid-soluble, amino acid nitrogen in soil during nitrogen immobilization and mineralization. Soil Biol. Biochem. 5:825–840.

Lee, K.E. 1991. The diversity of soil organisms. p. 72–89. *In* D.L. Hawksworth (ed.) The biodiversity of microorganisms and invertebrates: Its role in sustainable agriculture. CASA-FA Rep. Ser. 4. Redwood Press Ltd., England.

Lundgren, B. 1981. Fluorescein diacetate as a stain of metabolically active bacteria in soil. Oikos 36:17–22.

Martens, D.A., J.B. Johanson, and W.T. Frankenberger, Jr. 1992. Production and persistence of soil enzymes with repeated additions of organic residues. Soil Sci. 153:53–61.

McCarty, G.W., D.R. Shogren, and J.M. Bremner. 1992. Regulation of urease production in soil by microbial assimilation of nitrogen. Biol. Fert. Soils 12:261–264.

Medzon, E.L., and M.L. Brady. 1969. Direct measurement of acetylesterase in living protist cells. Bacteriol. 97:402–415.

Perry A.D.M., M.P. Amaranthus, J.G. Borchers, S.L. Borchers, and R.E. Brainerd. 1989. Bootstrapping in ecosystems. Bioscience 39:230–237.

Pillai, S.D., K.L Josephson, R.L. Bailey, C.P. Gerba, and I.L Pepper. 1991. Rapid method for processing soil samples for polymerase chain reaction amplification of specific gene sequences. Appl. Environ. Microbiol. 57:2283–2286.

Ross, D.J. 1973. Some enzyme and respiratory activities of tropical soils from New Hebrides. Soil Biol. Biochem. 5:559–567.

Rotman, B., and B.W. Papermaster. 1966. Membrane properties of living mammalian cells as studied by enzymatic hydrolysis of fluorogenic ester. Proc. Nat. Acad. Sci., U.S.A. 55:134–141.

Schlesinger, W.H. 1977. Carbon balance in terrestrial detritus. Annu. Rev. Ecol. System. 8:51–81.

Schnürer, J., and T. Rosswall. 1982. Fluorescein diacetate hydrolysis as a measure of total microbial activity in soil and litter. Appl. Environ. Microb. 43:1256–1261.

Skujins, J. 1967. Enzymes in soil. p. 371–414. *In* A.D. McLaren and G.H. Peterson (ed.) Soil biochemistry. Marcel Dekker, New York.

Skujins, J. 1973. Dehydrogenase: An indicator of biological activities in arid soils. Bull. Ecol. Res. Commun. NFR 17:235–241.

Skujins, J. 1976. Extracellular enzymes in soil. CRC Crit. Rev. Microbiol. 4:383–421.

Skujins, J. 1978. History of abiontic soil enzyme research. p. 1–49. *In* R.G. Burns (ed.) Soil enzymes. Academic Press, London.

Söderström, B.E. 1977. Vital staining of fungi in pure cultures and in soil with fluorescein diacetate. Soil Biol. Biochem. 9:59–63.

Speir, T.W. 1977. Studies on a climosequence of soils in tussock grasslands. 11. Urease, phosphatase and sulfatase activities of topsoils and their relationships with other properties including plant available sulfur. N.Z. J. Sci. 20:159–166.

Speir, T.W., and D.J. Ross. 1976. Studies on a climosequence of soils in tussock grasslands. 9. Influence of age of *Chionochloa rigida* on enzyme activities. N.Z. J. Sci. 19:389–396.

Stevenson, I.L. 1959. Dehydrogenase activity in soils. Can. J. Microbiol. 5:229–235.

Tabatabai, M.A. 1994. Soil enzymes. p. 775–833. *In* R.W. Weaver et al. (ed.) Methods of soil analysis. Part 2. SSSA Book Ser. 5. SSSA and ASA, Madison, WI.

Tabatabai, M.A., and J.M. Bremner. 1969. Use of *p*-nitrophenyl phosphate for assay of soil phosphatase activity. Soil Biol. Biochem. 1:301–307.

Tabatabai, M.A., and J.M. Bremner. 1970. Arylsulfatase activity of soils. Soil Sci. Soc. Am. Proc. 34:225–229.

Tabatabai, M.A., and J.M. Bremner. 1972. Assay of urease activity in soils. Soil Biol. Biochem. 4:479–487.

Turco, R.F., A.C. Kennedy, and M.D. Jawson. 1994. Microbial indicators of soil quality. p 73–90. *In* J.W. Doran et al. (ed.) Defining soil quality for a sustainable environment. SSSA Spec. Publ. 35. SSSA and ASA, Madison, WI.

Veibel, S. 1950. ß-glucosidase. p. 583–620. *In* J.B. Sumner and K. Myrgack (ed.) The enzymes. Vol. 1. Part 1. Academic Press, New York.

White, D.C., W.M. Davis, J.S. Nickels, J.S. King, and R.J. Bobbie. 1979. Determination of the sedimentary microbial biomass by extractable lipid phosphate. Oecologia. 40:51–62.

White, D.C., and F.E. Frerman 1967. Extraction, characterization and cellular localization of the lipids of *Staphylococcus aureus*. J. Bacteriol. 94:1854–1867.

Zak, J.C., M.R. Wilig, D.L. Moorhead, and H.G. Wildman. 1994. Functional diversity of microbial communities: A quantitative approach. Soil Biol. Biochem. 26:1101–1108.

Zelles, L., Q.Y. Bai, T. Beck, and F. Beese. 1992. Signature fatty acids in phospholipids and lipopolysaccharides as indicators of microbial biomass and community structure in agricultural soils. Soil Biol. Biochem. 24:317–323.

16 Soil Invertebrates as Indicators of Soil Quality

John M. Blair
Kansas State University
Manhattan, Kansas

Patrick J. Bohlen
Institute of Ecosystem Studies
Millbrook, New York

Diana W. Freckman
Colorado State University
Ft. Collins, Colorado

Soil quality, though difficult to define precisely, can be thought of as the ability of a soil to sustain biological productivity, maintain environmental quality, and promote plant, animal, and human health (Doran & Parkin, 1994). These basic functions of soil are dependent on the structural and functional integrity of a soil, and the impacts of management practices or unintentional disturbances. Soil is a heterogeneous mix of living and nonliving components, including a very complex assemblage of organisms and their by-products. The living components of soil, including microbes, plant roots, and soil invertebrates, have extremely important effects on soil characteristics relevant to soil quality. It seems to us essential, therefore, that the assessment of soil quality includes biological, as well as chemical and physical, properties of soils. Although most biological measurements to date have focused on microbial populations or activity, there is growing awareness of the importance of soil invertebrates as vital components of soils and as potential indicators of soil quality. Recent reviews of soil invertebrate ecology confirm that soil invertebrates affect soil structure, alter patterns of microbial activity and influence soil organic matter dynamics and nutrient cycling (Verhoef & Brussaard, 1990; Schulze & Mooney, 1994; Didden et al., 1994; de Ruiter et al., 1994; Heal et al., 1996). Furthermore, results from many recent studies point to the potential utility of using soil invertebrates as indicators of physical or chemical disturbances to soils (Wallwork, 1988; Bongers, 1990; Paoletti et al., 1991; Stork & Eggleton, 1992; Freckman & Ettema, 1993; Parmelee et al., 1993; Linden et al., 1994). In this chapter, we review some of the fundamental ways in which invertebrates affect soil structure and function, and contribute to the quality of soils. We then suggest the use of two major groups of soil invertebrates—

Copyright © 1996 Soil Science Society of America, 677 S. Segoe Rd., Madison, WI 53711, USA.
Methods for Assessing Soil Quality, SSSA Special Publication 49.

earthworms and nematodes—as potential indicators of soil quality, and provide rationale for the suitability of these two groups as bioindicators of soil quality. We also describe methods for sampling nematode and earthworm populations, and discuss interpretation of the results of these population estimates within the context of soil quality.

FUNCTIONAL GROUPS OF SOIL INVERTEBRATES

The abundance and diversity of soil invertebrates has been investigated in a variety of natural and managed ecosystems, although true species numbers are often unknown (Swift & Anderson, 1994). Soil invertebrate numbers are often large (Groombridge, 1992), and they comprise a significant portion of the belowground food webs in most terrestrial ecosystems (Petersen & Luxton, 1982). The incredible diversity of soil invertebrates is only beginning to be fully appreciated, and remains one of the great unknowns in the realm of biodiversity (Andre et al., 1994; Freckman, 1994). Many species of soil invertebrates have yet to be described, and the biology of those that have been described is often poorly known. Thus, the diversity of invertebrate life in the soil remains both a source of fascination and a challenge to those studying soil ecology.

Because it is, at the present time, impossible to identify all members of the soil invertebrate community, soil ecologists often group together various taxa or functional groups of invertebrates. Several schemes for grouping soil invertebrates have been proposed, some of which rely on taxonomic relationships or functional similarities. The major taxa comprising the soil invertebrate community also can be grouped according to average body width (Fig. 16–1). This scheme has appeal because there are some major differences in the way that organisms of different size both respond to, and interact with, the soil environment (Table 16–1).

The soil micro-fauna, which includes protozoa, rotifers, and nematodes, are microscopic aquatic organisms, restricted to life in the films of water around soil particles, or water-filled pore spaces. The movement of these organisms in soils depends on soil texture, available pore space, and the distribution of water. Their activities do little to directly alter soil structure. Nematodes are an important component of the micro-fauna, and are among the most abundant invertebrates in most soils. The various species of nematodes occurring in soils encompass a variety of feeding strategies (Freckman & Baldwin, 1990). Some species feed on plant roots, and have been relatively well-studied because of their economic impacts on crop plants; however, there also are many free-living species of soil nematodes that feed on soil microbes (bacteria or fungi) or which prey on small invertebrates, including other nematodes. The microbial-feeding nematodes may be the most important consumers of bacteria and fungi in many soil communities (Yeates, 1979), and their interactions with microbial decomposers affect decomposition and nutrient cycling processes (Anderson et al., 1981; Coleman et al., 1984; Freckman, 1988).

The soil meso-fauna includes the Enchytraeidae, a group of small, segmented worms related to earthworms, and a diverse assemblage of microarthro-

SOIL INVERTEBRATES

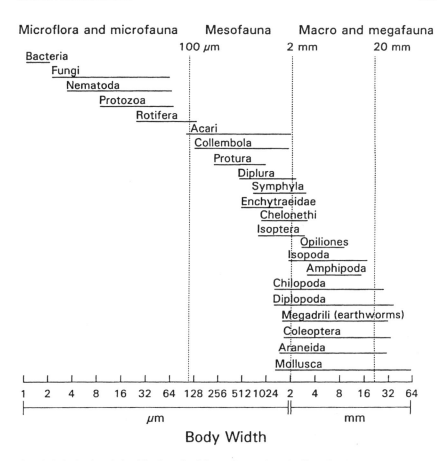

Fig. 16–1. A size-based classification of soil invertebrates (from Swift et al., 1979).

Table 16–1. Influences of soil biota on soil processes in ecosystems (from Hendrix et al., 1990).

	Nutrient cycling	Soil structure
Microflora	Catabolize organic matter Mineralize and immobilize nutrients	Produce organic compounds that bind aggregates Hyphae entangle particles onto aggregates
Microfauna	Regulate bacterial and fungal populations Alter nutrient turnover	May affect aggregate structure through interactions with microflora
Mesofauna	Regulate fungal and microfaunal populations Alter nutrient turnover Fragment plant residues	Produce fecal pellets Create biopores Promote humification
Macrofauna	Fragment plant residues Stimulate microbial activity	Mix organic and mineral particles Redistribute organic matter and microorganisms Create biopores Promote humification Produce fecal pellets

pods that usually comprise the bulk of the soil meso-fauna. The term *microarthropod* refers to an arbitrary grouping of the small arthropods that includes the mites (Acari), Collembola, Symphyla, Protura, Diplura, Pauropoda, small centipedes and millipedes, and small insects from several orders. Mites and Collembola generally account for 90 to 95% of total microarthropod numbers (Blair et al., 1994). The feeding habits of the microarthropods are exceptionally diverse and they typically include representatives of most trophic groups within the belowground food web. Microarthropods are not limited by availability of liquid water, and are free to move about in, and on, the soil within the constraints of available pore space. Microarthropods play a major role by grazing on microbial decomposers, which can alter both the composition and the activity of the microbial community (Hanlon & Anderson, 1979; Newell, 1984a,b), and by comminution, or physical fragmentation, of organic matter. Through their interactions with microbes and with other invertebrates, microarthropods can accelerate decomposition and nutrient cycling processes (Seastedt, 1984; Moore et al., 1988). Their direct influence on soil structure is limited, although they may be important in microaggregate formation in some soils.

The macro-fauna includes larger insects and other arthropods, as well as earthworms. These larger invertebrates are not limited to moving about in existing soil pores, but can burrow through the soil, creating large pores that increase soil aeration and water infiltration. Earthworms also alter soil structure by mixing soil with organic matter and depositing feces that can become stable soil aggregates (Tomlin et al., 1995). Earthworms can comprise a significant portion of the total biomass of invertebrates in some soils. For example, earthworms made up 97% of the total soil invertebrate biomass in winter wheat fields under integrated management in the Netherlands (Didden et al., 1994), and 96% of the soil invertebrate biomass in no-tillage agroecosystems in Georgia (Hendrix et al., 1986). Through their influence on soil structure and interactions with microbes and other decomposers, earthworms can alter the rates and patterns of nutrient cycling in terrestrial ecosystems (Blair et al., 1995).

RELEVANCE OF NEMATODES AND EARTHWORMS FOR ASSESSING SOIL QUALITY

The assumption in using soil invertebrates to assess soil quality is that changes in soil properties or management will alter species richness or species abundance in the decomposer food web, and that these changes will have consequences for soil processes, such as decomposition or nutrient mineralization. Although many groups of soil invertebrates may respond to soil disturbances, we have chosen to focus on nematodes and earthworms as potential indicator organisms. Both nematodes and earthworms respond to soil disturbances, and significantly influence soil processes, and therefore can serve as useful indicator species when assessing the effects of various land management practices or anthropogenic impacts on soil quality. We also believe that using both microfaunal (nematodes) and macrofaunal (earthworms) groups to assess soil quality has some advantages over using only one or the other. The two groups of invertebrates participate in different levels of the soil food web, influence nutrient cycling and soil structural changes at varying scales, and reflect different levels of disturbance in the soil physical and chemical environment. The presence and diversity of these two groups can indicate soil quality and fertility. Nematodes reflect soil change at a microsite level (soil pores, soil water, microbial populations), while earthworms reflect changes in these factors, as well as other larger scale factors (physical soil disturbance, reductions in amount of organic matter). Additionally, earthworms are not present in all soils (arid ecosystems, recently glaciated areas), and nematodes may be the most appropriate indicators in these soils.

Nematodes recently have been examined as indicators of changes in soil quality because they increase nutrient turnover and indirectly influence decomposition through their feeding on microbial decomposers. Many studies indicate that the abundance and composition of free-living soil nematodes are related to the status of the soil microbial community, and can provide additional information about the microbial community and processes (microbial biomass, fungal/bacterial ratios, N mineralization rates). Nematodes also appear to respond quickly in predictive ways to disturbances in ecosystems (Wasilewska, 1989; Freckman & Ettema, 1993; Ettema & Bongers, 1993; Yeates & Bird, 1994). Reasons for the interest in nematodes as indicators have been outlined by Freckman (1988), Bongers (1990), Wasilewska (1991) and Linden et al. (1994). These include: (i) the tremendous diversity of soil nematodes and their participation in many ecosystem functions at different levels of the soil food web; (ii) the presence of nematodes in soils of every terrestrial ecosystem, including extreme habitats such as Antarctica, which facilitates comparisons across many soils; (iii) the rapid response of nematodes to changes in their food resource base, because of their small size and short generation times; (iv) the relative stability of nematode populations, so that changes in population size or in nematode community structure can be used to infer soil disturbance; (v) the ability of most nematode species to survive adverse environmental extremes (freezing, droughts); (vi) the aquatic habitat of nematodes (living in soil water films) allows them to respond to changes in soil water quantity and quality at the microscale, (vii) the limited movement of nematodes in soil allows disturbance to be related to a particular source–point (but see Griffiths & Caul, 1993); and (viii) nematode trophic groups can be identified, and their varying life histories and reproductive capabilities can be used to indicate soil disturbance (Bongers, 1990).

Earthworms also have great potential as indicators of soil quality. Earthworm population densities can be related to soil organic matter levels (Hendrix et al., 1992), soil physical disturbances such as tillage (Lee, 1985) and potentially harmful chemicals (Edwards & Bohlen, 1992, 1995a). The population density of earthworms tends to increase with increasing organic matter inputs and decrease with soil disturbance. Through their feeding and burrowing activities, earthworms significantly alter soil structure and hydrologic properties (Tomlin et al., 1995), and make substantial contributions to nutrient mineralization (Didden et al., 1994). For these reasons, the possibility of manipulating earthworm populations, or their activities, to maintain soil quality in agroecosystems has been noted (Lavelle et al., 1989). Earthworms also influence other important biological indicators of soil quality, particularly the soil microbial biomass (Blair et al., 1995, 1996; Bohlen & Edwards, 1995). Earthworms are large and relatively easy to sample. Many common species can be identified with minimal training, and species can be placed in different functional groups that relate to their potential influence on soil properties. Many earthworm species also leave physical signs of their activity, such as surface casts, middens and burrows, which allow for a preliminary visual assessment of their importance in a particular system. As noted above, a shortcoming of using earthworms as indicators of soil quality is that they are not present in all soils, and some soils can attain good physical structure and fertility in the absence of earthworms (Linden et al., 1994). Despite these shortcomings, earthworms remain one of the most important members of the soil fauna, possessing many features that make them useful biological indicators of soil quality.

SAMPLING SOIL INVERTEBRATES TO ASSESS SOIL QUALITY

The methods chosen to assess soil quality using nematodes and earthworms are dependent on the goals of the study, knowledge of the soil invertebrate community in a particular habitat, and the anticipated use of the results. Any site information relating to the soil habitat of the animals, such as soil survey data, GIS maps of soil texture, temperature, hydrology and/or soil moisture, crop history, and soil chemistry, can provide a basis for establishing a sampling plan and a set of methods and procedures for the study. An initial preliminary sampling at the site for the purpose of testing specific extraction procedures and estimating the extent of taxon diversity and abundance is useful in estimating the level of taxonomic resolution and time commitment necessary for the study.

METHODS FOR ASSESSING SOIL NEMATODES

Temporal considerations

Nematodes can be used as indicators of soil quality without collecting prior baseline data on species or populations, although longer-term data can be helpful in interpreting results. Whether to sample a site once or on a temporal basis depends on the goals of the study. Plant parasitic nematode damage thresholds are frequently determined on the basis of two samplings during a growing season, an initial population and final population, times of sampling that generally follow host plant phenology and abiotic parameters that influence nematode activity and population growth (Seinhorst, 1970). Timing of sampling also should be considered relative to decomposition processes (i.e., time since incorporation of plant material into soil) since there is often a succession of different nematode species and functional groups as decomposition proceeds. A one time sampling for nonparasitic nematodes should occur when soil is sufficiently warm for both decomposition processes and plant root growth.

Scale

Sampling designs are dependent on the spatial scale of the soil being examined. Linear transects within or across rows of an agricultural field are frequently used, whereas for tree crops or forests, samples are often taken at the trunk and at the canopy drip line. In unmanaged ecosystems, the sampling design should represent the vegetative and landform heterogeneity of the system aboveground, which will reflect, to some degree, the variability of the nematode species and abundance belowground. Consideration of the relationship of belowground soil processes to aboveground primary productivity and diversity are important. Stolgren (1994) and Stolgren et al. (1995) have reviewed sampling plans for inventories of aboveground vegetation and vertebrate species as related to soils and geology maps, and noted problems and suggestions for establishing baselines for ecological monitoring. Similar syntheses for soil invertebrates would be useful for soil species inventories and for assessment of soil quality. Nematode species diversity varies with depth and is influenced by soil physical and chemical properties, plant root architecture, and buried organic matter, all of which must be considered in relation to the goals of the study.

Nematodes occur in all ecosystems, but their distribution is aggregated, making precise quantification a problem (Schmitt et al., 1990; Goodell & Ferris, 1981; Ferris et al., 1990). Information on the spatial distribution of plant parasitic nematodes indicates that frequency distributions typically follow a negative binomial distribution (Goodell & Ferris, 1981; Ferris et al., 1990).

There is less information on the spatial distribution of nonparasitic trophic groups, or on the recommended number of samples for these groups. Nonparasitic nematode groups were strongly spatially patterned at subhectare scales in an agroecosystem with soil that was relatively homogeneous, compared with natural ecosystems (Robertson & Freckman, 1995). In the agricultural site, edaphic factors (bulk density, texture, pH, C availability and N availability) explained <30% of the variance of the free-living nematodes (Robertson & Freckman, 1995). This spatial patterning also has been noted in agroecosystems for biological processes in soil, e.g., soil respiration (Aiken et al., 1991) and N availability and N gas loss (Robertson et al., 1993; Folorunso & Rolston, 1985). Nematode distribution often is related more to plant root distribution than to soil moisture or edaphic factors (Ingham et al., 1985; Freckman & Virginia, 1989; Robertson & Freckman, 1995).

RECOMMENDED NEMATODE SAMPLING PROCEDURES

Soil Sampling

The number of samples to be analyzed for soil quality assessment should consider labor, time, and costs. A single soil core can easily be weighed, mixed carefully and subdivided for soil moisture content (required to calculate numbers per gram dry weight of soil), and the nematodes extracted and counted. Because the spatial distribution of nematodes is highly aggregated, and to decrease the variability between soil samples, there should be a balance between the number of replicate soil cores and the number of samples processed. The number of replicate soil cores must be high (we recommend 5 or more), or many soil cores (again, we recommend more than 5) must be composited into one sample with many replicates. Johnson et al. (1972) found that 96% of the nematode species in a forest plot 18 m in diam. were collected from one sample composed of 16 soil cores. The Society of Nematologists (Barker, 1978) established guidelines for sampling nematode communities in agricultural systems. The depth of the sample is generally based on plant root architecture. For example, shallow rooted plants may be sampled from 0 to 10 cm, 10 to 20 cm, etc., whereas deep rooted plants such as grapes (*Vitis* sp.) may be sampled in 30-cm increments to a depth of several meters. In soils where there are no plants, such as the Antarctic Dry Valleys, and in other ecological studies comparisons of nematode populations are often based on 10-cm depth increments. Regardless of the depth of the samples, the number of cores (2 cm diam.) to be composited into one sample should be 10 cores for plots <5 m^2; 20 cores for plots 5 to 100 m^2; and 30+ cores for >100 m^2 plots. Total volume of soil per sample should be >500 g, or about a liter of soil per sample. Sample number should be 5 or greater.

Soil samples should be put into plastic bags, placed in an ice chest, and kept cool (<15°C). Exposure of collected soil to temperatures of 30 to 40°C, even for a short time, can kill many species. Samples should be refrigerated (5–15°C) immediately upon return to the laboratory. Many nematodes do not survive long storage, and samples should be processed within 1 wk of collection for reliable estimates of populations at the time of sampling. All samples should be mixed gently and thoroughly by hand prior to extraction. A 20-g soil sample is the minimum recommended for extraction.

Extraction Methods

No known method extracts all nematodes from soils. Viglierchio and Schmitt, (1983a,b) and Viglierchio and Yamashita (1983) in a series of papers on nematode extraction, noted that all methods tested were <50% efficient in removing nematodes from soil. Even extraction of known quantities of nematodes suspended in water as a control, using a sieving technique such as Cobb's sieving, varied from 51 to 88% depending on the nematode species. Differences in soil types, nematode density and size, and operator accounted for some of the variability. Preliminary tests of the efficiency of extraction techniques are recommended prior to beginning a study. The choice of extraction method varies among specific laboratories, but in general, elutriation, sugar centrifugation, sieving, or misting techniques are preferred over Baermann funnel methods for several reasons (Barker, 1985). Barker (1985) summarizes extraction techniques and efficiencies for several species and sizes of nematodes. These methods generally extract many sizes of nematodes, and a general diversity of nematodes, whereas the Baermann funnel method, popular with soil ecologists, is more variable and eliminates some of the larger worms from the analysis (Viglierchio & Schmitt, 1983a). Nevertheless, the Baermann funnel is used in many studies because it is more economical and less labor intensive. The general methods for nematode extraction using Baermann funnels are outlined in Freckman and Baldwin (1990). Here are a few hints on using this technique:

1. To assure the quality of samples and extraction, begin the study by mixing the soil gently, splitting the sample; send a duplicate subsample to a colleague for comparison.
2. This method extracts motile, free-living species and life-stages.
3. Higher extraction efficiencies are obtained from sandier soils than from clay soils.
4. Efficiency decreases with a variety of factors; a few causes of lower efficiency include: O_2 depletion within the funnel, dirty and pitted glassware, quality of the tissue paper, how tissue paper is layered on the funnel, excess sample volume (see Viglierchio & Schmitt, 1983a).
5. Types of tissue and aeration are important, and nematodes need to rapidly move out of the soil and into the funnel stem. Spreading out the sample on the tissue paper will increase the surface area from which nematodes can be extracted.
6. Most nematodes are recovered after 1 d. Often the total extraction time is 2 to 3 d, with collection of the extracted nematodes every 24 h (because of potential O_2 deficits in the funnel water). Longer extraction times (>3 d) allow the eggs of bacterial-feeding species to hatch, and an overestimation of these nematodes' importance can result. For example, in ecological studies of microbial contributions to soil processes or of microbial succession, you would not want to overestimate the proportion of bacterial feeders.

Interpretation of Nematode Results

Nematode taxonomic identification at the functional–trophic level is based on morphology and known feeding habits (Freckman & Baldwin, 1990; Yeates et al., 1993). Microscopic identification of many free-living nematodes to species requires experience and expertise, although identification to family level can be learned through a class and practice (Freckman & Baldwin, 1990; Yeates et al., 1993). More detailed analyses of nematode community structure may become possible with the advent of molecular techniques of identification (van der Knaap et al., 1993). Several diversity indices that use nematodes as indicators of soil quality are presently being tested to determine which index is the most rigorous for a range of soils and disturbances (Yeates & Bird, 1994; Freckman & Ettema, 1993; Neher & Campbell, 1994). Obviously, the more nematodes identified at any taxonomic level, the better the results, particularly since nematodes are patchily distributed in soil. Tests by Williams and Winslow (1955) and Peters (1952) indicate that the percentage of total nematodes counted per sample will influence results of nematode assays. For example, Williams and Winslow (1955) found a minimum of 100 *Heterodera* juveniles and eggs had to be counted to estimate with 90% accuracy, the actual extracted density. Seinhorst (1970) recommended counting all nematodes if the total extracted from a sample appeared to be < 200. Enumeration of only a small percentage of the extracted nematodes should be avoided, since nematode indices for soil quality assessments are frequently based on species or family richness and abundance. General recommendations of accepted rigor are: identification of 100% of the nematodes per sample for analyses based on trophic group level; identification of >25% of the nematodes extracted, or at least 100 nematodes per sample, for family or higher taxonomic groupings. Results for soil are generally presented as nematode numbers per soil dry weight, and with bulk density (see Sarrantonio et al., 1996, this publication; Arshad et al., 1996, this publication), nematode densities can be calculated per unit area (i.e., m^{-2}) to a specified sample depth.

A major question for soil quality samples is, what is the minimum level of nematode identification needed to assess changes in soils? Parmelee et al. (1995) noted that a trophic–functional categorization of nematodes based on esophageal morphology gave results comparable with a more detailed taxonomic analysis (also see Bohlen & Edwards, 1994). Other recent research has focussed on questions of redundancy of species in soils, emphasizing a more detailed species identification. The Maturity Index (MI; Bongers, 1990) was proposed for the family taxonomic level of nonparasitic nematode feeding groups as an index of soil disturbance. This index incorporates the life histories and feeding habits of nematodes, ranking them on a scale of 1 to 5, with taxa having rapid reproductive rates and high tolerance to disturbance ranked as 1, and species having slow reproductive rates and greater sensitivity to disturbance ranked as 5. Less disturbed soils would, therefore, generally have a higher MI index value. Freckman and Ettema (1993) found that of the diversity indices tested, a multivariate analysis and the MI best predicted disturbance in agroecosystem. Others have noted the potential utility of the MI (Ettema & Bongers, 1993; Neher & Campbell, 1994; Neher et al., 1995) and the MI's limitations (Bernard, 1992; Yeates & Bird, 1994). Yeates and Bird (1994) examined management regimes in Australia and found the MI did not describe the level of soil disturbance with a single sampling. They found species richness and the Shannon index best described changes in nematode composition related to soil management. The MI, as any index, appears to be useful for some quality assessments under some conditions. Further research and testing of indices using nematodes will be required to determine if a particular index of minimal taxonomic resolution can be used universally for predicting soil disturbance. Additional appraisals of nematodes as indicators of soil quality are discussed in Gupta and Yeates (1997).

METHODS FOR ASSESSING EARTHWORMS

Accurate quantitative sampling of earthworm populations or biomass is difficult without some knowledge of the basic life history of the populations being sampled. The depth distribution of individual worms varies with different species and between different life stages within a species. Three major ecological groupings of earthworms have been defined, based primarily on feeding and burrowing strategies (Bouché, 1977). *Epigeic* species live in or near the surface litter, and feed primarily on coarse particulate organic matter. They are typically small, and have high metabolic and reproductive rates as adaptations to the highly variable environmental conditions at the soil surface. *Endogeic* species live within the soil profile and feed primarily on soil and associated organic matter (geophages). They generally inhabit temporary burrow systems which are filled with cast material as the earthworms move through the soil. *Anecic* species, such as the familiar *Lumbricus terrestris*, live in more or less permanent vertical burrow systems which may extend up to 2 m into the soil profile. They feed primarily on surface litter that they pull into their burrows, and also may create *middens* at the burrow entrance, consisting of a mixture of cast soil and partially incorporated surface litter. Although these ecological categories are very broad, they indicate that different methodologies of sampling are required for different ecological groupings of earthworms. There also is seasonal variation in depth distribution for many species, with some species adopting resting stages deeper in the soil to escape unfavorable conditions at certain times of year. Whatever method is used to sample populations, small individuals are easily overlooked, leading to inaccurate estimates of population density. It is therefore critical to have some knowledge of the species present, their life cycles and seasonal variation, before deciding upon specific methods for sampling earthworm populations.

There are two main categories of methods for sampling earthworm populations: physical and behavioral. Physical methods involve removing a known volume of soil and sorting through it manually or mechanically. Behavioral methods rely upon the response of earthworms to an irritant, usually a solution of dilute formalin, applied to the soil. The irritant causes the earthworms to come to the surface where they can be collected manually. Different methods work better for different earthworm species and the two methods, physical and behavioral, can often be combined successfully to provide more accurate results than could be attained by using either method alone. Before choosing a particular approach, preliminary studies should be done to assess the relative differences of the various methodologies and their applicability to the particular site under study. Thorough assessments of earthworm populations can be labor intensive, and the methods selected will necessarily be a compromise between the level of accuracy desired and the labor and time available. There are several excellent reviews of methods for sampling earthworm populations (Satchell, 1969; Bouché, 1972; Bouché & Gardner, 1984; Lee, 1985; Edwards & Bohlen, 1995b). We provide here a summary of methods that can be adapted successfully to the vast majority of sampling situations.

RECOMMENDED EARTHWORM SAMPLING PROCEDURES

Handsorting

All physical methods for assessing earthworm populations involve removing a known volume of soil from which the earthworms can be manually or mechanically collected. The soil can simply be dug with a spade or removed with a mechanical device, such as a hydraulic coring tool (e.g., Berry & Karlen, 1993). When digging with a spade, it can be difficult to remove replicable volumes of soil, a problem that can be overcome by driving a steel quadrat into the ground, to mark the edges of the sampling unit. Typical dimensions for the surface area of soil removed are 25 by 25 cm and we recommend that sample units of this size be used. Smaller sampling areas may be inadequate when population densities are low. Also, the proportion of physically damaged worms in the sample is greater when very small sampling units are used. Larger quadrats decrease sampling efficiency when soils are handsorted, because of the great volumes of soil that must be sorted (Zicsi, 1958). The depth of the soil removed must match the vertical distribution of the species present. This can be determined by preliminary sampling, and may vary with time of year (e.g., Persson & Lohm, 1977; Baker et al., 1992). We recommend sampling depths of 20 to 30 cm. Soil can be removed incrementally if more detailed data on depth distribution are desired.

Earthworms can be separated from the soil either by handsorting or washing–sieving. Handsorting can be done in the field by placing the soil on plastic sheets and carefully sorting through it or by returning soil to the laboratory. The soil should be handsorted immediately or stored at 4°C, because any worms killed during digging will decay rapidly. Handsorting is the simplest method but is laborious. Handsorting does not require any special apparatus, and does not require that large amounts of soil be removed from the field.

One disadvantage of handsorting is that it is not effective at recovering very small specimens. Small worms, variously defined as worms smaller than 200 mg fresh weight (Satchell, 1963; Persson & Lohm, 1977) or <2 cm in length (Reynolds, 1973), are easily overlooked during handsorting. In most cases, small worms are not a large portion of total earthworm biomass, but they may contribute significantly to earthworm numbers. Handsorting can be supplemented with washing and sieving to correct for very small specimens or collect earthworm cocoons. Washing and wet-sieving can be accomplished using a standard soil sieve (Parmelee & Crossley, 1988) or more elaborate mechanical washing–sieving devices (e.g., Edwards et al., 1970; Bouché & Beugnot, 1972). We recommend that standard soil sieves be used with the mesh size selected to prevent small earthworms or cocoons from washing through. Soils with a large clay content may need to be soaked in a 0.5% metaphosphate solution to facilitate sieving. Formalin may be added to the solution to preserve the worms. Another situation in which handsorting is ineffective and washing–sieving may be preferred is when sampling turf or pasture grass mats with dense fibrous root systems. A combination of handsorting and washing–sieving should be used for turf mats.

Physical sampling methods are by far the best for an accurate population assessment of shallow-dwelling earthworm species such as those belonging to the lumbricid genera *Aporrectodea* and *Octolasion* or the native American megascolecid genus *Diplocardia*. However, species such as *Lumbricus terrestris*, which form permanent burrows that may extend up to 2 m deep, are not adequately sampled by physical methods. Small immature specimens of this species may be recovered by handsorting, but adults, which can comprise the majority of earthworm biomass at some sites, are rarely recovered by digging or coring methods. Behavioral sampling methods must be used for an adequate assessment of populations of species such as *L. terrestris*.

Behavioral Methods

The most widely used behavioral sampling method is the formalin expulsion technique, first described by Raw (1959). This method consists of applying a dilute formalin solution (0.25% or approximately 14 mL of 37% formalin in 2 L water) to a known area of soil. Earthworms are irritated by the formalin solution and expelled onto the soil surface where they can be easily collected. Typically, 2 L of dilute formalin are applied to a 50 by 50 cm quadrat (0.25 m^2), and amounts added to different sized areas should be adjusted proportionately. The solution should be applied gradually to prevent it from pooling and running off the selected quadrat. Quadrat frames made from strips of aluminum flashing or sheet metal can be pressed a few centimeters into the ground to mark the boundaries of the quadrat and prevent the extractant from running off. Earthworms may take several minutes to emerge from the soil and collection should continue until worms cease emerging. The advantage of formalin expulsion is that it is much less labor intensive than physical sampling methods.

The formalin expulsion method does have some serious limitations. It is most effective for species that have vertical burrows that open to the surface, and is less effective for horizontally burrowing species. It is totally ineffective at recovering worms in resting stages. The method is not effective at low soil temperatures (below 4–8°C), or when the soil is either very wet or dry. Temperature optima may vary at different locations and under different environmental conditions. Because of these limitations, the formalin method works best when used during times of years when populations are most active, typically spring and fall in temperate regions. Under these conditions, the method is excellent for deep-dwelling species and may even provide a good relative measure of populations of those species that are not recovered as well by formalin as by handsorting (Bouché & Gardner, 1984; Bohlen et al., 1995a,b). Another problem with formalin extraction is that it is difficult to determine the exact volume of soil being sampled, because there is no way to determine the flow paths of the formalin. Nonetheless, it is by far the best method for sampling *L. terrestris* and similar species. If formalin can not be used due to objections over its toxicity, a nontoxic alternative is to use a solution of mustard flour (0.33%; Gunn, 1992), although this method is not as well-studied or standardized as the formalin method and cannot be recommended for routine scientific sampling.

Combination of Physical and Behavioral Methods

In situations where both deep-dwelling and shallow-dwelling species coexist, we recommend a combination of physical and behavioral methods for the most accurate assessment of earthworm populations. We recommend that formalin solution be applied to the bottom of the pit dug to obtain samples for handsorting. Worms emerging in the bottom of the pit can be combined with those collected by handsorting (Martin, 1976; Barnes & Ellis, 1979). We have used this technique in our studies of earthworm populations in corn (*Zea mays* L.) agroecosystems on a silt loam soil in Ohio, and have found it to be very satisfactory (Bohlen et al., 1995b).

Interpretation and Analysis of Earthworm Data

Data for earthworm populations are expressed as total biomass or number of earthworms per unit area, generally as grams or number per square meter. The data can be presented for the total earthworm community, each species within the community or for functional groups (e.g., deep burrowers or shallow burrowers) or size classes (e.g., small immature worms or mature adults). Biomass can be expressed as fresh weight, oven dry weight or ash-free-dry-weight. Ash-free-dry-mass (AFDM) is preferred because it allows for the most accurate comparison of data from different times or studies. It is determined by burning the oven-dried worms (60°C to constant weight), or a ground subsample, in a muffle furnace at 500°C for 4 h (g AFDM = g dry wt. − g ash wt.). This corrects for varying mass of soil in the earthworms' intestines, which can account for 50 to 70% of earthworm dry mass. A difficulty that arises in analyzing data from physical sampling methods is that each sample usually contains body parts, in addition to whole individuals. This is not a problem for estimates of earthworm biomass per unit area, but it complicates the accurate determination of numbers of individuals in a sample. An arbitrary method for overcoming this problem is to count as an individual any fragment of a whole worm that contains the anterior portion (head). Body fragments lacking a head are included in total biomass but not counted as individuals.

An essential component of interpreting data on earthworm populations is to determine an appropriate baseline for comparison. Results from particular sites can be compared to values in the literature for similar sites; however, differences in site history or edaphic factors can significantly affect earthworm numbers, even in similar sites. Therefore, data from controlled experiments with replicated plots provide the best basis for determining cause and effect relationships between different management practices and earthworm population density, biomass or species composition. An alternate approach is to sample populations across a landscape at sites representing different ecosystem types, management practices, landscape positions, or soil textural classes and relate results to differences among the sites (Hendrix et al., 1992). Factors controlling earthworm abundance can be difficult to determine. For example, Bohlen et al. (1995a) compared earthworm populations from two agricultural fields on similar soil types that had been managed identically for 50 yr and found that the fields differed greatly in numbers, biomass and species composition of earthworms. The proximity of source populations and the presence of barriers to colonization can strongly influence the development of earthworm communities at a particular site. For this reason, it is useful to have information on the distribution of different species across the landscape, which should be considered before designing a sampling plan. Comparisons also can be made over time to determine seasonal dynamics or long-term changes in earthworm populations. This approach is particularly useful for assessing the response of populations to changes in management or tillage practices.

SUMMARY

We have indicated the rationale and potential utility of using soil invertebrates as indicators of soil quality, or as indicators of change in soil quality due to external factors (management practices, chemical pollutants, climate change, and others). Because of the variability in soil invertebrate populations, and the variety of natural factors that can influence their abundances, the greatest potential use of soil invertebrates may be in assessing changes in soil quality resulting from a change in management practice, or as biological indicators of a directional environmental change. In these instances, comparisons with nearby nonimpacted sites, or documentation of changes in invertebrate populations over time can provide information on soil function. We provided some general recommendations regarding sampling procedures for both nematodes and earthworms; however, it is important to emphasize that the details of a given sampling–monitoring regime will depend on the questions being addressed and site-specific characteristics of the soils, vegetation, and invertebrate communities. In fact, we strongly suggest that a preliminary sampling be done to characterize the soil invertebrate community of a particular site as a prelude to determining a site-specific sampling plan. Although we focussed on the use of nematodes and earthworms as particularly useful indicator organisms, we also should point out that other soil invertebrate groups may be useful in assessing changes in soil quality (i.e., Blair & Crossley, 1988; van Straalen et al., 1988; Parmelee et al., 1993). Although considerable research remains to be done on both the biodiversity and functional significance of soil invertebrates, it is essential that the assessment and monitoring of soil quality include the soil fauna.

REFERENCES

Aiken, R.M., M.D. Jawson, K. Grahammer, and A.D. Polymenopoulos. 1991. Positional, spatially correlated and random components of variability in carbon dioxide efflux. J. Environ. Qual. 20:301–308.

Anderson, R.V., D.C. Coleman, C.V. Cole, and E.T. Elliott. 1981. Effects of the nematodes *Acrobeloides* sp. and *Mesodiplogaster lheritieri* on substrate utilization and nitrogen and phosphorus mineralization in soil. Ecology 62:549–555.

Andre, H.M, M.-I. Noti, and P. Lebrun. 1994. The soil fauna: the other last biotic frontier. Biodiversity Conserv. 3:45–56.

Arshad, M., B. Lowery, and B. Grossman. 1996. Physical tests for monitoring soil quality. p. 123–141. *In* J.W. Doran and A.J. Jones (ed.) Methods for assessing soil quality. SSSA Spec. Publ. 49. SSSA, Madison, WI.

Baker, G.H., V.J. Barrett, R. Grey-Gardner, and J.C. Buckerfield. 1992. The life history and abundance of the introduced earthworms *Aporrectodea trapezoides* and *A. caliginosa* (Annelida: Lumbricidae) in pasture soils in the Mount Lofty Ranges, South Australia. Aust. J. Ecol. 17:177–188.

Barker, K.R. 1978. Determining nematode population responses to control agents. p. 114–125. *In* E.I. Zehr (ed.) Methods for evaluating plant fungicides/nematicides and bactericides. Am. Phytopathol. Soc., St. Paul, MN.

Barker, K.R. 1985. Nematode extraction and bioassays. p. 19–38. *In* K.R. Barker et al. (ed.) An advanced treatise on meloidogyne. Vol. 2. Methodology. N.C. State Univ. Press, Raleigh.

Barnes, B.T., and F.B. Ellis. 1979. Effects of different methods of cultivation and direct drilling and disposal of straw residues, on populations of earthworms. J. Soil Sci. 30:669–679.

Bernard, E.C. 1992. Soil nematode biodiversity. Biol. Fert. Soils 14:99–103.

Berry, E.C., and D.L. Karlen. 1993. Comparison of alternative farming systems. II. Earthworm population density and species diversity. Am. J. Alter. Agric. 8:21–26.

Blair, J.M., and D.A. Crossley, Jr. 1988. Litter decomposition, nitrogen dynamics and litter microarthropods in a southern Appalachian hardwood forest eight years following clearcutting. J. Appl. Ecol. 25:683–698.

Blair, J.M., R.W. Parmelee, M.F. Allen, D.A. McCartney, and B.R. Stinner. 1996. Changes in soil N pools in response to earthworm population manipulations under different agroecosystem treatments. Soil Biol. Biochem. (In press).

Blair, J.M., R.W. Parmelee, and P. Lavelle. 1995. Influences of earthworms on biogeochemistry. p. 127–158. *In* P.F. Hendrix (ed.) Earthworm ecology and biogeography in North America. Lewis Publ., Boca Raton, FL.

Blair, J.M., R.W. Parmelee, and R.L. Wyman. 1994. A comparison of forest floor invertebrate communities of four forest types in the northeastern U.S. Pedobiologia 38:146–160.

Bohlen, P.J., and C.A. Edwards. 1994. The response of nematode trophic groups to organic and inorganic nutrient inputs in agroecosystems. p. 235–244. *In* J.W. Doran et al. (ed.) Defining soil quality for a sustainable environment. SSSA Spec. Publ. 35. SSSA, Madison, WI.

Bohlen P.J., and C.A. Edwards. 1995. Earthworm effects on N dynamics and soil respiration in microcosms receiving organic and inorganic nutrients. Soil Biol. Biochem. 27:341–348.

Bohlen, P.J., W.M. Edwards, and C.A. Edwards. 1995a. Earthworm community structure and diversity in experimental agricultural watersheds in Northeastern Ohio. Plant Soil 164:536–543.

Bohlen, P.J., R.W. Parmelee, J.M. Blair, C.A. Edwards, and B.R. Stinner. 1995b. Efficacy of methods for manipulating earthworm populations in large-scale field experiments in agroecosystems. Soil Biol. Biochem. 27:993–999.

Bongers, T. 1990. The maturity index: An ecological measure of environmental disturbance based on nematode species composition. Oecologia 83:14–19.

Bouché, M.B. 1972. Lombriciens de France. Ecologie et systématique. INRA Publ. 72-2. Inst. Natl. Recherches Agric., Paris.

Bouché, M. B. 1977. Stratégies lombriciennes. Soil organisms as components of ecosystems. Ecol. Bull. (Stockholm) 25:122–132.

Bouché, M.B., and R.H. Gardner. 1984. Earthworm functions: VIII. Population estimation techniques. Rev. Ecol. Biol. du Sol 21:37–63.

Bouché, M. B., and M. Beugnot. 1972. Contribution à l'approche méthodologique de l'étude des biocénoses: II. L'extraction des macroéléments du sol par lavage-tamisage. Ann. Zool. Ecol. Anim. 4:5.37–544.

Coleman, D.C., R.V. Anderson, C.V. Cole, J.F. McClellan, L.E. Woods, J.A. Trofymow, and E.T. Elliott. 1984. Roles of protozoa and nematodes in nutrient cycling. p. 17–28. *In* D.M. Kroll (ed.) Microbial–plant interactions. ASA Spec. Publ. 47. ASA, Madison, WI.

de Ruiter, P.C., J. Bloem, L.A. Bouwman, W.A.M. Didden, G.H.J. Hoenderbloom, B. Lebbink, J.Y.C. Marinissen, J.A. de Vos, M.J. Vreeken-Buijs, K.B. Zwart, and L. Brussaard. 1994. Simulation of dynamics in nitrogen mineralization in the belowground food webs of two arable farming systems. Agric. Ecosys. Environ. 51:199–208.

Didden, W.A.M., J.C.Y. Marinissen, M.J. Vreeken-Buijs, S.L.G.E. Burgers, R. de Fluiter, M. Guers, and L. Brussaard. 1994. Soil meso- and macrofauna in two agricultural systems: Factors affecting population dynamics and evaluation of their role in carbon and nitrogen dynamics. Agric. Ecosys. Environ. 51:171–186.

Doran, J.W., and T.B. Parkin. 1994. Defining and assessing soil quality. p. 3–21. *In* J.W. Doran et al. (ed.) Defining soil quality for a sustainable environment. SSSA Spec. Publ. 35. SSSA, Madison, WI.

Edwards, C.A., and P.J. Bohlen. 1992. The effects of toxic chemicals on earthworms. Rev. Environ. Contam. Toxic. 125:24–99.

Edwards, C.A., and P.J. Bohlen. 1995a. The effects of contaminants on the structure and function of soil communities. Acta Zool. Fennica 196:284–289.

Edwards, C.A., and P.J. Bohlen. 1995b. Biology and Ecology of Earthworms. Chapman & Hall, London.

Edwards, C.A., A.E. Whiting, and G.W. Heath. 1970. A mechanized washing method for separation of invertebrates from soil. Pedobiologia 10:141–148.

Ettema, C.H., and T. Bongers. 1993. Characterization of nematode colonization and succession in disturbed soil using the Maturity Index. Biol. Fert. Soils 16:79–85.

Ferris, H., T.A. Mullens, and K.E. Ford. 1990. Stability and characteristics of spatial description parameters for nematode populations. J. Nematol. 22:427–439.

Folorunso, O.A., and D.E. Rolston. 1985. Spatial and spectral relationships between field-measured denitrification gas fluxes and soil properties. Soil Sci. Soc. Am. J. 49:1087–1093.

Freckman, D.W. 1988. Bacterivorous nematodes and organic-matter decomposition. Agric. Ecosyst. Environ. 24:195–217.

Freckman, D.W. 1994. Life in the soil: Soil biodiversity: Its importance to ecosystem processes. Colorado State Univ., Fort Collins, CO.

Freckman, D.W., and J.G. Baldwin. 1990. Nematoda. p. 155–200. *In* D.L. Dindal (ed.) Soil biology guide. Wiley Interscience, New York.

Freckman, D.W., and C.H. Ettema. 1993. Assessing nematode communities in agroecosystems of varying human intervention. Agric. Ecosyst. Environ. 45:239–261.

Freckman, D.W., and R.A. Virginia. 1989. Plant feeding nematodes in deep-rooting desert ecosystems. Ecology 70:1665–1678.

Goodell, P.B., and H. Ferris. 1981. Sample optimization for five plant-parasite nematodes in an alfalfa field. J. Nematol. 13:304–313.

Groombridge, B. 1992. Global Biodiversity: Status of the Earth's Living Resources. Chapman & Hall, London.

Griffiths, B.S., and S. Caul. 1993. Migration of bacterial-feeding nematodes, but not protozoa, to decomposing grass residues. Biol. Fert. Soils 15:201–207.

Gunn, A. 1992. The use of mustard to estimate earthworm populations. Pedobiologia 36:65–67.

Gupta, V.V.S.R., and G.W. Yeates. 1997. Soil microfauna as indicators of soil health. *In* C.E. Pankhurst et al. (ed.) Bioindicators of soil heath. CAB Int., Wallingford (In press).

Hanlon, R.D.G., and J.M. Anderson. 1979. The effects of Collembola grazing on microbial activity in decomposing leaf litter. Oecologia 38:93–100.

Heal, O.W., S. Struwe, and A. Kjoller. 1996. Diversity of soil biota and ecosystem function. p. 385–402. *In* B. Walker and W. Streffen (ed.) Global change and terrestrial ecosystems. IGBP. Vol. 1. Cambridge Univ. Press, Cambridge, England.

Hendrix, P.F., D.A. Crossley, Jr., J.M. Blair, and D.C. Coleman. 1990. Soil biota as components of sustainable agroecosystems. p. 637–654. *In* C.A. Edwards et al. (ed.) Sustainable agricultural systems. Soil and Water Conserv. Soc., Ankeny, IA.

Hendrix, P.F., B.R. Mueller, R.R. Bruce, G.W. Langdale, and R.W. Parmelee. 1992. Abundance and distribution of earthworms in relation to landscape factors on the Georgia Piedmont, U.S.A. Soil Biol. Biochem. 24:1357–1361.

Hendrix, P.F., R.W. Parmelee, D.A. Crossley, Jr., D.C. Coleman, E.P. Odum, and P.M. Groffman. 1986. Detritus food webs in conventional and no-tillage agroecosystems. Bioscience 36:374–380.

Ingham, R.E., J.A. Trofymow, E.R. Ingham, and D.C. Coleman. 1985. Interactions of bacteria, fungi, and their nematode grazers: Effects on nutrient cycling and plant growth. Ecol. Monogr. 55:119–140.

Johnson, S.R., V.R. Ferris, and J.M. Ferris. 1972. Nematode community structure of forest woodlots: I. Relationships based on similarity coefficients of nematode species. J. Nematol. 4:175–182.

Lavelle, P, I. Barois, A. Martin, Z. Zaidi, and R. Schaefer. 1989. Management of earthworm populations in agro-ecosystems: A possible way to maintain soil quality? p. 109–122. *In* M. Clarholm and L. Bergström (ed.) Ecology of rable land. Kluwer Academic Publ., the Netherlands.

Lee, K.E. 1985. Earthworms: Their ecology and relationships with soils and land use. Academic Press, New York.

Linden, D.R., P.F. Hendrix, D.C. Coleman, and P.C.J. van Vliet. 1994. Faunal indicators of soil quality. p. 91–106. *In* J.W. Doran et al. (ed.) Defining soil quality for a sustainable environment. SSSA Spec. Publ. 35. SSSA, Madison, WI.

Martin, N.A. 1976. Effect of four insecticides on the pasture ecosystem. V. Earthworms (Oligochaeta: Lumbricidae) and Arthropoda extracted by wet sieving and salt flotation. N.Z. J. Agric. Res. 1:175–176.

Moore, J.C., D.E. Walter, and H.W. Hunt. 1988. Arthropod regulation of micro- and mesobiota in below-ground detrital food webs. Ann. Rev. Entomol. 33:419–439.

Neher, D.A., and C.L. Campbell. 1994. Nematode communities and microbial biomass in soils with annual and perennial crops. Appl. Soil Ecol. 1:17–28.

Neher, D.A., S.L. Peck, J.O. Rawlings, and C.L. Campbell. 1995. Measures of nematode community structure and sources of variability among and within agricultural fields. p. 187–201. *In* H.P. Collins et al. (ed.) The significance and regulation of soil biodiversity. Kluwer Academic Publ., the Netherlands.

Newell, K. 1984a. Interaction between two decomposer basidiomycetes and a collembolan under Sitka spruce: Distribution, abundance and selective grazing. Soil Biol. Biochem. 16:227–233.

Newell, K. 1984b. Interaction between two decomposer basidiomycetes and a collembolan under Sitka spruce: grazing and its potential effects on fungal distribution and litter decomposition. Soil Biol. Biochem. 16:235–239.

Paoletti, M.G., M.R. Favretto, B.R. Stinner, F.F. Purrington, and J.E. Bater. 1991. Invertebrates as bioindicators of soil use. Agric. Ecosys. Environ. 34:341–362.

Parmelee, R.W., P.J. Bohlen, and C.A. Edwards. 1995. Analysis of ematode trophic structure in agroecosystems: functional groups vs. high resolution taxonomy. p. 203–207. *In* H.P. Collins et al.

(ed.) The significance and regulation of soil biodiversity. Kluwer Academic Publ., the Netherlands.

Parmelee, R.W., and D.A. Crossley, Jr. 1988. Earthworm production and role in the nitrogen cycle of a no-tillage agroecosystem on the Georgia Piedmont. Pedobiologia 32:353–361.

Parmelee, R.W., R.S. Wentsel, C.T. Phillips, M. Simini, and R.T. Checkai. 1993. A soil microcosm for testing the effects of chemical pollutants on soil fauna communities and trophic structure. Environ. Toxicol. Chem. 12:1477–1486.

Persson, T., and U. Lohm. 1977. Energetical significance of the annelids and arthropods in a Swedish grassland soil. Ecol. Bull. (Stockholm) 23:1–211.

Peters, B.G. 1952. Toxicity tests with vinegar eelworm: 1. Counting and culturing. J. Helmin. 26:97–110.

Petersen, H., and M. Luxton. 1982. A comparative analysis of soil fauna populations and their role in decomposition processes. Oikos 39:287–388.

Raw, F. 1959. Estimating earthworm populations by using formalin. Nature (London) 184:1661–1662.

Reynolds, J.W. 1973. Earthworm (Annelida:Oligochaeta) ecology and systematics. p. 95–120. *In* D. Dindal (ed.) Proc. Soil Microcommunities Conf. 1st, Syracuse, NY. U.S. Atomic Energy Commission, Washington, DC.

Robertson, G.P., J.R. Crum, and B.G. Ellis. 1993. The spatial variability of soil resources following long-term disturbance. Oecologia 96:451–456.

Robertson, G.P., and D.W. Freckman. 1995. The spatial distribution of nematode trophic groups across a cultivated ecosystem. Ecology 76:1425–1432.

Sarrantonio, J.W. Doran, M.A. Liebig, and J.J. Halovorson. 1996. On-farm assessment of soil quality and health. p. 83–105. *In* J.W. Doran and A.J. Jones (ed.) Methods for assessing soil quality. SSSA Spec. Publ. 49. SSSA, Madison, WI.

Satchell, J.E. 1963. Nitrogen turnover by a woodland population of *Lumbricus terrestris*. p. 60–66. *In* J. Doeksan and J. van der Drift (ed.) Soil organisms. North Holland Publ. Company, Amsterdam.

Satchell, J.E. 1969. Methods of sampling earthworm populations. Pedobiologia 9:20–25.

Schulze, E.-D., and H.A. Mooney. 1994. Biodiversity and ecosystem function. Springer-Verlag, New York.

Schmitt, D.P., K.R. Barker, J.P. Noe, and S.R. Koenning. 1990. Repeated sampling to determine the precision of estimating nematode population densities. J. Nematol. 22:552–559.

Seastedt, T.R. 1984. The role of microarthropods in decomposition and mineralization processes. Ann. Rev. Entomol. 29:25–46.

Seinhorst, J.W. 1970. Dynamics of populations of plant parasitic nematodes. Ann. Rev. Phytopath. 8:131–136.

Stolgren, T.J. 1994. Planning long-term vegetation studies at landscape scales. p. 209–241. *In* J.H. Steele and T.M. Powell (ed.) Ecological time series. Chapman & Hall, New York.

Stolgren, T.J., J.F. Quinn, M. Ruggiero, and G.S. Waggoner. 1995. Status of biotic inventories in U.S. national parks. Biol. Conserv. 71:97–106.

Stork, N.E., and P. Eggleton. 1992. Invertebrates as determinants and indicators of soil quality. Am. J. Altern. Agric. 7:39–47.

Swift, M.J., and J.M. Anderson. 1994. Biodiversity and ecosystem function in agricultural ecosystems. p. 15–42. *In* E.-D. Schulze and H.A. Mooney (ed.) Biodiversity and ecosystem function. Springer-Verlag, New York.

Swift, M.J., O.W. Heal, and J.M. Anderson. 1979. Decomposition in terrestrial ecosystems. Univ. of California Press, Berkeley, CA.

Tomlin, A.D., M.J. Shipitalo, W.M. Edwards, and R. Protz. 1995. Earthworms and their influence on soil structure and infiltration. p. 159–183. *In* P.F. Hendrix (ed.) Earthworm ecology and biogeography in North America. Lewis Publ., Boca Raton, FL.

van der Knaap, E., R.J. Rodriguez, and D.W. Freckman. 1993. Differentiation of bacterial-feeding nematodes in soil ecological studies by means of arbitrarily primed PCR. Soil Biol. Biochem. 25:1141–1151.

van Straalen, N.M., H.S. Kraak, and C.A.J. Denneman. 1988. Soil arthropods as indicators of acidification and forest decline in the Veluwe area, The Netherlands. Pedobiologia 32:47–55.

Verhoef, H.A., and L. Brussaard. 1990. Decomposition and nitrogen mineralization in natural and agroecosystems: The contribution of soil animals. Biogeochem. 11:175–211.

Viglierchio, D.R., and R.V. Schmitt. 1983a. On the methodology of nematode extraction from field samples: Baermann funnel modifications. J. Nematol. 15:438–444.

Viglierchio, D.R., and R.V. Schmitt. 1983b. On the methodology of nematode extraction from field samples: comparison of methods for soil extraction. J. Nematol. 15:450–454.

Viglierchio, D.R., and T.T. Yamashita. 1983. On the methodology of nematode extraction from field samples: Density flotation techniques. J. Nematol. 15:445–449.

Wallwork, J.A. 1988. The soil fauna as bioindicators. p. 203–215. *In* Biologia ambiental: Actas del Congreso de Biologia Ambiental (II Congreso Mundial Vasco). San Sebastian, Spain. 1987. Servicio Editorial, National Basque Univ., Paris.

Wasilewska, L. 1989. Impact of human activities on nematode communities in terrestrial ecosystems. p. 123–132. *In* M. Clarholm and L. Bergström (ed.) Ecology of arable land. Kluwer Academic Publ., the Netherlands.

Wasilewska, L. 1991. Long term changes in communities of soil nematodes on fen peat meadows due to the time since their drainage. Ekologia Polska 39:59–104.

Williams, T.D., and R.D. Winslow. 1955. A synopsis of some laboratory techniques used in the quantitative recovery of cyst-forming and other nematodes from soil. p. 375–384. *In* D.K. McE. Kevan (ed.) Soil zoology: Proc. of Nottingham Second Easter School in Agricultural Science. Academic Press, New York.

Yeates, G.W. 1979. Soil nematodes in terrestrial ecosystems. J. Nematol. 11:213–229.

Yeates, G.W., and A.F. Bird. 1994. Some observations on the influence of agricultural practices on the nematode fauna of some South Australian soils. Fund. Appl. Nematol. 17:133–145.

Yeates, G.W., T. Bongers, R.G.M.D. Goede, D.W. Freckman, and S.S. Georgieva. 1993. Feeding habits in soil nematode families and genera- an outline for soil ecologists. J. Nematol. 25:315–331.

Zicsi, A. 1958. Determination of number and size of sampling unit for estimating lumbricid populations of arable soils. p. 68–71. *In* P.W. Murphy (ed.) Progress in soil zoology. Butterworth, London.

17 Tests for Risk Assessment of Root Infection by Plant Pathogens

Ariena H. C. van Bruggen and Niklaus J. Grünwald

University of California
Davis, California

Soil quality is partially determined by the risk of root infection by plant pathogens. This risk is dependent on the population density of plant pathogens in soil and the ability of the biological community to suppress pathogens and disease development. Disease suppression is the phenomenon that less disease is incited than would be expected in the presence of a susceptible host and a virulent plant pathogen, under conditions normally conducive for infection (Hornby, 1983). Soil conduciveness also is called soil receptivity, in particular in the french literature (Alabouvette, 1986).

Some level of disease suppression is present in all soils, which can be demonstrated by comparing disease severity in natural field soil and sterilized soil both artificially infested with a pathogen; however, the level of suppression varies for different soils and management regimes. In general, a soil supporting a perennial natural vegetation such as a pasture or forest ecosystem is more suppressive to pathogens than soil in an agroecosystem (Alabouvette, 1986; Ko, 1982; Ko & Kao, 1989; Chakraborty & Warcup, 1983). Similarly, organically managed soils are often more suppressive to root pathogens than conventionally managed soils (van Bruggen, 1995); however, lower disease levels in organically managed fields do not necessarily imply that disease suppression is operational. Pathogens may by chance not have been introduced into disease-free soils. One would have to conduct controlled experiments to prove that disease is suppressed in a particular soil.

Disease suppression can take place at various stages in the life cycle of a pathogen: a reduction in survival of resting structures and mycelium in host debris, inhibition of germination of spores or resting structures (called fungistasis), a reduction in growth in the rhizosphere and on the root surface, or inhibition of penetration into the root (Schneider, 1982). The soil suppressiveness tests described below address either the saprophytic phase or the parasitic phase of plant pathogens or both. Soil quality tests in relation to conduciveness to plant disease development will entail two kinds of tests: (i) to determine the inoculum density of plant pathogens in terms of the numbers of propagules that are currently present in the soil, and (ii) to determine the disease potential in terms of conduciveness of the soil to pathogen growth, infection, and disease develop-

---

Copyright © 1996 Soil Science Society of America, 677 S. Segoe Rd., Madison, WI 53711, USA.
Methods for Assessing Soil Quality, SSSA Special Publication 49.

ment. This chapter will be focused on disease suppression by competition for nutrients and general antibiosis. The micro- and mesofauna also play a role as predators of plant pathogens (Chakraborty & Warcup, 1983; Curl et al., 1988), but in this chapter we will focus on the microflora.

INOCULUM DENSITY OF PLANT PATHOGENS

The inoculum density of a plant pathogen in soil is the result of buildup and decline processes over the past years, and thus may reflect soil receptivity to this pathogen. Soil-borne fungal plant pathogens survive either as specialized survival structures like sclerotia, chlamydospores, or oospores in soil or as mycelium or fruiting bodies associated with infected plant debris. Soil-borne bacterial plant pathogens survive as dormant cells in soil or as active cells in association with plant debris or in the rhizosphere of nonhosts. Survival structures such as sclerotia can survive for many decades, and build-up to large populations when a host is planted frequently. The presence of such pathogens in soil often determines if certain crops can be grown in a particular field or not, and therefore has a major impact on soil quality.

Since pathogen populations are dynamic, the time of sampling is of paramount importance for estimating inoculum density. For comparison of different soils with respect to pathogen populations, soil samples should be collected at the same time in relation to the cropping season. The highest numbers are usually encountered towards the end of the growing season of a susceptible crop, while the lowest numbers are found for survival structures in the off-season (Rush et al., 1992).

For some pathogens there may be a positive correlation between inoculum density as determined before planting and inoculum potential in terms of capacity to cause disease, but for many soilborne pathogens, in particular fungi with some competitive saprophytic ability, there often is no direct relation between estimated inoculum density and expected disease severity (Kinsbursky & Weinhold, 1988; Keinath & Fravel, 1992). In those cases, techniques using dead or live baits (sensitive plants) that measure inoculum potential rather than inoculum density may result in better correlations with disease intensity than the numbers of propagules present at one moment in time.

Methods for the isolation and quantification of plant pathogens from soil and for measurement of inoculum potential have been summarized in two useful books by Johnson and Curl (1972) and Singleton et al. (1992).

Extraction and Quantification of Propagules from Soil

The most commonly used technique for isolation of fungi and bacteria from soil is the soil-dilution plate technique (Rush et al., 1992). A soil sample is added to sterile water or 0.2% water agar and shaken for about 30 min. Ten-fold serial dilutions are made with sterile water or 0.2% water agar, and an aliquot (0.1 or 1 mL) of the last three dilutions is spread over a solid agar plate or dispensed in a petridish to which cooled, molten (45°C) agar medium is added, respectively. The agar medium needs to be selective for the pathogen to be quantified. Semiselective media have been developed for many soil-borne plant pathogens, usually by adding various antimicrobial chemicals (Johnson & Curl, 1972; Singleton et al., 1992); however, none of the media are completely selective for the intended use, and individual colonies will need to be identified by microscopic or biochemical techniques for fungi and bacteria, respectively. Finally, the number of colony forming units (CFU) per weight or volume of soil is calculated.

For many diseases the initial inoculum density in soil may be too low to be detected by dilution plating. Yet, these inoculum densities may cause significant disease, for example damping-off caused by *Pythium ultimum* (Lifshitz & Hancock, 1981). Pathogens with a good competitive saprophytic ability such as *Pythium* species could be quantified by first amending soil samples with a food base such as dried leaf fragments followed by dilution plating of amended soil in order to obtain a good correlation between inoculum intensity and expected disease severity (Lifshitz & Hancock, 1981). On the other hand, if inoculum levels are high enough to obtain significant relationships between inoculum density and disease severity, dilution plating of soil may be sufficient. This was for example the case with *Verticillium dahliae* in commercial potato (*Solanum tuberosum* L.) fields (Nicot & Rouse, 1987b).

The disadvantage of the dilution plate method is that profusely sporulating and fast-growing fungi are dominant on the plates while nonsporulating, slow-growing pathogens are generally not recovered. An alternative method for non-sporulating or slow-growing fungi is the Warcup plate method, wherein a small soil sample (about 0.1 g) is crushed in a drop of water in a petridish to which cooled molten agar medium is added (Rush et al., 1992). Fungal colonies that grow to the surface can be identified and counted. Instead of dispersing soil particles in the agar as is done with the Warcup method, particles of air-dried soil also have been distributed over an agar surface using an Anderson air sampler (Butterfield & DeVay, 1977; DeVay et al., 1982). The Anderson sampler consists of two sieve plates (pore sizes 1.18 and 0.81 mm), a sampling tower, and an air suction tube. Petri plates with selective agar medium are placed under the sieve plates, and soil particles are sucked through the sieve plates onto the agar. Colonies of the pathogen under study are counted after an incubation period. This method has been used for quantification of microsclerotia of *Verticillium dahliae* (causing wilt diseases) and oospores of *Pythium* species (causing damping-off; Butterfield & DeVay, 1977; DeVay et al., 1982; Nicot & Rouse, 1987a). The inoculum density of *Pythium* as determined by this method correlates well with disease severity in the field, as long as the inoculum level is high enough for detection (DeVay et al., 1982). The number of microsclerotia of *V. dahliae* as determined by the Anderson sampler technique, however, was not correlated to field symptoms of Verticillium wilt of cotton (*Gossypium hirsutum* L.; Butterfield & DeVay, 1977).

For fungi that produce sclerotia, these survival structures also can be separated from soil by wet-sieving and/or flotation methods. These methods were developed many decades ago, for example for *V. dahliae*, *Sclerotium* species (causing Southern blight or root rot on many crops), *Phymatotrichum omnivorum* (causing Texas root rot of cotton), and *Sclerotinia* species (causing watery soft-rots in many crops), using screens with opening sizes adapted to the size of the sclerotia (Singleton et al., 1992). To facilitate flotation sucrose solutions can be used. Sclerotia are counted directly under a microscope. *Rhizoctonia solani* associated with plant debris also is separated from soil by wet-screening and flotation on water. Debris particles are then transferred to a selective medium and *R. solani* colonies are counted under a microscope (Weinhold, 1977). The correlations between sclerotial counts at the beginning of the season and disease severity at the end of the season have generally been very good. The advantage of these techniques is that large soil samples can be processed allowing for quantification of low inoculum levels; however, when the wet-sieving technique was compared with the soil dilution plating and Anderson sampler techniques for quantification of *Verticillium dahliae* in artificially infested sand, the wet sieving technique had the greatest bias (only 64% of the microsclerotia recovered) and was most time-consuming. The Anderson sampler was unbiased and time-efficient but had a lower precision than the other techniques (Nicot & Rouse, 1987a).

Disease Tests with Natural Inoculum

The potential for root rot in a field can sometimes be predicted by simple greenhouse tests with soil collected from that field. The best known example of such a predictive test is the indexing test for root rot of pea caused by *Aphanomyces euteiches*. Ten pea (*Pisum sativum* L.) seeds are sown in potted soil carefully maintained at moderate soil moisture levels to allow germination, and then exposed to high soil moisture to promote infection by *A. euteiches*. After one month a disease severity index is calculated for the seedlings in each pot. This disease severity index correlated well with disease severity in the field. If the disease severity index is less than half of the maximum, planting of peas is considered safe for that field. Otherwise, planting of peas is not considered safe (Parke & Grau, 1992). Similar bioassays can be used for a complex of pea pathogens including *Fusarium solani* f.sp. *pisi*, *Thielaviopsis basicola* and *Aphanomyces euteiches* (Oyarzun, 1993).

Baiting Techniques

Baiting techniques have been used to detect propagules of plant pathogens that have some competitive saprophytic ability, such as *Pythium* and *Rhizoctonia* species. Baits such as autoclaved stem or tuber sections, leaves, or seeds are mixed with soil or a soil extract and the proportion of baits infested is determined after a certain incubation period. Stanghellini and Kronland (1985) placed water agar blocks on top of disks of potato tubers on soil in petriplates to bait *Pythium aphanidermatum* (causing seedling damping-off). They obtained a linear relationship between the probit-transformed number of baits colonized and the log-transformed oospore density as determined by dilution plating on a selective medium (Stanghellini & Kronland, 1985). We modified this method by using 24-well tissue culture plates and a selective medium (containing 150 mg rose bengal, 200 mg streptomycin sulfate, and 15 g water agar per liter) instead of water agar on top of the potato disks (Fig. 17–1). This allows us to process large num-

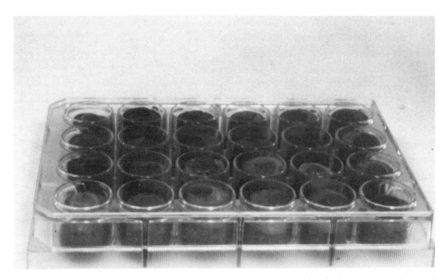

Fig. 17–1. Baiting technique for *Pythium aphanidermatum* using a 24-well tissue culture plate. 350 µL of soil are added to each well, covered with a disk of potato bait and topped with a disk of water agar amended with selective inhibitors. After 48 h of incubation at 28°C the water agar disks are transferred to a selective medium and incubated for 24 h at 35°.

bers of soil samples with 24 replications per sample, and prevents growth of *Pythium* from one unit of inoculum into more than one disk.

Baiting techniques also have been developed for *Phytophthora* species, for example *P. parasitica*, the causal agent of tomato [*Lycopersicon esculentum* (L.) Karsten] root rot (Neher et al., 1993), even though *Phytophthora* species generally don't have a high competitive saprophytic ability; however, by floating the baits on water containing the test soil, zoospores that swim towards the baits have a competitive edge over nonzoosporic fungi (Neher et al., 1993).

Obligate parasites such as *Polymyxa betae*, the vector of beet necrotic yellow vein virus (BNYVV), can only be isolated from soil by baiting with a living host plant (Tuitert, 1990).

Baiting techniques are convenient and sensitive but not truly quantitative, because the probability of contact between the propagule and the bait may vary depending on the amount of soil used and the environmental conditions of the test since the pathogen needs to grow into the bait. Thus, the efficiency of baiting techniques can only be compared when these techniques are used in conjunction with dilution end point or most probable number methods.

Dilution End Point and Most Probable Number Methods

For both of these methods natural field soil is mixed with sterilized soil or sand at different ratios. With the dilution end point method two-fold dilutions are made of natural field soil up to about seven dilutions, while with the most probable number method usually 10-fold dilutions and a larger number of replications are being used (Johnson & Curl, 1972). Use of the most probable number technique to quantify soil-borne pathogens was recently reviewed by Adams and Welham (1995). A bait such as a surface-sterilized fruit or leaf is inserted in the soil-mix (saturated with water if one wants to test for *Phytophthora* species) so that all propagules in the soil have a chance to reach the bait. With the dilution endpoint method the last dilution that gives infection is a measure of infective units per volume of soil. The number of infected baits in a dilution series can be used to calculate the most probable number of infective units per volume of soil from a most probable number table or computer program (Pfender et al., 1981; Tuitert, 1990). Computer programs allow more flexibility in the extent and number of the dilutions and replications (Pfender et al., 1981; Tuitert, 1990). MPN methods also have been used with living baits (host plants), in particular for quantification of biotrophic pathogens or vectors such as *Polymyxa betae* (Ciafardini & Marotta, 1989; Tuitert, 1990). In Tuitert's assay (1990), infested soil is serially diluted with sterile sand and dispersed in small pots in which beet seedlings are planted. The soil is kept moist, and after 6 wk the roots are examined microscopically for *P. betae* and plant tissue tested for BNYVV by ELISA (see under immunological techniques). In a similar most probable number bioassay for BNYVV (Ciafardini & Marotta, 1989), suspensions of soil infested with *P. betae* are mixed with suspensions of autoclaved soil and transferred to wells of tissue culture plates filled with sterile sand. Beet seedlings in glass rings are then placed on top of the sand in the wells so that the roots grow throughout the bottom layer of the sand–soil mix. After incubation inside a box to limit evaporation, all roots are checked for infection by *P. betae*. Most probable number methods are more accurate than traditional bioassay methods and show a good relationship with the incidence of BNYVV in the field (Ciafardini & Marotta, 1989; Tuitert & Hofmeester, 1994).

Immunological Techniques

Monoclonal antibodies have been produced against several soil-borne plant pathogens (Werres & Steffens, 1994), but techniques for quantification of fungi and bacteria in soil are still rare.

Of the immunological techniques, the enzyme-linked immunosorbent assay (ELISA) is most commonly used, in particular the double antibody sandwich or DAS ELISA. With this method, antibodies are used to coat the wells of polystyrene microtitre plates. The antigen (in a soil extract) is added to the wells and becomes bound to the immobilized antibodies. Secondary antibodies to which an enzyme has been coupled are then added to the plate. These also bind to the antigen, forming a *sandwich*, with the antigen in the center. Next, the substrate for the enzyme [for example hydrogen peroxide for horseradish peroxidase] is added to the wells and its degradation by the enzyme results in a color-change that can be quantified by an ELISA-plate spectrophotometer. A problem with DAS-ELISA is that the detection threshold is relatively high, for example $>10^4$ bacterial cells per milliliter of soil extract.

To lower the detection threshold, several variations of the ELISA technique have been developed. Chemoluminescence-based ELISA allows quantification of bacterial cells in a soil suspension down to 100 cells/mL which amounts to 10 cells per ELISA plate well (Schloter et al., 1992). As in the standard ELISA, the secondary antibody carries horseradish peroxidase that reacts with hydrogen peroxide to produce oxygen radicals. The radicals are detected by adding 3-aminophthalacidhydrazid (Luminol), which produces light after oxidation. The light intensity is quantified with a luminometer.

An alternative ELISA method is the immunomagnetic capture technique in which magnetic polystyrene beads are coated with monoclonal antibodies against a certain pathogen and mixed in a soil suspension. The beads are then pulled to the wall of the container with a magnet, washed in buffer, and resuspended in a suspension with an enzyme-linked polyclonal antibody against the same pathogen. The beads are washed again and resuspended in substrate buffer. A color change of the substrate can be measured with a spectrophotometer. Alternatively, the extracted beads can be plated on a selective medium. This last method was used, for example, to quantify spores of *Streptomyces lividans* (not a plant pathogen but related to the pathogen causing scab on potatoes) in soil down to a concentration of 5×10^2 cells/mL (Wipat et al., 1994).

Although monoclonal antibodies have been produced against several fungal plant pathogens, it is still relatively difficult to detect and quantify fungi directly in soil extracts; however, by using a biological amplification step *Rhizoctonia solani* AG4 can be detected reliably (Thornton et al., 1993). A soil-antigen extract is obtained by shaking a 5-g soil sample in 5 mL of semi-selective medium followed by centrifugation. In a diagnostic-ELISA, microtitre plate wells are coated with the soil-antigen extract, rinsed and blocked with buffer, and then incubated with monoclonal antibody followed by goat anti-mouse polyclonal antibody conjugated to peroxidase. After addition of substrate (hydrogen peroxide), the color change is quantified with an ELISA plate reader.

DAS-ELISA test kits are now available to detect *Phytophthora*, *Pythium*, and *Rhizoctonia* species in soil (in Werres & Steffens, 1994). These test kits are relatively easy to use, also for nonpathologists; however, nonpathogenic species of the same fungi also are often detected with these test kits, and the relationship between positive ELISA test results and risk of infection by the pathogens has not been established.

In conclusion, immunological techniques are promising for detection of plant pathogenic fungi and bacteria in soil, but quantification of plant pathogens in soil is still problematic. Moreover, the relationship between the inoculum density detected by immunological techniques and the risk of infection and disease development has not been established.

DNA Techniques

DNA and other organic molecules can be extracted from soil after lysing of microorganisms in soil (by using lysozyme to lyse bacterial cells and chitinase–cellulase mixtures to lyse fungal cells) followed by freeze–thawing procedures (Tsai & Olson, 1991). Alternatively, bacteria (but not fungi) can first be extracted from soil after dispersing the soil with sodium-saturated cation-exchange resin and then lysed followed by DNA extraction (Jacobsen & Rasmussen, 1992). The first method resulted in higher contamination of DNA with soil constituents such as tannins and humic acids compared with the second method (Left et al., 1995); however, several techniques have been developed to separate directly extracted DNA from humic acids, tannin, and other contaminating soil constituents by electrophoresis in low-melting-temperature agarose (Porteus & Armstrong, 1993), electrophoresis on polyvinylpyrrolidone amended agarose gels (Young et al., 1993), purification columns (Tsai & Olson, 1991, 1992), CsCl gradient centrifugation (Jacobsen & Rasmussen, 1992), or various precipitation steps, among others in glass milk (Smalla et al., 1993). Large amounts of extracted DNA can be used directly for DNA–DNA hybridization on a nylon membrane or nitrocellulose paper with a DNA probe for a particular pathogen. Alternatively, small amounts of purified, isolated DNA can be amplified by polymerase chain reaction (PCR) using universal prokaryotic or eukaryotic primers corresponding to 16S or 18S ribosomal RNA sequences (Porteus & Armstrong, 1993). After amplification, the DNA can be blotted onto a nylon membrane or nitrocellulose paper and hybridized to a specific labeled probe. More recently, DNA has been amplified using specific primers allowing detection of 500 bacterial cells per gram of soil, that were added to sterilized soil (Tsai & Olson, 1992).

Specific DNA probes and/or primers have been constructed for various soilborne plant pathogens, for example *Pseudomonas solanacearum* causal agent of bacterial wilt in various crops (Seal et al., 1992), *Pythium* species that cause damping-off or are used as biocontrol agents (Martin, 1991), *Phytophthora* species that cause root rot (Goodwin et al., 1990; Lee et al., 1993), *Gaeumannomyces graminis* causal agent of take-all of wheat (*Triticum aestivum* L.) and barley (*Hordeum vulgare* L.; Henson et al., 1993), and *Verticillium dahliae* causal agent of various wilt diseases and potato early dying (Li et al., 1994).

Although the detection levels using DNA extraction and PCR amplification techniques can be quite low, we need to realize that these techniques are currently only qualitative (Smalla et al., 1993). For quantification of inoculum densities one would need to combine DNA techniques with end-point dilution or MPN methods (Smalla et al., 1993). Moreover, we don't know if fungal DNA is extracted only from mycelium or also from dormant survival structures. Finally, DNA amplified by PCR also will be detected from nonviable and even degraded cells (Josephson et al., 1993). So far, DNA detection techniques have not been related to inoculum potential of plant pathogens in soil.

SUPPRESSIVENESS BY MICROORGANISMS

Suppressiveness tests can be performed with pathogens and their host plants or with pathogens in vitro if a correlation exists between the in vitro test and disease suppression on the host plant. In addition, general soil microbiological tests to measure microbial activity can be related to general disease suppression and then used as indicators for suppressiveness.

Bioassays with Host Plants and Pathogens

Results from bioassays with host plants and pathogens generally show the best correlation with disease suppression in the field. All tests for disease suppression in a certain soil involve addition of a pathogen to that soil, either at different inoculum densities or at one density added to both sterilized and nonsterilized field soil. Nonsterilized soil can be disturbed (due to sieving and mixing in of inoculum) or remain undisturbed. All these methods are very labor intensive.

Addition of Different Inoculum Densities

Five or six densities of inoculum of the pathogen under study are mixed in with natural field soils in pots (Alabouvette, 1986; Mandelbaum & Hadar, 1990; Oyarzun et al., 1994). Plants are grown for a predetermined period in the various soils. Disease severity is determined and plotted against inoculum level for each soil. The initial slope of a nonlinear regression line or the intercept with the x-axis of the regression line of transformed data is a measure of disease suppression: the smaller the slope or the more to the right the intercept the more suppressive is the soil (Mandelbaum & Hadar, 1990; Oyarzun et al., 1994; Wijetunga & Baker, 1979). To allow comparison of disease suppression in different seasons and with different soils, the tests need to be conducted under strictly standardized conditions (Oyarzun et al., 1994).

Comparison of Natural and Sterilized Soil

In order to be able to compare suppressiveness of different soils, usually a relative measure of suppressiveness of each soil type is calculated by comparing disease severity in natural field soil with that in the same soil after sterilization, both amended with propagules of the pathogen of interest. Sterilization can be accomplished by autoclaving at 120°C for 1 h on two consecutive days (to kill organisms with spores that survive the first autoclaving treatment), by fumigation with methylbromide, chloropicrin, or by gamma-radiation (Alabouvette, 1986). We recently tested soils from organic and conventional farms for suppression of corky root of tomatoes (caused by *Pyrenochaeta lycopersici*) using gamma-irradiated and nonirradiated soil artificially infested with the pathogen (Workneh & van Bruggen, 1994a). The relative increase in disease severity in sterilized soil compared with natural soil (a measure of disease suppression) was significantly higher for organically managed than conventionally managed soils, indicating that corky root was more suppressed in soil from organic than from conventional farms. Recently, we developed a method (unpublished) to test soils for suppressiveness against *R. solani* or *P. aphanidermatum* using 20-cm long polyvinyl chloride (PVC) pipes filled with infested (at two inoculum levels) or noninfested, autoclaved or nonautoclaved soil, standing in a plexiglass container with a layer of sand on the bottom (to maintain the water potential). Tomato seeds are planted in the soil, which is then covered by vermiculite to prevent drying out (Fig. 17–2). The autoclaved soil is exposed to microbial recolonization from the air for 1 wk before it is used in the test (to avoid toxic effects from autoclaved soil on seedlings).

Using Undisturbed Soil

The soil physical environment is a major factor affecting root disease expression, in particular soil porosity and waterholding capacity. Thus, soil struc-

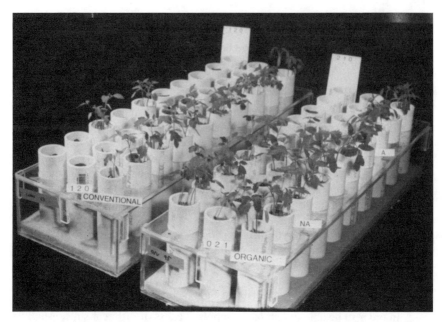

Fig. 17–2. Growth chamber bioassay with field soil from organic and conventional farms performed in plexiglass trays holding 4-cm wide polyvinyl chloride (PVC) pipes. Autoclaved (in back of trays) and nonautoclaved (in front), organic (right tray) or conventional (left tray) soils were either not inoculated, or inoculated at two different levels with *Rhizoctonia solani*. The higher degree of suppressiveness can be directly seen from the considerably better stand of tomato plants in the organic compared with the conventional soil.

ture is an integral part of soil quality with respect to disease suppression. For this reason it would be better to use undisturbed soil cores or soil blocks for disease suppressiveness tests; however, addition of inoculum of the pathogen is very difficult without disturbing the soil. Large survival structures such as sclerotia of *Sclerotinia* or *Sclerotium* species could simply be inserted into soil at a specific distance (for example 2 cm) from the root base of the first test plants in a row of seedlings. The percentage of plants becoming infected at different time intervals forms an indication of the relative receptivity of different soils to these pathogens. Pathogens with small survival structures that need to be in close proximity to the root to be able to infect the root cannot be added to soil easily; however, it is possible to transplant an infected seedling next to a row of healthy seedlings in a container with test soil. The pathogen can then grow from the infected seedling to healthy seedlings. The row length of infected seedlings after a specific time interval will be a measure of conduciveness of the soil to the spread of the pathogen (Workneh, 1993). This test is in part based on an assay for damping-off by *Rhizoctonia solani* in which mycelial disks of *R. solani* are inserted into various compost mixes in containers with rows of seedlings. The length of a row with infected seedlings is considered an indicator for receptivity of the mix for *Rhizoctonia* damping-off (Stephens et al., 1981).

Bioassays with Pathogens without Host Plants

Growth and Survival of Pathogens in or on Soil

Growth of a pathogen over a soil surface can be indicative of conduciveness of that soil for the pathogen and the disease it causes. For example, a simple assay was developed to assess soil for suppressiveness to *Pythium splendens* causing damping-off of cucumber (Ko & Kao, 1989). Sporangia were suspended in cucumber root extract to break the dormancy and then placed on the smooth surface of a small soil block. Spore germination was determined after 24 h of incubation. A similar test also was developed to study the influence of a food base on growth of *Phytophthora erythroseptica* (Johnson & Curl, 1972). Fungal growth through soil also can be measured using soil colonization tubes in which the pathogen is added at one end of the tube. Hyphae can be observed directly through the tube, or infested soil can be removed from the tube at regular distances from the inoculation point and plated onto an agar medium (Johnson & Curl, 1972).

Inhibition of spore germination, fungistasis, has been measured in various different ways, the most common methods being placement of fungal spores on soil to which different amounts of either nutrients or sterile, washed sand have been added. In these assays, the fungistatic effect is overcome by nutrient enrichment or by dilution at increasingly higher nutrient or sand concentrations (Wacker & Lockwood, 1991). The amount of amendment (nutrient or sand) that will result in 50% germination is indicative of the level of fungistasis for a particular pathogen - soil combination. Nash Smith (1977) developed a method to study germination of chlamydospores of *Fusarium oxysporum* f.sp. *tracheiphilum* (causal agent of cotton wilt) in the presence of cotton root tips growing in wilt suppressive or conducive soils. Cotton seedlings were grown in vermiculite in square petri dishes standing on one side; the taproot was guided through a whole in the lower side of each dish into a small container with test soil fortified with chlamydospores. At regular time intervals the root tips in the test soil were cut off and chlamydospores in adjacent soil were checked for germination and germtube lysis. Inhibition of germination and enhanced germtube lysis were positively correlated with wilt suppression in the field (Nash Smith, 1977).

Survival of pathogen structures in soil is one aspect of disease suppression. Several authors have buried survival structures or hyphae in natural soil and retrieved them after various incubation periods. For example, microsclerotia of *Verticillium dahliae* causing potato early dying disease were infiltrated on nylon membranes (pore size 0.048 mm) and buried in soils with different levels of suppressiveness. After 1 mo, the nylon membranes were retrieved and placed on a selective medium. Germination of microsclerotia was checked microscopically (Keinath & Fravel, 1992). Similarly, microsclerotia of *Pyrenochaeta lycopersici* were buried in soils with different levels of suppressiveness to corky root of tomato (Workneh & van Bruggen, 1994a). The viability of sclerotia as determined by dilution plating on a selective medium diminished faster in the more suppressive soils from organic farms than in conducive soils from conventional farms. Mandelbaum and Hadar (1990) buried nylon fabric (pore size 0.08 mm) enmeshed with hyphae of *Pythium aphanidermatum* in various soils and checked

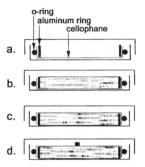

Fig. 17–3. Diagram of the agar-ring test procedure. The assembly consists of a metal ring onto which a moist cellophane membrane is strapped with an O-ring. (a) The ring and cellophane are placed in a glass-petri-dish facing downward. (b) After autoclaving the assembly, 15 mL of soil and approximately 20 mL of cooled, molten water-agar are poured into the ring, (c) the assembly is left over-night for cooling, the ring is turned with the cellophane facing up, and (d) a 3-mm agar-plug of a fungal culture is placed on the center of the cellophane.

the hyphae microscopically for lysis after 24 h. The extent of lysis corresponded to suppressiveness of the soils as measured in bioassays with cucumber seedlings.

Growth of Pathogens over a Membrane on Soil

An in vitro assay comparing relative fungal growth in autoclaved and natural soil can be used to determine soil receptivity to a phytopathogenic fungus (Davet, 1976; Williams & Willis, 1962; Van der Hoeven & Bollen, 1980). For this assay (Fig. 17–3), cellophane is strapped over an aluminum ring cut from irrigation pipes and held down by an o-ring (Buna-N, Nitrile ring). This unit is placed with the cellophane facing downwards inside a glass petri dish and autoclaved. Next, 15 mL of either autoclaved or nonautoclaved, natural soil are added to the dish inside of the aluminum ring, followed by 20 mL of 8% water agar. The unit is left upside down over night. The next morning, the cellophane surface is turned with the cellophane facing up, and an agar-disk of 3-mm diam. is cut with a sterilized cork-borer from a colony of a plant pathogenic fungus and placed in the center of the cellophane surface. After a predetermined incubation period (24 h for *Pythium* or *Rhizoctonia* species) under standardized conditions, the diameter of the fungal colony is measured in two perpendicular directions and averaged (Fig. 17–4). The diameter of the disk is subtracted from the average colony diameter to determine radial growth. We used five plates each of autoclaved and natural soil for each sampling unit. After averaging over these five plates a unitless index for the relative reduction in growth on nonsterilized versus sterilized soil is calculated as:

$$\text{Suppressiveness Index} = 1 - \frac{\text{(radial growth on unsterilized soil)}}{\text{(radial growth on sterilized soil)}}$$

The density of the mycelium growing over the cellophane surface may be affected by the level of disease suppression. Therefore, it is advisable to develop and calibrate a discrete scoring scale corresponding to hyphal density per unit area. The score for hyphal density can be multiplied by the radial growth for each colony, and an adjusted suppressiveness index can be calculated.

The agar ring test can be modified to account for availability of nutrients from root exudates under field conditions by adding axenically grown seedlings from surface sterilized seeds under the cellophane.

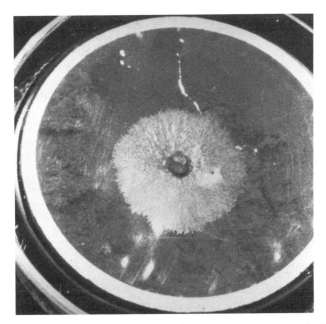

Fig. 17–4. Agar-ring test of a 24-h old colony of *Pythium aphanidermatum* on a cellophane surface covering an autoclaved soil–water agar mixture. The diameter of a similar colony growing over a nonautoclaved soil water agar mixture would typically be one-half to two-thirds of the diameter over an autoclaved mixture.

Indicator Tests for Microbial Activity

Various enzymatic activities have been shown to be correlated with disease suppression (Inbar & Chet, 1991, Inbar et al., 1991; Workneh et al., 1993). Thus, high enzymatic activities may be indicative of disease suppressiveness of a soil.

Fluorescein Diacetate Hydrolysis

FDA hydrolysis is a measure of general microbial activity indicating the degree of hydrolytic activity of various enzymes such as lipases, proteases, and esterases that are hydrolyzing FDA to fluorescein, which can be detected spectrophotometrically (Schnuerer & Rosswall, 1982). FDA hydrolysis has been reported to be a good indicator of suppressiveness of soils to several soil-borne plant pathogens. For instance, it has been found, that FDA hydrolysis is negatively correlated with severity of damping-off caused by *Pythium ultimum* (Inbar et al., 1991; Chen et al., 1988), of corky root of tomatoes caused by *Pyrenochaeta lycopersici* (Workneh et al., 1993; Workneh & van Bruggen, 1994b), and Phytophthora root rot of tomatoes caused by *Phytophthora parasitica* (Workneh et al., 1993). However, in some studies, microbial activity was not a reliable predictor of suppression of *Pythium* damping-off (Mandelbaum & Hadar, 1990).

Because the FDA hydrolysis assay has often been used as an indicator for disease suppression, a summary of this method is presented here. Immediately after sampling and sieving (2 to 4 mm mesh) of soil, 40-g subsamples are brought to a soil matric potential of −35 J/kg in a pressure plate apparatus for 48 h. Triplicate 5-g subsamples of soil are suspended in 20 mL 60 mM sodium phosphate buffer (pH 7.6). Next, 0.2 mL of an FDA solution (2 mg of FDA per mL of acetone) are added to each sample and incubated on a rotary shaker for exactly 30 min, at which point the reactions are stopped by adding 20 mL acetone. An additional 5-g subsample of soil is subjected to the same procedure, with the exception that acetone is added before adding the FDA to serve as a control. The next step involves removal of soil particles by centrifugation at 6 000 × g for 5 min and subsequent filtration of the supernatant through 2.5 µm glass-fiber filters. Absorbance is determined on a spectrophotometer at 490 nm after zeroing with the control sample. To account for adsorption of FDA by clay particles and organic matter in the soil, it is recommended to prepare a standard curve for each soil sample (Chen et al., 1988).

Other Enzymatic Assays

Enzymatic assays can either reflect general microbial activity (for example, dehydrogenase and catalase tests) or specific activities (for example, cellulase, chitinase, and xylanase assays). Indicator dyes are generally used to quantify the enzymatic activity colorimetrically. For example, tests for dehydrogenase activity are based on the principle that actively metabolizing cells transform tetrazolium salts into formazan, which can be quantified colorimetrically (Johnson & Curl, 1972). Further details for dehydrogenase assays are presented in Johnson and Curl (1972).

Recently, efficient colorimetric assays have been developed using microtitre plates to estimate the hydrolytic activities for endoacting cellulases, xylanases, chitinases, 1,3-β-glucanases and amylases of soil microbial communities (Wirth & Wolf, 1992). Carboxy-methyl (CM) substituted, watersoluble polysaccharide derivatives, labeled covalently with Remazol Brilliant Blue R (RBB) in the case of cellulose (CM-cellulose-RBB), starch (starch-RBB) and xylan (CM-xylan-RBB) or Remazol Brilliant Violet 5R (RBV) in the case of chitin (CM-chitin-RBV) are the substrates used for these assays. Dye-labeled substrate in aqueous solution and buffer are equilibrated in a water bath at 40°C. After the addition of 100 µL soil extract to the 350 µL wells in 96 well microtitre plates, the plates are incubated for 30 min to 6 h depending upon the activity of the enzyme under study. The reaction is stopped by adding 50 µL of 1 M HCL in the case of CM-chitin-RBV and 2 M HCL in the case of all other substrates to precipitate the substrates. After cooling the plates on ice and centrifugation, 175 µL of the supernatant are transferred to another microtitre plate with 175 µL wells and enzyme activity is measured spectrophotometrically at 550 nm for CM-chitin-RBV and 600 nm for all other substrates. Control wells are prepared analogously, without the addition of soil extract during the incubation (Wirth & Wolf, 1992). These enzymatic tests have not yet been related to disease suppressiveness.

CONCLUSION

All tests to assess the risk of infection by a root pathogen are very labor intensive and time-consuming, especially if a host plant is involved in the assay. For some pathogens such as *Pyrenochaeta lycopersici* a bioassay can take as long as 12 wk (6 wk to rear the inoculum and 6 wk for the disease assay). Moreover, the presence of a mixture of pathogens in a soil can obscure the symptoms of the test pathogen added to that soil. It would therefore be very useful to develop indicator tests for general disease suppression. Measuring microbial activity could be one of those tests, but FDA hydrolysis is not always closely correlated with disease suppression. Considering that not only the amount of organic matter in a soil but also its quality and stage of decomposition determines general disease suppression, it is important to characterize the composition of organic matter and the associated microflora in search of potential indicators for disease suppression. Moreover, fast and reliable techniques to detect root pathogens in soil also are needed to quantify endogenous pathogens and monitor growth and survival of introduced pathogens in soil.

ACKNOWLEDGMENTS

We are thankful to Gerrit Bollen for reviewing the manuscript, to Gerrie Tuitert for reading the section on baiting techniques and most probable number methods, and to Vic Claassen for checking the section on DNA techniques.

REFERENCES

Adams, M.J., and S.J. Welham. 1995. Use of the most probable number technique to quantify soilborne pathogens. Ann. Appl. Biol. 126:181–196.

Alabouvette, C. 1986. Fusarium wilt suppressive soil from the Chateaurenard region: Review of a 10 year study. Agronomie 6:273–284.

Butterfield, E.J. and J.E. DeVay. 1977. Reassessment of soil assays for *Verticillium dahliae*. Phytopathology 67:1073–1078.

Chakraborty, S., and J.H. Warcup. 1983. Soil amoebae and saprophytic survival of *Gaeumannomyces graminis tritici* in a suppressive pasture soil. Soil Biol. Biochem. 15:181–185.

Chen, W., H.A.J. Hoitink, A.F. Schmitthenner, and O.H. Tuovinen. 1988. The role of microbial activity in suppression of damping-off caused by *Pythium ultimum*. Phytopathology 78:314–322.

Ciafardini, G., and B. Marotta. 1989. Use of the most-probable-number technique to detect *Polymyxa betae* (Plasmodiophoromycetes) in soil. Appl. Environ. Microbiol. 55:1273–1278.

Curl, E.A., R. Lartey, and C.M. Peterson. 1988. Interactions between root pathogens and soil microarthropods. Agric. Ecosyst. Environ. 24:249–261.

Davet, P. 1976. Comportement sur divers substrats des champignons associés à la maladie des racines liégeuses de la tomate au liban. Ann. Phytopathol. 8:159–169.

DeVay, J.E., R.H. Garber, and D. Matheron. 1982. Role of Pythium species in the seedling disease complex of cotton in California. Plant Dis. 66:151–154.

Goodwin, P.H., J.T. English, D.A. Neher, J.M. Duniway, and B.C. Kirkpatrick. 1990. Detection of *Phytophthora parasitica* from soil and host tissue with a species-specific DNA probe. Phytopathology 80:277–281.

Henson, J.M., T. Goins, W. Grey, D.E. Mathre, and M.L. Elliott. 1993. Use of polymerase chain reaction to detect *Gaeumannomyces graminis* DNA in plants grown in artificially and naturally infested soil. Phytopathology 83:283–287.

Hornby, D. 1983. Suppressive soils. Ann. Rev. Phytopathol. 21:65–85.
Inbar, Y., M.J. Boehm, and H.A.J. Hoitink. 1991. Hydrolysis of fluorescein diacetate in sphagnum peat container media for predicting suppressiveness to damping-off caused by *Pythium ultimum*. Soil Biol. Biochem. 23:479–483.
Inbar, J., and I. Chet. 1991. Detection of chitinolytic activity in the rhizosphere using image analysis. Soil Biol. Biochem. 23:239–242.
Jacobsen, C.S., and O.F. Rasmussen. 1992. Development and application of a new method to extract bacterial DNA from soil based on separation of bacteria from soil with cation-exchange resin. Appl. Environ. Microbiol. 58:2458–2462.
Johnson, L.F., and E.A. Curl. 1972. Methods for research on the ecology of soil-borne plant pathogens. Burgess Publ. Co., Minneapolis, MN.
Josephson, K.L., C.P. Gerba, and I.L. Pepper. 1993. Polymerase chain reaction detection of nonviable bacterial pathogens. Appl. Environ. Microbiol. 59:3513–3515.
Keinath, A.P., and D.R. Fravel. 1992. Induction of soil suppressiveness to Verticillium wilt of potato by successive croppings. Am. Potato J. 69:503–514.
Kinsbursky, R.S., and A.R. Weinhold. 1988. Influence of soil on inoculum density-disease relationships of *Rhizoctonia solani*. Phytopathology 78:127–130.
Ko, W.-H. 1982. Biological control of *Phytophthora* root rot of papaya with virgin soil. Plant Dis. 66:446–448.
Ko, W.-H., and C.-W. Kao. 1989. Evidence for the role of calcium in reducing root disease incited by *Pythium* spp. p. 205–217. *In* A.W. Engelhard (ed.). Soilborne plant pathogens: Management of diseases with maro- and microelements. APS Press, St. Paul, MN.
Lee, S.B., T.J. White, and J.W. Taylor. 1993. Detection of *Phytophthora* species by oligonucleotide hybridization to amplified ribosomal DNA spacers. Phytopathology 83:177–181.
Left, L.G., J.R. Dana, J.V. McArthur, and L.J. Shimkets. 1995. Comparison of methods of DNA extraction from stream sediments. Appl. Environ. Microbiol. 61:1141–1143.
Li, K.-N., D.I. Rouse, and T.L. German. 1994. PCR primers that allow intergeneric differentiation of ascomycetes and their application to *Verticillium* spp. Appl. Environ. Microbiol. 60:4324–4331.
Lifshitz, R., and J.G. Hancock. 1981. An enrichment method to estimate potential seedling disease caused by low densities of *Pythium ultimum* inocula in soils. Plant Dis. 65:828–829.
Mandelbaum, R., and Y. Hadar. 1990. Effects of available carbon source on microbial activity and suppression of *Pythium aphanidermatum* in compost and peat container media. Phytopathology 80:794–804.
Martin, F.N. 1991. Selection of DNA probes specific for isolates or species in the genus *Pythium*. Phytopathology 81:742–746.
Nash Smith, S. 1977. Comparison of germination of pathogenic *Fusarium oxysporum* chlamydospores in host rhizosphere soils conducive and suppressive to wilts. Phytopathology 67:502–510.
Neher, D.A., C.D. McKeen, and J.M. Duniway. 1993. Relationships among Phytophthora root rot development, *P. parasitica* populations in soil, and yield of tomatoes under commercial field conditions. Plant Dis. 77:1106–1111.
Nicot, P.C., and D.I. Rouse. 1987a. Precision and bias of three quantitative soil assays for *Verticillium dahliae*. Phytopathology 77:875–881.
Nicot, P.C., and D.I. Rouse. 1987b. Relationship between soil inoculum density of *Verticillium dahliae* and systemic colonization of potato stems in commercial fields over time. Phytopathology 77:1436–1355.
Oyarzun, P.J. 1993. Bioassay to assess root rot in pea and effect of root rot on yield. Neth. J. Plant. Path. 99:61–75.
Oyarzun, P.J., G. Dijst, and P.W.T. Maas. 1994. Determination and analysis of soil receptivity to *Fusarium solani* f.sp. *pisi* causing dry root rot of peas. Phytopathology 84:834–842.
Parke, J.L., and C.R. Grau. 1992. Aphanomyces. p. 27–30. *In* L.L. Singleton et al. (ed.) Methods for research on soilborne phytopathogenic fungi. APS Press, St. Paul, MN.
Pfender, W.F., D.I. Rouse, and D.J. Hagedorn. 1981. A "most probable number" method for estimating inoculum density of *Aphanomyces euteiches* principal incitant of pea root rot in naturally infested soil. Phytopathology 71:1169–1172.
Porteus, L.A., and J.L. Armstrong. 1993. A simple mini-method to extract DNA directly from soil for use with polymerase chain reaction amplification. Current Microb. 27:115–118.
Rush, C.M., J.D. Mihail, and L.L. Singleton. 1992. Introduction. p. 3–6. *In* L.L. Singleton et al. (ed.). Methods for research on soilborne phytopathogenic fungi. APS Press, St. Paul, MN.

Schloter, M., W. Bode, A. Hartmann, and F. Beese. 1992. Sensitive chemoluminescence-based immunological quantification of bacteria in soil extracts with monoclonal antibodies. Soil Biol. Biochem. 24:399–403.

Schneider, R.W. (ed.) 1982. Suppressive soils and plant disease. APS Press, St. Paul, MN.

Schnuerer, J., and T. Rosswall. 1982. Fluorescein diacetate hydrolysis as a measure of total microbial activity in soil and litter. Appl. Environ. Microbiol. 43:1256–1261.

Seal, S.E., L.A. Jackson, and M.J. Daniels. 1992. Isolation of a *Pseudomonas solanacearum*-specific DNA probe by subtraction hybridization and construction of species-specific oligonucleotide primers for sensitive detection by the polymerase chain reaction. Appl. Environ. Microbiol. 58:3751–3758.

Singleton, L.L., J.D. Mihail, and C.M. Rush. 1992. Methods for research on soilborne phytopathogenic fungi. Am. Phytopathol. Soc., St. Paul, MN.

Smalla, K., N. Cresswell, L.C. Mendonca-Hagler, A. Wolters, and J.D. van Elsas. 1993. Rapid DNA extraction protocol from soil for polymerase chain reaction-mediated amplification. J. Appl. Bacteriol. 74:78–85.

Stanghellini, M.E., and W.C. Kronland. 1985. Bioassay for quantification of *Pythium aphanidermatum* in soil. Phytopathology 75:1242–1245.

Stephens, C.T., L.J. Herr, H.A.J. Hoitink, and A.F. Schmitthenner. 1981. Suppression of *Rhizoctonia* damping-off by composted hardwood bark medium. Plant Dis. 65:796–797.

Thornton, C.R., F.M. Dewey, and C.A. Gilligan. 1993. Development of monoclonal antibody-based immunological assays for the detection of live propagules of *Rhizoctonia solani* in soil. Plant Pathol. 42:763–773.

Tsai, Y.-L., and B.H. Olson. 1991. Rapid method for direct extraction of DNA from soil and sediments. Appl. Environ. Microbiol. 57:1070–1074.

Tsai, Y.-L., and B.H. Olson. 1992. Detection of low numbers of bacterial cells in soils and sediments by polymerase chain reaction. Appl. Environ. Microbiol. 58:754–757.

Tuitert, G. 1990. Assessment of the inoculum potential of *Polymyxa betae* and beet necrotic yellow vein virus (BNYVV) in soil using the most probable number method. Neth. J. Plant Pathol. 96:331–341.

Tuitert, G., and Y. Hofmeester. 1994. Epidemiology of beet necrotic yellow vein virus in sugar beet at different initial inoculum levels in the presence or absence of irrigation. Europ. J. Plant Pathol. 100:19–53.

van Bruggen, A.H.C. 1995. Plant disease severity in high-input compared to reduced-input and organic farming systems. Plant Dis. 79:976–984.

Van der Hoeven, E.P., and G.J. Bollen. 1980. Effect of benomyl on soil fungi associated with rye: 1. Effect on the incidence of sharp eyespot caused by *Rhizoctonia cerealis*. Neth. J. Plant Pathol. 86:163–180.

Wacker, T.L., and J.L. Lockwood. 1991. A comparison of two assay methods for assessing fungistasis in soils. Soil Biol. Biochem. 23:411–414.

Weinhold, A.R. 1977. Population of *Rhizoctonia solani* in agricultural soils determined by a screening procedure. Phytopathology 67:566–569.

Werres, S., and C. Steffens. 1994. Immunological techniques used with fungal plant pathogens: Aspects of antigens, antibodies and assays for diagnosis. Ann. Appl. Biol. 125:615–643.

Williams, L.E., and G.M. Willis. 1962. Agar-ring method for in vitro studies of fungistatic activity. Phytopathology 52:368–369.

Wipat, A., E.M.H. Wellington, and V.A. Saunders. 1994. Monoclonal antibodies for *Streptomyces lividans* and their use for immunomagnetic capture of spores from soil. Microbiology (UK) 140:2067–2076.

Wirth, S.J., and G.A. Wolf. 1992. Micro-plate colorimetric assay for endo-acting cellulase, xylanase, chitinase, 1,3-β-glucanase and amylase extracted from forest soil horizons. Soil Biol. Biochem. 24(6):511–519.

Wijetunga, C., and R. Baker. 1979. Modeling of phenomena associated with soil suppressive to *Rhizodonia solani* after being plated to successive crops of radishes. Phytopathology 69:1287–1293.

Workneh, F. 1993. Comparison of severity of corky root (*Pyrenochaeta lycopersici*) and Phytophthora root rot (*Phytophthora parasitica*) on tomato and associated soil and plant variables on organic and conventional farms. Ph.D. thesis. Univ. of California, Davis.

Workneh, F., and A.H.C. van Bruggen. 1994a. Suppression of corky root of tomatoes in organically managed soil associated with soil microbial activity and nitrogen status of soil and tomato tissue. Phytopathology 84:688–694.

Workneh, F., and A.H.C. van Bruggen. 1994b. Microbial density, composition, and diversity in organically and conventionally managed rhizosphere soil in relation to suppression of corky root of tomatoes. Appl. Soil Ecol. 1:219–230.

Workneh, F., A.H.C. van Bruggen, L.E. Drinkwater, and C. Shennan. 1993. Variables associated with corky root and Phytophthora root rot of tomatoes in organic and conventional farms. Phytopathology 83:581–589.

Young, C.C., R.L. Burghoff, L.G. Keim, V. Minak-Bernero, J.R. Lute, and S.M. Hinton. 1993. Polyvinylpyrrolidone-agarose gel electrophoresis purification of polymerase chain reaction-amplifiable DNA from soils. Appl. Environ. Microbiol. 59:1972–1974.

18 Assessing Organic Chemical Contaminants in Soil

Thomas B. Moorman

USDA-ARS, National Soil Tilth Laboratory
Ames, Iowa

Assessments of soil quality attempt to estimate the extent and state of the soil resource. The presence and bioactivity of organic chemical pollutants can affect soil quality and the movement of these compounds from soil into surface or groundwater represents an additional hazard to human and animal health. Soil contains a large number of naturally occurring organic chemicals, some of which may be injurious to plants, animals, or humans (e.g., allelopathic agents or fungal toxins); however, the greatest concern is over the presence of potentially toxic or carcinogenic compounds that are of anthropogenic origin (xenobiotics). Xenobiotic compounds are of particular concern because of their large-scale production and relatively long persistence in the environment. Some of these compounds, such as pesticides, are directly applied to soils during agricultural operations. In the 10 states comprising the American Midwest, 23 herbicides (7.71 × 10^7 kg) and eight insecticides (5.94 × 10^6 kg) were applied in 1994 to 25.3 million ha (NASS, 1994). Typically, within a smaller region the number of compounds applied to soils and plants is much less. For instance, five herbicides account for 92% of the 16.3 million kg of herbicide applied to 4.86 million ha in Iowa (NASS, 1994). Other chemicals are inadvertently introduced to soil from industrial activities, including aliphatic and aromatic hydrocarbons (gasoline, oils, or solvents), chlorinated aromatic and aliphatic compounds, polynuclear aromatic hydrocarbons (PAH), and polychlorinated biphenyls (PCB). Leisinger (1983) estimated that 65 000 chemicals are in everyday use, but only 114 toxic organic substances are designated priority pollutants by the U.S. Environmental Protection Agency (USEPA). Finally, organic chemicals are introduced into soil during the disposal of waste materials, primarily sewage sludges.

Depending upon the chemical properties and concentration, soil properties, and environmental factors these substances persist in the soil where they may affect human and animal health, plants and soil microorganisms. Research on the effects of agricultural pesticides began in the 1950s with concerns over accumulation of toxic compounds and their possible effects on soil microorganisms and soil fertility (Moorman, 1994a). Parallel concerns have arisen concerning chemicals of industrial origin. The ecotoxicological effects of a chemical compound in the soil are governed by the inherent toxicity of the compound, the concentration in the soil environment and the duration of exposure. Chemicals persisting in the

---

Copyright © 1996 Soil Science Society of America, 677 S. Segoe Rd., Madison, WI 53711, USA.
Methods for Assessing Soil Quality, SSSA Special Publication 49.

soil also may move into plants and the food chain or into surface water and groundwater. This chapter provides information on basic methods for assessing the movement and persistence of organic chemicals in soil, including sampling, analyses for pesticides and other organic compounds, and interpretation of these results within the context of soil quality. Due to the large number of organic compounds in soil and the heterogeneous nature of these compounds it is not possible to specify even a small set of standard methods. Instead, this chapter will cover key information for the design and execution of a specific plan for soil assessment.

PLANNING AN ASSESSMENT

Assessing the potential movement and persistence of organic chemicals in soils within the context of soil quality involves three distinct steps: planning, measurement (or data gathering), and interpretation. The first step in planning an assessment is to identify the needs, scope, and goals of the project and to determine the required accuracy of estimates obtained from the data. Important questions may include:

- Are specific compounds suspected to be present at the site and at what levels?
- Do the residues exceed certain limits?
- Can residues move from the soil into surface water or groundwater?
- Can residues be taken up by crops?
- How long will the residues persist?
- Are the source(s) of the contaminants known and are they point sources or non-point sources?
- What is the scale (plot, field, watershed, and region) of the assessment?
- What is the impact of pollutants on soil?

SAMPLING METHODS

Inferences about the quantity of an organic compound present at a particular site are made from samples taken from the site or area of interest. Designing an effective sampling plan requires the following steps:

1. Determine the number of samples that can be taken and analyzed. This determination is generally made by evaluating the resources (time, money, equipment, and people) available for the sample preparation and analytical determinations.
2. Define a study area. A study area may consist of a field or parcel of land, a farm, or watershed. The sources and potential movement of contaminants should be considered, in order to define a study area that is an appropriate size.
3. Determine temporal aspects of sampling plan. Sampling may be dictated to detect the maximum concentration of contaminants, such as organ-

ic compounds introduced in sewage sludge or pesticide residues after spraying. Sampling patterns in time should take into account the temporal pattern of inputs into the soil, if these are known. Repeated sampling can be used to estimate the rate of dissipation of a compound, although estimates of this type are influenced by environmental conditions.

4. Sampling locations. Three common designs for sampling an area are simple random sampling, stratified random sampling, and systematic sampling. The assumptions and statistical analysis related to these sampling plans are summarized by Dick et al. (1996, this publication) and discussed in more detail by Gilbert (1987). Application of survey and geostatistical techniques to environmental monitoring are given by Flatman and Yfantis (1984). The source or potential sources of contaminants and the degree of spatial variability resulting from soils or from transport processes (runoff or leaching) should be considered in the selection of a sampling design. It also may be extremely valuable to locate areas within the study area, or as close by as possible, that are not impacted (uncontaminated). These areas should be similar in terms of soils, vegetation, and land use and can serve as a source of soil for establishing extraction and analysis efficiency and for preparing spiked and analyte-free control samples.

5. Soil sampling. The choice of composite or individual samples can depend upon the experimental design, resources for sampling, and sample size requirements. Compositing is the process of taking equally sized soil samples from an area and combining them into a single sample. Statistical aspects of composite sampling and sampling depth increments in soils are given elsewhere (Gilbert, 1987). In addition to statistical concerns, the sampling procedure should be designed to reduce the risks of sample contamination and/or loss of the compound of interest. Samples can become cross-contaminated with chemicals retained on sampling equipment or during handling in the field. Materials for sample containers should be chosen to prevent loss of the compound through volatilization or sorption, which are possible with nonpolar compounds and plastic containers. The possibility that extraneous compounds can be introduced in the sample during the sampling process needs to be evaluated. If sampling gear is to be used again, effective cleaning procedures must be devised. Sample storage also should be considered, with freezing being a common method of preservation. If samples are air dried, then the loss of compounds during drying and the possibility of cross-contamination should be evaluated.

ANALYSIS METHODS

The quantitative determination of organic compounds in soil and water is the result of an effective extraction procedure followed by concentration and finally analysis.

Chemical Extraction and Analysis

Extraction

Extraction of the analyte (compound of interest) from soil is required for either instrumental or biochemical analysis. For many organic compounds that are moderately to poorly soluble in water (PAH, PCB, many pesticides, or chlorinated hydrocarbons) extraction is accomplished using a nonpolar or moderately polar solvent. Solubility of the compound in the extractant and low amounts of co-extractive compounds, the efficiency of the extraction and reproducibility are key factors to obtaining reliable data (Dao et al., 1983). Acetone and methanol are two solvents widely used as extractants for pesticides, while hexane has been used for less soluble compounds, such as hexachlorobenzene. Following extraction, concentration of the extract may be required before analysis. The need for concentration depends upon the concentration of contaminants present in the soil and the sensitivity of the analytical method.

Chromatography and Detection

The most common methods of instrumental analysis are gas chromatography (GC) and high performance liquid chromatography (HPLC). In both procedures the analyte is separated from other compounds in the sample extract, detected, and quantified. Factors that influence the choice of the method include physical and chemical restraints imposed by the analyte, desired sensitivity of the method, possible interferences, and the ease and speed of the analysis. In GC, sample extracts are injected into a heated stream of gas that volatilizes the analyte. The gas stream passes through a column consisting of a solid support covered by a liquid coating. The relative affinity of the compounds for the liquid phase determines the order that different compounds elute from the column (retention time). The most common GC detectors for measurement of organics in environmental samples are the electron capture detector (EC), which is well suited to the measurement of halogenated substances, and the NP (alkali flame) detector that is well suited to measurement of N and P containing compounds. In HPLC, the compounds are carried in a liquid stream (water, buffers, or solvents) through a column that retains the different compounds in a differential manner. In both systems detectors at the end of the column measure the compounds of interest. The principal detector for HPLC is the ultraviolet (UV) spectrophotometer, which is well suited to detection of a large variety of compounds. Fluorescence spectrophotometers also are used for compounds that fluoresce after exposure to certain wavelengths of light and can provide sensitivity and selectivity (e.g., Mueller & Moorman, 1991). In general, detection by GC is more sensitive than that for HPLC (Koskinen, 1988). This advantage may be offset by somewhat easier operations and greater versatility of the HPLC. Nonvolatile compounds or compounds that are not thermally stable are analyzed by HPLC, rather than GC. In some instances, the analyte can be chemically modified by derivitization to produce an analyte that is more thermostable or more easily detectable (e.g., Ahmad & Crawford, 1990; Hayakawa et al., 1995).

Quantification

Regardless of the analytical instrumentation, quantification is made using standards containing known quantities of the analyte to establish the identity of a compound and the amount present in a sample. As the analyte passes through the detector a characteristic peak is produced having an area that is proportional to the concentration of the analyte. Compound identity is established by comparing retention times of peaks produced from sample extracts to those of the standards. Most commonly used is the external standard method in which a series of determinations are made and the instrument response (peak area) is related to concentration by linear or curvilinear regression. Concentrations of unknowns are calculated using a rearrangement of this regression equation. The range of standard concentrations should cover the expected range of concentrations in the sample extracts and the standards should be prepared in a solution similar to the sample matrix. Other information that standards may be used for are the determination of precision, bias, and accuracy. Precision indicates the variability of replicate analyses of a single standard concentration (reproducibility), while bias refers to the deviation of measured values from a known standard value (overestimation or underestimation). Accuracy refers to the combination of high precision and low bias (Gilbert, 1987).

Additional confirmation is sometimes desired to establish that peaks obtained from unknown samples are in fact the compound of interest. Presently, mass selective detectors can be used following conventional GC or HPLC detectors. These small, postcolumn spectrometers confirm the identity of the compound through analysis of the mass fragment patterns. For HPLC, the use of diode array detectors allows the absorbance spectrum of a peak to be compared against the absorbance spectrum of any authentic sample. This second confirmation tool may eliminate some *false positive* detections, but it is less rigorous than mass spectral analysis.

Quality Control

In the design of analysis program for a laboratory (or even a contract laboratory) certain steps can be taken to ensure the highest quality data. These steps include (i) the inclusion of replicate spiked blanks; and (ii) inclusion of replicate controls (soil similar to the samples, but without the compound(s) of interest). Collecting soil similar to that in the study area that is free of contamination and spiking it with known quantities of the analytes of interest at appropriate concentrations will confirm that extraction and chromatography procedures are effective in terms of extraction efficiency and reproducibility. If the analyses are conducted over a long period of time, then periodic analysis of spiked samples provides a measure of laboratory performance over that period of time. Processing untreated soils should provide a check for any contamination during handling in the laboratory and co-extractive compounds. It also is important to establish the extraction efficiency and the minimum detectable amount. Care should be taken to convert the minimum detectable amount into units that are comparable to the units of concentration used for the data. Analysis of replicate subsamples (or sample *splits*) are another means of testing reproducibility.

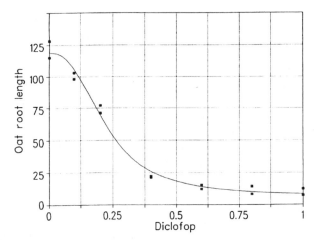

Fig. 18–1. An oat root growth bioassay for the herbicide diclofop. Oat root lengths (mm) are shown in response to different doses of diclofop (mg/kg soil). A logistic dose-response regression (solid line) results in an I_{50} (50% inhibition) value of 0.221 ± 0.014 mg/kgsoil (data from Hsiao & Smith, 1983).

Bioassay

Bioassay procedures have long been used to assay for concentrations of toxic materials in extracts of soil and water. The most commonly used bioassay procedures are those used to measure herbicide concentrations in soil. In these procedures herbicide-free soil similar to the soil containing the unknown herbicide residues is treated with a range of known concentrations of herbicide and planted with a sensitive species. The growth of plants in the treated soils is plotted against concentration, as shown in Fig. 18–1. In this example, the herbicide diclofop was assayed by measuring oat (*Avena sativa* L.) root lengths after 5 d growth in 250 g soil (Hsiao & Smith, 1983). Herbicide concentrations in soil are determined by comparison to a dose-response curve (regression curve of plant growth vs. concentration). In practice, a set of standards (plants exposed to a range of known concentrations) must be prepared each time the bioassay is performed. One disadvantage of bioassays is that accurate measurements are possible only in a limited range of concentrations. Figure 18–1 shows that the oat bioassay is relatively insensitive at very low and high concentrations of diclofop. The limit of detection is dependent upon the test organism, but can approach limits of detection of instrumental methods. Hsaio and Smith (1983) reported a detection limit of 1 ng chlorsulfuron g^{-1} soil by corn (*Zea mays* L.) root length bioassay, which is comparable to that obtained by chemical extraction and GC analysis (Ahmad & Crawford, 1990). Another disadvantage of bioassays is the potential influence of any differences between the soils used to prepare the standards and the soils for bioassay. Compounds that do not produce rapid effects in a test organism may be difficult to analyze by bioassay; however, bioassay techniques are low cost and require little specialized equipment. They may be an ideal tool for farmers concerned about herbicide carry-over in production fields.

Immunoassay

Immunoassay methods use antibodies capable of specific binding to the compound of interest to quantify the concentration in an environmental sample matrix. Immunoassays are available for a wide variety of compounds including more than 35 herbicides and insecticides, the PCBs phenanthrene, and benzo[a]-pyrene, trinitrotoluene (TNT), pentachlorophenol and other compounds. Lists of available antibodies, details of antibody production and methods testing are given elsewhere (Van Emon & Lopez-Avila, 1992; Meulenberg et al., 1995). The immunoassay works by introducing a sample (typically water or dilute solvent-water mixtures) into wells of microtiter plates, previously coated with a hapten–protein mixture. A known quantity of antibody is introduced, which binds to both the analyte and the hapten. Binding to the hapten is inversely proportional to the solution concentration of the analyte. After rinsing to remove analyte-antibody complex and any excess antibody, an enzyme-linked secondary antibody is introduced. Secondary antibodies bind to the remaining antibody-hapten complex. After rinsing, an enzyme substrate is introduced, which is converted to a chromogenic compound by the bound enzyme. The chromogen is measured by absorbance using a plate reader. Several alternatives to the plate reader are commercially available at less cost and the use of these techniques in the field is becoming practical (Van Emon & Gerlach, 1995). Typically, very good agreement is found between immunoassay and instrumental methods. For instance, Schlaeppi et al. (1989) reported strong correlation (r^2 of 0.91) between immunoassay and HPLC with similar limits of detection.

Practical application of immunoassay requires the necessary equipment and procedures to control problems arising in immunoassay analysis. One obstacle to their use for soil is the need for extraction of analytes from the soil matrix. This can be accomplished using methods similar to those for chemical analysis. Following dilution to reduce solvent concentrations, the immunoassay technique can proceed. A second problem is that of specificity. Antibodies are capable of cross-reacting with structurally similar compounds, leading to false positives. Commercial manufacturers of immunoassay reagents are aware of these possible cross-reactions and can often provide information. It is important to develop a quality assurance program similar to that described for instrumental analysis. Van Emon and Gerlach (1995) reported that of 62 water and soil samples analyzed by field-portable immunoassay kits, 44 provided data in an acceptable range, 14 false positives were recorded, and there was a failure to detect analytes in 4 of the samples.

ENVIRONMENTAL FATE AND IMPACTS

This section describes basic procedures to evaluate persistence of organics in soil, sorption and transport in soil, and the effects of contaminants on soil organisms. Transport of organic chemicals in soil is controlled by the interaction of the chemicals with selected soil attributes and the flow of water in the soil. Two processes that can be estimated or measured are of critical importance: sorption and degradation. For each of these processes brief methodology is presented for measuring the process. In addition, methods are presented for estimating parameters that describe the processes and for integrating them for purposes of soil quality evaluations.

Sorption

Sorption is a term generally indicating the uptake or binding of a solute by the soil without indicating the mechanism of binding. Adsorption is a term often used to refer to the binding of solutes to soil surfaces and desorption to indicate the reverse reaction. Both adsorption and desorption processes are governed by the reversible binding of the organic solute to the soil, although it is recognized that the strength and reversibility of the binding varies according to the soil properties and the properties of the organic solute (Hassett & Banwart, 1989). Sorption is inversely related to the relative mobility of moderately polar and nonpolar (nonionic) organic chemicals in soil.

Adsorption is measured most commonly in a laboratory procedure that involves the measurement of a partition coefficient (K_d) defined as:

$$K_d = C_s / C_w$$

where C_s is the sorbed concentration (mg/kg soil) and C_w is the concentration in water (mg/L). Typically, small amounts of dry soil (3 to 5 g) are placed in a glass centrifuge tube and the organic compound is added at an appropriate concentration in a solution of 0.01 M $CaCl_2$. After equilibration (gentle shaking) for 24 to 48 h the slurry is centrifuged to remove the soil and the solution concentration is measured. The amount sorbed is calculated from the change in solution concentration. Preliminary studies are recommended to demonstrate mass balance, stability of the compound over the time of the assay, and the approach to near equilibrium conditions. If a well-mixed and sieved (2 mm) soil is used, three replicate determinations are usually sufficient. If this determination is performed at different initial concentrations then a Freundlich adsorption isotherm can be calculated:

$$K_f = C_s / C_w^{(1/n)}$$

where K_f is a measure of the strength of adsorption and ($1/n$) describes the degree of curvature of the isotherm (the variation in K_f across the range of concentrations). Normally, these parameters are estimated by linear regression from the log transformed form of the equation [log (C_s) = log (K_f) + ($1/n$) × log (C_w)]. Further details on experimental methodology are reported by McCall et al. (1981). Another useful parameter is the sorption coefficient normalized for soil organic C content, K_{oc}.

$$K_{oc} = K_d / f_{oc}$$

The term f_{oc} is organic C content expressed on a fractional basis. For many nonpolar organic compounds, sorption is positively correlated with organic C content and the K_d values can be estimated from K_{oc} and the organic C content of a particular soil. If the K_{oc} value is derived from a sufficient range of soils, it is broadly applicable to a wide range of soils and sediments. Alternatively, a K_{oc} determined using a small number of samples from a specific area can be used with increased confidence for estimation in similar soils or sediments. For moderately sorbed compounds, such as triazine and sulfonylurea herbicides, pH and clay content affect adsorption, which reduces the accuracy of estimating K_d from K_{oc}. The term K_{om} is analogous to K_{oc}, except that organic matter is used in the calculation rather than organic C. Additional information on the use of octanol-water coefficents and other techniques for estimating sorption are described by Hassett and Banwart (1989).

Degradation and Persistence

Degradation and persistence are two related terms used to describe the loss of an organic compound. Dissipation refers to the loss of a compound from a soil volume, such as the loss of atrazine from the tillage layer within a particular field. Persistence refers to the tendency of a compound to persist within some defined soil volume. The apparent dissipation or persistence of compounds is affected by losses in runoff, leaching and volatilization. Degradation refers to the chemical or biological breakdown of a compound and thus is a component of the dissipation process.

The determination of a dissipation rate in the field is accomplished by sampling over time, assuming that all sources of contaminant are known and quantified. This approach has been commonly used to determine the field dissipation rates for pesticides. In many instances the dissipation of a pesticide in the field appears to follow first-order kinetics with respect to concentration; however, environmental conditions and biological and chemical attributes of the soil affect the dissipation rate of pesticides (Nash, 1988; Moorman, 1994b). The time scale of sampling should be considered carefully. Long-term studies are better suited to the study of slow processes or processes that are highly sensitive to environmental conditions.

A second approach is to determine a degradation rate in laboratory studies and to use the rate to predict degradation or leaching through the use of simple or complex models. In this procedure, the organic chemical is added at known concentrations to replicate soil samples maintained at constant temperature and moisture conditions. The standard conditions of 25°C and 33 kPa water potential were recommended for standard incubation studies (Laskowski et al., 1983). Periodically samples are extracted, analyzed, and the rate of degradation calculated. Studies with soils incubated under standard conditions appear to be useful and accurate for comparative purposes. For instance, Mueller et al. (1992) showed that soil depth greatly reduced microbial biomass and fluometuron degradation rate. The high cost and variability associated in obtaining field measured half-lives has led to a reliance on laboratory derived half-lives for predictive purposes. Frehse and Anderson (1983) compared the half-life from field studies to the half-life from laboratory studies and obtained a linear correlation. Field-derived half-lives of insecticides and herbicides were less than those from the laboratory, as might be expected from due to the leaching and volatilization losses under field conditions; however, some fungicides had greater half-lives in the field than the laboratory.

Several sources of information are available that provide ranges of sorption coefficients and soil half-lives, as well as measures of water solubility, volatility and other information. Information of this type relating to pesticides has been compiled by Wauchope et al. (1992), Augustijn-Beckers et al. (1994), and Montgomery (1993). Similar information on priority pollutants and industrial solvents also has been compiled (Howard, 1989; Howard, 1990).

DATA USE AND INTERPRETATION

One aspect of soil quality is the ability of soil to retain and detoxify organic chemicals. Chemical loading arises from different sources: pesticide use, sewage sludge applications, air-borne deposition, spills, land-farming of waste materials, and still other sources. Only pesticide use occurs over sufficiently broad areas that it can be considered a non-point source and thus has impact at the watershed and national scale. Sewage sludge disposal and land-farming of organic wastes occur at the field scale, but are still point sources in that only relatively small areas are used for these practices. Air-borne deposition of organics from industrial sites, incinerators, refineries, and power plants may cover large areas of ground, but the pattern of chemical loading (and thus hazard and impact) are essentially those of point sources. The central questions relating to all of these activities and soil quality are:

- Are toxic organics present in concentrations that affect soil productivity, plant growth, and animal health or human health?
- What soils are more vulnerable or less vulnerable to the planned introduction of organic chemicals?

Survey Assessments

The first question seeks to determine the present state of soil quality with respect to organic chemical content in soil. Examples of this type of assessment might include determining the concentration of pesticides in agricultural fields or determining the content of dioxin at a site. After delineation of representative scales and study areas and an initial determination of representative analytes a sampling scheme is designed and executed. These types of studies are essentially monitoring surveys of affected or potentially affected areas. When monitoring is conducted over time some inferences on processes can be made.

A national survey of organic contaminants of soil was reported by Webber and Singh (1995), who conducted a survey of contaminants (122 compounds) from eight benchmark agricultural sites and six production fields in Canada. This study reported data for a number of pesticides, PCB, PAH, chlorinated benzenes, and other compounds. The effectiveness of their interpretation was affected by the lack of recognized scales to judge the potential hazard of these compounds in soil. While this type of survey provides a broad geographic coverage, there was

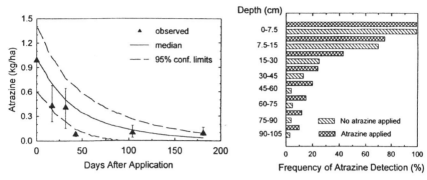

Fig. 18–2. Concentration of atrazine after application in the top 120 cm of a soil profile in central Iowa corn field (left). The solid line represents median atrazine concentration and 95% confidence limits (dashed lines) predicted by a probabilistic model based on field-scale variation in soil properties. The median measured concentration (triangles) and interquartile range (error bars) also are shown (Miller et al., 1995). The frequency of atrazine detections based on core sampling and GC analysis declines with depth (right). Combined frequency distributions are shown for three fields over 1992 and 1993 and includes crop years with and without atrazine application (Hatfield & Baker, 1995).

no assessment of temporal patterns or the sources of the nonagricultural compounds.

Smaller scale assessments of organic compounds in soil are far more common. Temporal trends and contaminant sources need to be considered in the design of some studies. Miller et al. (1995) reported on a study to monitor herbicide residues that was designed to determine the persistence and vertical movement of herbicides in field soils. Three fields were sampled and atrazine residues determined in years when atrazine was applied and years when no atrazine was applied. The time-course of atrazine dissipation at one of these field sites is shown in Fig 18–2. Atrazine residues were measurable in nearly all samples of surface soil, even in years when atrazine was not applied (Fig. 18–2). The residues in years with no application were below levels that cause phytotoxicity to the rotational crop. In subsurface samples, a large number of samples did not contain detectable atrazine concentrations. Concentrations, based on samples above the detection limit of 5 ppb, decreased with depth. Preferential movement of the herbicide in soil could account for the observed pattern of detects and nondetects. The monitoring data show that atrazine does reach the lower depths of soil and that it is relatively persistent. Tile-monitoring studies in this area also show movement of atrazine from soil profiles, even in years when it is not applied (Hatfield & Baker, 1995).

Long-term studies are effective in revealing the incremental effects of practices on soil quality. Wang et al. (1995) reported on a project to examine the effect of sewage sludge applications on the accumulation of chlorobenzenes in soil. It is important to account for all possible sources of a contaminant in order to accurately assess its dissipation in soil. Chlorobenzene contents of the applied sludges were determined and total applied loads were calculated. Sludge-treated and control (no sludge applied) plots were sampled by removing nine cores from each plot. Data were reported as the mean, standard deviation, median, and range for samples taken during a number of years. The large difference between mean and median concentration is indicative of a nonnormal distribution, which is not uncommon. The chlorobenzene contents in the control plots were attributed to air-borne deposition. Differences between the treated and control areas established that 10% of the total chlorobenzene applied to soil in 25 sewage sludge applications (1942 to 1961) remained in soil in 1991. Measurable differences were found 30 yr after sludge applications ceased. In another report Wang and Jones (1994) demonstrated that the uptake of chlorobenzenes by plants (carrots, *Daucus carota* L.) was correlated with soil concentration, but <1% of the total chlorobenzene was taken up.

Predicting Contaminant Movement

Assessing the quality of the soil with respect to contaminant movement (i.e., predicting vulnerability) is fundamentally different from surveys which assess the amount of a contaminant or a suite of contaminants. In general, the approach to predicting vulnerability is to use spatially distributed soil attribute data and climatic data to predict chemical leaching or runoff. The predictions are made using simple or complex models. Simulations are performed using inputs for relatively homogeneous land areas. Both model inputs (soil attributes) and model results can be organized with the aid of geographic information systems (GIS).

Large scale applications of this methodology have used the AF model (Rao et al., 1985) and DRASTIC. The AF model computes the attenuation factor (AF), which is the fraction of applied mass of contaminant at some specified depth (water table). The model's data requirements related to soils are the depth to the water table, annual recharge, average organic C (f_{oc}) content or organic C content for specified depths, water content at field capacity and bulk density. Data requirements specific for a contaminant are the K_{oc}, the first-order degradation rate constant, and Henry's constant. The model can be applied to the entire profile or applied to layers of soil of some specified depth over an area that is assumed to be relatively uniform. Kleveno et al. (1992) compared the results of AF to those from the more detailed, mechanistic, PRZM model. Both models agreed on the relative leachability of 10 pesticides in six of seven different simulation scenarios, but Kleveno et al. (1992) concluded that AF was insufficient for predicting quantities of pesticides entering groundwater. Khan and Liang (1989) used the AF model and soil data obtained from soil surveys to map the relative leachability of soils on Oahu island. Later, Loague et al. (1990) performed a series of analyses that demonstrated that soil variability at the Order level was significant and that uncertainty in the data lead to uncertainty in the AF model (Loague et al., 1990). Uncertainty with organic C content and K_{oc} were greater than the other parameters in the AF model.

There are several alternative methodologies for assessing potential pesticide leaching. Evans and Myers (1990) outlined an approach that utilized DRASTIC, a procedure for ranking soil vulnerability, with a GIS-based map of soil and topography to produce an assessment of soil vulnerability for a 100 square mile area in Delaware. The DRASTIC procedure assigns numerical scores for depth to groundwater, net recharge, aquifer material characteristics and hydraulic conductivity, soil texture, vadose zone attributes, and topography, which are aggregated into a composite score indicative of vulnerability. Kellogg et al. (1994) described an assessment based on a screening procedure that rates soils and chemicals on a one to four scale describing decreasing leachability. The leachability ratings were based on numerous simulations combining different soil attributes and pesticide properties (half-life, K_{oc}, and Henry's constant) using the GLEAMS model. The vulnerability of soils were mapped using data from the Natural Resources Inventory (NRI) to produce a national assessment of potentially vulnerable soils. The studies described above represent assessments that attempt to differentiate regions of high vulnerability from those having low vulnerability based, at least in part, on soil attributes. Integration of this information with other aspects of soil quality, such as erodibility or potential for plant productivity, will increase our ability to make appropriate land management decisions.

Soil Quality and Risk Assessment

Contaminants in soils generate two levels of concern: the hazard to human health and the effect of soil contaminants on soil function in the agricultural or natural ecosystem. The ability of a soil to bind, detoxify, and degrade contaminants will minimize impacts and thus represent a component of soil quality. Risk assessment concerns evaluation of the impacts of contaminants in terrestrial and aquatic ecosystems (Suter et al.,1993). Risk assessment provides a science-based method for the establishment of guidelines or regulations that define increasing levels of risk associated with contaminants in soil. Unlike contaminants in water resources, national guidelines addressing soil contamination have not been established in the USA. Presently, guidelines for organic chemicals in soil are based on the risk from exposure to a particular contaminant in soil on a site-specific basis. Emphasis is placed on determining the potential risk of contaminant movement from soils into water resources or food crops. Beck et al. (1995) have reviewed Dutch and Canadian regulatory approaches to organic contaminants in soils. Target and intervention values for selected organic contaminants have been established for some compounds in these countries. Target values represent a no-effect threshold concentration and concentrations above the intervention value indicate a need for active remediation. Risk-based limits on contaminant levels in soil should be considered as a starting point for soil quality assessments.

An example of the risk assessment process is discussed by Blacker and Goodman (1994), who reported on the soil sampling and data analysis methods used by the USEPA to assess dioxin contamination at Times Beach, MO. Dioxin was present as a contaminant in waste oil applied to control road dust. Dioxin was assumed to be redistributed by wind and water transport of the soil. The goal of the survey was to identify areas where dioxin content exceeded 1 ng/g soil. This concentration was based on an expected increase in cancer risk of 10^{-6} from exposure to soil containing 1 ng dioxin/g soil (1 ppb). The investigators considered the increased cost of smaller cell sizes and increased sampling intensity versus the effects of larger cell sizes, which increase the risk of not remediating hot spots or remediating soil that was actually below the 1 ng/g level. A sampling program was eventually devised that used three composite samples obtained from 15.2 × 30.4 m grid cells. The cell size was chosen to represent a residential lot. The three samples were extracted and analyzed to determine if concentration exceeded 1 ng/g ($P < 0.05$). Overlaying of kriged dioxin concentrations on to a map with elevation contours showed the effects of surface runoff on dioxin transport. While this study was designed to guide bioremediation efforts, it illustrates the decisions concerning sampling design and resource allocation that become increasingly evident at larger scales. It also illustrates the need to establish effective criteria for interpreting the hazard associated with concentration of a particular contaminant.

In addition to the immediate risks to human health, the impacts of organic chemical contamination on the functions of soil need to be considered. Populations of soil microorganisms and microfauna, nutrient cycling, and plant-microorganism interactions in soil may be affected by the toxicity of these compounds in soil. Existing information indicates most pesticides applied at agronomically relevant concentrations cause only short-term effects (Domsch et al., 1983; Moorman, 1994a); however, high concentrations, such as those present at spill sites, may affect microbial biomass, respiration, and N transformations in soils. Integrated methods for assessing soil quality may have considerable use in evaluating the success of soil remediation programs and in assessing long-term soil responses to contaminants.

REFERENCES

Ahmad, I., and G. Crawford. 1990. Trace analysis of the herbicide chlorsulfuron in soil by gas chromatography-electron capture detection. J. Agric. Food Chem. 38:138–141.

Augustijn-Beckers, P.W.M., A.G. Hornsby, and R.D. Wauchope. 1994. The SCS/ARS/CES pesticide properties database for environmental decision making: II. Additional compounds. Rev. Environ. Contam. Toxicol. 137:1–82.

Blacker, S., and D. Goodman. 1994. Risk-based decision making case study: application at a Superfund cleanup. Environ. Sci. Technol. 28:471–477.

Beck, A.J., S.C. Wilson, R.E. Alcock, and K.C. Jones. 1995. Kinetic constraints on the loss of organic chemicals from contaminated soils: implications for soil quality limits. Critical Rev. Environ. Sci. Technol. 25:1–43.

Dao, T.H., T.L. Lavy, and J. Dragun. 1983. Rationale of the solvent selection for soil extraction of pesticide residues. Residue Rev. 87:91–104.

Dick, R.P., D.R. Thomas, and J.J. Halvorson. 1996. Standardized methods, sampling, and sample pretreatment. p. 107–121. In J.W. Doran and A.J. Jones (ed.) Methods for assessing soil quality. SSSA Spec. Publ. 49. SSSA, Madison, WI.

Domsch, K.H., G. Jagnow, and T.-H. Anderson. 1983. An ecological concept for the assessment of side-effects of agrochemicals on soil microorganisms. Residue Rev. 86:66–105.

Evans, B.M., and W.L. Myers. 1990. A GIS-based approach to evaluating regional groundwater pollution potential with DRASTIC. J. Soil Water Conserv. 45:242–245.

Flatman, G.F., and A.A. Yfantis. 1984. The survey and the census. Environ. Monit. Assess. 4:335–349.

Frehse, H., and J.P.E. Anderson. 1983. Pesticide residues in soil - Problems between concept and concern. p. 23–32. In J. Miyamato (ed.) IUPAC Pesticide chemistry: Human welfare and the environment. Pergammon Press, Oxford.

Gilbert, R.O. 1987. Statistical methods for environmental sampling. Van Norstrand Reinhold, New York.

Hassett, J.J., and W.L. Banwart. 1989. The sorption of nonpolar organics by soils and sediments. p. 31–44. In B.L. Sawhney and K. Brown (ed.) Reactions and movement of organic chemicals in soils. SSSA Special Publ. 22. SSSA, Madison, WI.

Hatfield, J.L., and J.L. Baker. 1995. Report of the Iowa management systems evaluation areas project (MSEA). USDA-ARS, Natl. Soil Tilth Lab., Ames, IA

Hayakawa, K., T. Murahashi, M. Butoh, and M. Miyazaki. 1995. Determination of 1,3-, 1,6-, and 1,8-dinitropyrenes and 1-nitropyrene in urban air by high performance liquid chromatography using chemiluminescence detection. Environ. Sci. Technol. 29:928–932.

Howard, P.H. 1989. Handbook of environmental fate and exposure data for organic chemicals. Vol. 1. Large production and priority pollutants. Lewis Publ., Boca Raton, FL.

Howard, P.H. 1990. Handbook of environmental fate and exposure data for organic chemicals, Vol. 2. Solvents. Lewis Publ., Boca Raton, FL.

Hsiao, A.I., and A.E. Smith. 1983. A root bioassay procedure for the determination of chlorsuluron, diclofop acid, and sethoxydim residues in soil. Weed Res. 23:231–236.

Kellogg, R.L., M.S. Maizel, and D.W. Goss. 1994. The potential for leaching of agrichemicals used in crop production: A national perspective. J. Soil Water Conserv. 49:294–298.

Khan, M.A., and T. Liang. 1989. Mapping pesticide contamination potential. Environ. Manage. 13:233–242.

Kleveno, J.J., K. Loague, and R.E. Green. 1992. Evaluation of a pesticide mobility index: Impact of recharge variation and soil profile heterogeneity. J. Contam. Hydrol. 11:83–99.

Koskinen, W.C., 1988. Analysis of pesticides in soil and water. p. 70–78. In D.W. Nelson and R.H. Dowdy (ed.) Methods for ground water quality studies. Proc. Natl. Workshop, Arlington, VA.. Univ. Nebraska, Lincoln.

Laskowski, D.A., R.L. Swann, P.J. McCall, and H.D. Bidlack. 1983. Soil degradation studies. Residue Rev. 85:139–158.

Leisinger, T. 1983. Microorganisms and xenobiotic compounds. Experientia 39:1183–1191.

Loague, K., R.E. Green, T.W. Giambelluca, T.C. Liang, and R.S. Yost. 1990. Impact of uncertainty in soil, climatic, and chemical information in a pesticide leaching assessment. J. Contam. Hydrol. 5:171–194.

Meulenberg, E.P., W.H. Mulder, and P.G. Stoks. 1995. Immunoassays for pesticides. Environ. Sci. Technol. 29:553–561.

McCall, P.J., D.A. Laskowski, R.L. Swann, and H.J. Dishberger. 1981. Measurement of sorption coefficients and their use in environmental fate analysis. p. 89–109. In Test protocols for environmental fate and movement of toxicants. Symp. Proc., Assoc. Off. Anal. Chem., Washington, DC. 1980. Assoc. Off. Anal. Chem., Washington, DC.

Miller, J.G., D.B. Jaynes, and T.B. Moorman. 1995. Prediction of atrazine persistence in a central Iowa field. p. 109–118. In C. Heatwole (ed.) Water quality modeling. Proc. Int. Symp., Orlando, FL. Am. Soc. Agric. Eng., St. Joseph, MO.

Montgomery, J.H. 1993. Agrochemicals desk reference: Environmental data. Lewis Publ., Boca Raton, FL.

Moorman, T.B. 1994a. Effects of herbicides on the ecology and activity of soil and rhizosphere microorganisms. Rev. Weed Sci. 6:151-176.

Moorman, T.B. 1994b. Pesticide degradation by soil microorganisms: environmental, ecological, and management effects. p. 121–165. In J.L. Hatfield and B.A. Stewart (ed.) Advances in soil science. Soil biology: Effects on soil quality. Lewis Publ., Boca Raton, FL.

Mueller, T.C., and T.B. Moorman. 1991. Liquid chormatographic determination of fluometuron and metabolites in soil. J. Assoc. Off. Anal. Chem. 74:671–673.

Mueller, T.C., T.B. Moorman, and C.E. Snipes. 1992. Effect of concentration, sorption, and microbial biomass on degradation of the herbicide fluometuron in surface and subsurface soils. J. Agric. Food Chem. 12:2517–2522.

NASS. 1994. Agricultural chemical usage: 1993 field crops summary. Natl. Agric. Statistics Serv., USDA, Washington, DC.

Nash, R.G. 1988. Dissipation from soil. p. 131–169. In R. Grover (ed.) Environmental chemistry of herbicides. Vol. I. CRC Press, Boca Raton, FL.

Rao, P.S.C., A.G. Hornsby, and R.E. Jessup. 1985. Indices for ranking the potential for pesticide contamination of groundwater. Soil Crop Sci. Fl. Proc. 44:1–8.

Schlaeppi, J.-M., W. Föry, and K. Ramsteiner. 1989. Hydroxyatrazine and atrazine determination in soil and water by enzyme-linked immunosrbent assay using specific monoclonal antibodies. J. Agric. Food Chem. 37:1532–1538.

Suter, G.W., L.W. Barnthouse, S.M. Bartell, T. Mill, D. Mackay, and S. Patterson. 1993. Ecological risk assessment. Lewis Publ., Boca Raton, FL.

Van Emon, J.M., and C.L. Gerlach. 1995. A status report on field-portable immunoassay. Environ. Sci. Technol. 29:312–317.

Van Emon, J.M., and V. Lopez-Avila. 1992. Immunochemical methods for environmental analysis. Anal. Chem. 64:79–88.

Wang, M.-J., and K.C. Jones. 1994. Uptake of chlorobenzenes by carrots from spiked and sewage sludge-amended soil. Environ. Sci. Technol. 28:1260–1267.

Wang, M.-J., S.P. McGrath, and K.C. Jones. 1995. Chlorobenzenes in field soil with a history of multiple sewage sludges applications. Environ. Sci. Technol. 29:356–362.

Wauchope, R.D., T.M. Buttler, A.G. Hornsby, P.W.M. Augustijn-Beckers, and J.P. Burt. 1992. The SCS/ARS/CES pesticide properties database for environmental decision making. Rev. Environ. Contam. Toxicol. 123:1–164.

Webber, M.D., and S.S. Singh. 1995. Contamination of agricultural soils. p. 87–96. *In* D.F. Acton and L.J. Gregorich (ed.) The health of our soils: Toward sustainable agriculture in Canada. Publ.1906/E. Agric. and Agri-Food Canada, Ottawa.

19 Soil Quality in Central Michigan: Rotations with High and Low Diversity of Crops and Manure[1]

E. Franco-Vizcaíno

Michigan State University
East Lansing, Michigan

Cover crop, intercrop and manure residues can help protect the soil from erosion, as well as increase organic matter content and water retention, and the efficiency of N use (Karlen et al., 1992). Crop diversity may improve soil quality by increasing the amount, quality and diversity of residues returned to the soil, and by lengthening the time that roots are actively growing in the soil.

Many believe that the robustness of agricultural systems can be improved by imitating natural ecosystems; however, little information has been gathered on crop and residue diversity or its impact on soil quality. Much opportunity exists in Michigan for increasing crop diversity within the traditional corn-based (*Zea mays* L.) system. Knowledge of physical, chemical, and biological measures of soil quality will serve as a base for recommendations and accelerated adoption of increased crop diversity.

APPROACH

Farmers worked cooperatively with researchers to select matched field sites for comparisons of low and high diversity cropping systems. Cropping history, including manure application, was obtained for the 4-yr period prior to soil evaluation. Final site selection for paired comparisons was based on cropping history as well as similarity in topography, aspect, soil type, and distance between paired sites (Table 19–1). A corn field immediately adjacent to the Living Field Laboratory (Kellogg Biological Station, Hickory Corners, MI) was used as a control for the study.

The number of residue sources was determined by considering each crop, cover crop species and manure application. For example, continuous corn for 5

---

[1] Research supported by the C.S. Mott Chair for Sustainable Agriculture, Dep. of Crop and Soil Sciences, Michigan State University, East Lansing.

Copyright © 1996 Soil Science Society of America, 677 S. Segoe Rd., Madison, WI 53711, USA. *Methods for Assessing Soil Quality*, SSSA Special Publication 49.

Table 19–1. Landscape and soil characteristics, and 1989 to 1993 history of cropping and manuring of study sites in south central Michigan.

Comparison	Landscape	Distance between study sites ~ m	Soil series % slope	Determined texture % gravel-sand silt-clay†	Cropping	Cover crops	Manure ~ Mg ha⁻¹	Residue diversity
Control	Nearly level	40	Kalamazoo sl 0–2%	cl (1-26-44-30)	A A A A C	— — — — —	— — — — —	2
Control	Nearly level		Kalamazoo sl 0–2%	l (15-42-33-25)	A A A A C	— — — — —	— — — — —	2
1 high	S shoulder, small knoll	200	Spinks ls 0–6%	ls (2-86-7-7)	Tr S C S C	— — — — —	— — — 25 —	4
1 low	S shoulder, small knoll		Spinks ls 0–6%	ls (2-87-7-6)	C C C C C	— — — — —	— — 25 25 —	2
2 high	Nearly level bottom	200	Capac l 0–3%	cl (1-40-24-35)	C C S W C	cl — — cl cl	25 25 — — —	4
2 low	Nearly level bottom		Capac l 0–3%	scl (1-53-18-29)	C C C C C	— — — — —	25 25 25 25 25	2
3 high	Nearly level	100	Capac l 0–3%	scl (16-49-21-29)	C C S W C	cl — — cl cl	— — — — —	4
3 low	Nearly level		Capac l 0–3%	scl (4-51-23-36)	C S W C C	— — — — —	— — 25 — —	3
4 high	Rolling, midslope	100	Marlette fsl 2–6%	sl (10-67-23-10)	C S C S C	cl — — — —	25 25 25 25 25	4
4 low	Rolling, midslope		Marlette fsl 2–6%	sl (6-66-24-11)	C C C C C	— — — — —	— — — — —	2
5 high	Nearly level	1000	Capac l 0–3%	scl (3-45-27-28)	C S W C C	— — cl cl cl	25 25 25 25 25	4
5 low	Nearly level		Capac l 0–3%	scl (3-53-25-22)	A A A C C	— — — — —	— — — — 25	3
6 high	Small undulations	150	Ithaca l 0–3%	scl (8-46-22-31)	Cu W C S C	— — — — —	25 25 25 25 —	5
6 low	Small undulations		Ithaca l 0–3%	scl (1-52-21-27)	C C S C C	— — — — —	— — — 25 —	2
7 high	Nearly level	2000	Kalamazoo sl 0–2%	sl (1-58-25-17)	C C W C C	— — — — cl	25 — 25 — —	4
7 low	Small undulations		Kalamazoo sl 2–6%	sl (6-56-26-18)	C C C C C	— — — — —	— — — — —	1
8 high	Small undulations	100	Capac l 0–3%	cl (6-39-33-29)	C B Cu W C	— — — cl —	25 25 25 — —	6
8 low	Small undulations		Capac l 0–3%	sl (6-54-28-18)	C C C C C	— — — — —	25 25 25 — —	2
9 high	S shoulder, small knoll	4000	Marlette fsl 2–6%	sl (17-64-20-16)	A W C S C	og — v — v	— — — 12.5 —	7
9 low	S shoulder, small knoll		Marlette fsl 2–6%	sl (10-58-23-19)	W F C S C	— — cl — —	— — — — —	4

† Bulked samples, sand + silt + clay = 100%. C = Corn, S = Soybeans, A = Alfalfa, W = Wheat, Tr - Triticale, Cu = cucumbers, F = Fallow, cl = clover, og = orchard grass, v = vetch. Manure was generally from on-farm dairy or hog operations, and it was assumed that the type applied did not change from year to year. Capac: fine-loamy, mixed, mesic Aeric Ochraqualfs. Ithaca: fine, mixed, mesic, Glossaquic Hapludalfs. Kalamazoo: fine-loamy, mixed, mesic Typic Hapludalfs. Marlette: fine-loamy, mixed, mesic Haplic Glossudalfs. Spinks: sandy, mixed, mesic Psammentic Hapludalfs.

yr with no cover crops or manure was counted as one source; whereas, a rotation that included corn, wheat (*Triticum aestivum* L.), and soybeans [*Glycine max* (L.) Merr.], clover as a cover crop, and manure applied every other year counted as five. Selection of a pair of fields for comparison required that the pair have a minimum difference of two residue sources. Of the field pairs selected, it was later determined that two (Pairs 2 and 5) did not differ by two residue sources because farm records and/or farmer recollection were incorrect.

Each field plot was approximately 0.01 ha in size. Six sampling stations were established in each field plot in nontrafficked interrows of corn having few obvious disturbances such as fertilizer bands. Soil properties evaluated at the 0- to 20-m depth included bulk density; percentage of gravel (>2mm); particle-size analysis by the hydrometer method; water holding capacity (30 kPa to 1.5 MPa); penetrometer resistance ($n = 6$ per station); A-horizon and rooting depth ($n = 2$ per station); infiltration rate (time required for 2.5 cm of water to infiltrate); inorganic N ($NO_3^+ + NO_2^- + NH_4^+$) by colorimetry; pH; mineralizable N by anaerobic incubation; total C by high temperature combustion; total N by the Kjeldahl procedure; extractable P by the Bray method and soil respiration and microbial biomass. Infiltration and respiration measurements were made in the early morning and in early afternoon approximately 4 to 6 h after the first irrigation. In-field soil quality measurements were conducted using the methods described by Sarrantonio et al. (1996, this publication). Corn yield was determined by harvesting four, 6.8-m rows within each plot.

A paired comparison t-test was used to test differences in soil properties within a field pair and across the nine pairs. Differences in soil properties between rotations with high and low residue diversity were tested by two-way ANOVA using a procedure that treated sampling station values as replicates. Three field pairs that differed greatly in the number of years manure was applied were considered separately. Infiltration data were transformed to the $\log_{10}$ and microbial biomass C to the square-root form in order to obtain a normal distribution for analysis.

Soil quality is typically assessed by measuring a number of soil and crop properties. In our study, 21 different properties were measured on each field. Currently, there is not a specific threshold value assigned to each property that separates a higher quality soil from a lower quality soil. It is likely that such threshold values would, at a minimum, be soil specific. In addition, it is likely that a single soil could have some measured properties that indicate a higher soil quality while having other measured properties that indicate lower soil quality. For example, soil A, as compared with soil B, could have higher yield and soil respiration as indicators of higher soil quality but higher penetration resistance and lower water holding capacity. Therefore, to facilitate our discussion, we assumed that the following characteristics were associated with higher soil quality:

- corn yield was higher or rooting was deeper,
- topsoil was deeper or had a lower bulk density or resistance to penetration,
- topsoil had a higher water holding capacity or faster infiltration,
- topsoil had a higher concentration of total C, total N or mineralizable N but lower extractable N or P,

- topsoil had a lower ratio of C to N,
- topsoil had a higher ratio of mineralizable N to total C,
- topsoil had a higher microbial biomass, or higher microbial respiration rate (but lower specific respiratory activity),
- topsoil had a higher ratio of microbial biomass to total C, or higher soil respiration rate.

RESULTS AND DISCUSSION

Soils among all field pairs were of medium texture and bulk density, slightly acid to neutral and fertile (Table 19–2). Few patterns could be discerned from comparisons of <individual> low and high-diversity pairs; however, improved soil quality, as indicated by enhanced soil property values, tended to be associated with high residue diversity plots. For example, of 63 significant differences in soil properties (excluding pH) that occurred in the nine comparisons, 38 were in the direction of improved soil quality in the high diversity plots versus 25 in the low diversity plots. And in a majority of comparisons (Pairs 2, 3, 4, 6, 7, and 9) a higher number of significant differences occurred in the high diversity plots than the low diversity plots.

High and variable concentrations of extractable N in some plots (Pairs 1, 3, and 5) may have been due to sampling near fertilizer bands. The high concentrations of extractable P were generally associated with long-term manuring (Pairs 1, 4, 7, and 8). Later interviews with farmers indicated that at three locations (Pairs 2, 4, and 5), the low diversity plots had, in fact, received yearly applications of manure during the 5-yr study period. Similar variability in soil properties of the control plots also was observed.

Combining results from all nine comparisons showed that corn yield and total and mineralizable N were higher in fields receiving a higher diversity of residues (Table 19–3). The variation between high and low residue diversity plots was 25% or less as indicated by the ratios. Among the 22 soil properties evaluated, 9 differed by <5% among high and low diversity plots. But all 13 soil properties showing an increase of 5% or greater in the direction of improved soil quality occurred in the high diversity plots.

For paired low and high diversity plots having similar manure histories (Pairs 1, 3, 6, 7, 8, and 9), significant improvement occurred in resistance to penetration, total and mineralizable N, soil respiration after irrigation, and ratio of mineralizable N to total C (Table 19–3). But in addition, 14 of 16 improvements of 5% or greater in soil quality properties were in the high diversity plots. Increased extractable N and P were the two soil properties that showed an increase of 5% or greater in the low diversity plots. Variation between these high and low diversity plots receiving similar manuring was as high as 57%.

For the comparisons with dissimilar manuring (Pairs 2, 4, and 5), where manure was applied much more frequently to the low diversity than the high diversity plots, soil fertility was similar among pairs except for extractable phosphorus (Table 19–3). Moreover, of 16 improvements of 5% or greater in soil

Table 19–2. Paired comparisons of soil physical, chemical, and biological properties in corn fields with high or low diversity of residues returned to the soil during 1989–1993. Values are means ($n = 6$) for the 0- to 20-cm soil layer unless noted otherwise.

Comparison	Residue diversity	pH	Bulk density	Penetration resistance	Water holding capacity	Topsoil depth	Corn rooting depth	Infiltration rate Initial†	Infiltration rate After irrigation§	Corn yield¶	Total C	Total N
			g cm^{-3}	kg cm^{-2}		cm		cm min^{-1}			Mg ha^{-1}	
Control	Low	5.2	1.05	0.9	1.90#	nd	24.8	3.7††	0.4	nd††#	29.1	2.84**
Control	Low	5.4	1.25**	1.0	1.25	nd	25.8	0.2	0.9	nd††#	26.1	2.46
1	High	6.5#	1.38	wet nd	0.98	25.9	23.3	0.4	0.4	5.56	19.3	1.33
1	Low	5.8	1.43#	wet nd	1.02	28.5#	23.3	0.4	0.3	7.35#	21.7	1.73*
2	High	6.9	1.01	0.89	2.29	26.3	21.8	23.2	9.4	10.8*	48.4	6.01
2	Low	7.1*	1.03	0.46	1.69	31.4*	22.0	11.4	7.2	8.9	45.5	4.62
3	High	6.3	1.06	0.67	1.20	30.0	24.8	17.5*	3.4*	11.1**	34.2	3.66
3	Low	6.5	1.29*	2.40*	1.34	29.8	22.6	0.6	0.1	5.84	35.3	3.43
4	High	5.9	1.26	1.86	2.48	25.1	19.7	1.7	0.9	10.7	34.4#	2.86
4	Low	5.7	1.25	1.06	2.20	22.8	22.8#	3.7	1.7	10.1	27.9	2.58
5	High	6.4	1.15	1.88	1.52	28.6	23.2	7.8	3.0	11.1*	31.7	2.96
5	Low	6.3	1.29	1.84	1.51	28.1	25.4**	12.2	7.4	6.48	36.2	3.67*
6	High	6.1	1.25	0.36	1.10	25.3	25.1	1.0	0.3	10.3	32.7	3.61
6	Low	5.8	1.23	0.83**	1.49*	23.8	23.8	1.3	0.3	9.74	34.4	3.20
7	High	5.6	1.44	0.93	1.99*	29.8*	27.8**	0.3	0.1	8.23	35.3**	3.36**
7	Low	6.1	1.40	1.48	1.38	26.7	21.8	0.3	0.1	7.15	21.2	2.11
8	High	5.8	1.29**	1.08*	1.72	26.1	22.6	0.9	0.1	10.3	38.2	3.49
8	Low	6.8	1.15	0.59	2.36#	24.4	20.9	3.3	0.2	10.7	34.4	3.23
9	High	5.8	1.27	1.03	2.18*	21.6	20.1	1.8	0.3	5.48	21.2	2.09
9	Low	5.9	1.20	0.72	1.00	23.9**	23.3**	1.2	0.4	8.22*	21.2	2.10

(continued on next page)

Table 19–2. Continued.

Comparison	Residue diversity	Extractable N	Mineralizable N	Extractable P	Microbial biomass C	Soil respiration		Microbial respiration	Specific respiratory activity	Mineralizable N/ total C	C/N	C_{mic}/ C_{titak}
						initial	after irrigation					
		kg ha^{-1}				kg CO_2-C		ha^{-1}d^{-1}	††	mgNg^{-1}C	w/w	%
Control	Low	46.9	28.9	108	935	40.9	24.7	16.9	18.0	0.99	10.3	3.22
Control	Low	44.5	29.2	119	947	33.4	18.6	22.9	24.6	1.12	10.6	3.72
1	High	29.1	14.4#	506#	720	17.2	10.8	23.9	36.1	0.75#	14.7	3.77
1	Low	191*	3.30	436	811	17.3	8.83	33.1**	50.2	0.15	12.6	3.81
2	High	50.0*	36.6	157	1338	23.8	10.6	21.7*	16.9#	0.74#	8.4	2.77
2	Low	36.9	30.1	118	1153	30.4	15.8#	8.90	7.77	0.67	9.9*	2.58
3	High	243#	27.0	92.2*	1054	42.4	18.0	11.4	10.5	0.79	9.4	3.17**
3	Low	14.3	27.4	77.1	974	29.7	13.9	10.3	10.6	0.78	10.3*	2.79
4	High	24.6	40.0	252	749	32.8	10.5	5.87	7.37	1.23	11.9	2.37
4	Low	29.6	48.6	990**	877	27.9	6.68	9.75	10.9#	1.71	10.8	3.15
5	High	39.4	19.0	120	1081	22.2	3.08	13.4	11.8	0.60	10.8	3.42
5	Low	207	30.7	119	1133	34.8	32.4**	13.2	12.0	0.84	9.9	3.16
6	High	63.5	24.0	176*	664	39.9	6.21	9.50	18.0	0.75	9.1	2.02
6	Low	52.0	18.8	126	617	60.3#	6.86	9.17	15.5	0.56	10.7**	1.85
7	High	29.9	49.9*	567*	909**	27.7	29.6*	22.7	24.8	1.44	10.5	2.58
7	Low	22.8	30.1	135	567	31.1	15.0	10.2	19.1	1.43	10.0	2.69
8	High	81.4#	26.2	170	848	37.1	15.4	13.8	16.0	0.68	10.9	2.23
8	Low	31.5	25.8	355**	1000†	27.1	11.4	16.4	16.8	0.75	10.7	2.93†
9	High	28.7	29.0**	58.8	757†	6.42	12.5	7.96	10.7	1.39**	10.1	3.31
9	Low	94.9*	18.3	171**	579	45.7	11.0	5.36	9.67	0.87	10.1	2.80

†,*,** Significant at the 0.1, 0.05, and 0.01 levels, respectively (symbols arbitrarily placed on larger value); nd = not determined.
‡ Falling head, 2.5 cm H2O.
§ 4–6 h after initial irrigation.
¶ n = 4.
\# yield averaged 8.53 Mg/ha at adjacent experimental field trials under the same management.
†† mg CO_2–C g^{-1} Cmic d^{-1}.

Table 19–3. Comparison of soil properties in corn-based rotations with high or low diversity of residues returned to the soil, analyzed by two-way ANOVA procedures using subsamples as replicates.

Soil property	All nine comparisons			Similar manuring Comparisons 1,3,6,7,8,9			Dissimilar manuring Comparisons 2,4,5		
	High diversity	Low diversity	Ratio	High diversity	Low diversity	Ratio	High diversity	Low diversity	Ratio
Bulk density, g cm^{-3}	1.24	1.25	0.99	1.28	1.28	1.00	1.14	1.19	0.96
Penetration resistance, kg cm^{-2}	0.97	1.04	0.93	0.68	1.00*	0.68	1.54†	1.12	1.38
Corn rooting depth, cm	23.1	22.9	1.01	23.9†	22.6	1.06	21.5	23.4*	0.92
Topsoil depth, cm	26.5	26.6	1.00	26.4	26.2	1.01	26.7	27.4	0.97
Water holding capacity, cm	1.72	1.54	1.12	1.53	1.41	1.09	2.10	1.80	1.17
Infiltration rate, cm min^{-1}	2.21	1.77	1.25	1.26	0.84	1.50	6.81	7.98	0.85
Infiltration rate after irrigation, cm min^{-1}	0.68	0.58	1.17	0.33	0.21	1.57	2.94	4.47	0.66
pH	6.0	6.0	1.00	5.9	6.0	0.98	6.2	6.1	1.02
Total C, Mg ha^{-1}	32.8†	30.9	1.06	30.2†	28.0	1.08	38.2	36.5	1.05
Total N, Mg ha^{-1}	3.27*	2.96	1.10	2.92**	2.63	1.11	3.95	3.62	1.09
C:N ratio	10.6	10.6	1.00	10.8	10.7	1.01	10.4	10.2	1.02
Extractable N, kg ha^{-1}	65.5	75.5	0.87	79.2	67.7	1.17	38.0	91.1	0.42
Mineralizable N, kg ha^{-1}	29.6*	25.9	1.14	28.5***	20.7	1.38	31.9	36.4	0.87
Extractable P, kg ha^{-1}	233	281	0.83	262	217	1.21	177	409**	0.43
Soil respiration, kg C ha^{-1} d^{-1}	34.8	33.5	1.04	38.0	34.7	1.10	27.8	31.0	0.90
Soil respiration after irrigation	13.2	13.6	0.97	15.5*	11.0	1.41	8.22	18.3**	0.45
Microbial biomass C, kg ha^{-1}	900	855	1.05	822	758	1.08	1060	1050	1.01
Microbial respiration, kg C ha^{-1} d^{-1}	14.5	12.9	1.12	14.9	14.1	1.06	13.7	10.4	1.32
Specific microbial respiration, mg g^{-1} d^{-1}	16.9	16.8	1.01	19.4	20.4	0.95	12.0	10.1	1.19
C$_{microbial}$/C$_{total}$, %	2.87	2.86	1.00	2.88	2.81	1.02	2.86	2.94	0.97
Mineralizable N/C$_{total}$ mg^{-1} g^{-1}	0.93	0.87	1.07	0.97**	0.77	1.26	0.86	1.06	0.81
Corn yield, Mg ha^{-1}	9.29*	8.28	1.12	8.51	8.17	1.04	10.9*	8.51	1.28

†, *, **, *** Significantly different at the 0.1, 0.05, 0.01, and 0.001 levels, respectively (symbols arbitrarily placed on larger value).

properties, 9 were in the direction of improved soil quality in the low diversity plots vs. 7 in the high diversity plots. This suggests that prolonged manuring, as evidenced by high extractable phosphorus, may be detrimental to economic yields. But higher residue diversity seemed to counteract the adverse effects of prolonged manuring. Variation between high and low diversity plots receiving dissimilar manuring was as high as 68%.

Water infiltration rates and soil respiration were lower for all plots after irrigation as compared to preirrigation (Table 19–2). The postirrigation assessment serves as a measure of the soil's response to near-optimal conditions of moisture and aeration; however, the 4- to 6-h drainage period between infiltration measurements was not adequate to allow for the soil to fully drain to field capacity. Thus, water-filled pore space exceeded 60% and, as discussed by Parkin et al. (1996, this publication), CO_2 diffusion rates were markedly reduced.

These results suggest that increased diversity of residues was associated with an improvement in soil quality which was expressed as improved soil tilth, nutritional status, and biological activity. The increase in total, but not extractable or mineralizable N suggests that the improvement in nutritional status resulted from an increase in the pool of organic N in the soil. Results presented here are consistent with those of Reganold et al. (1993).

Indices of cropping diversity and manuring frequency also were devised and correlated with soil properties (e.g., microbial biomass C, $r = 0.57$). This assessment revealed that improvements in soil quality could not be associated simply with diversity in rotations, cover crops or manure applications, but rather with diversity and frequency in all three sources of residue.

Although factors such as quantity and quality of residues could not be controlled, the cropping histories do not suggest that these factors bias the results. It is likely that replacing corn with either wheat or soybeans would result in lower quantities, and equal or slightly higher quality, of residues returned to the soil. Thus, results reported here indicate that a higher diversity of C inputs from crop residues can lead to improved soil quality after a single rotation cycle.

ACKNOWLEDGMENTS

This research was supported by the C.S. Mott Chair for Sustainable Agriculture. I thank Tom Willson, Hugh Smeltekop, Todd Martin, Christie McGrath, Brian Cook, Elaine Parker, and Curtis Beard for help with field and laboratory work, and Rich Leep and Jack Knorek for help interviewing farmers.

REFERENCES

Karlen, D.L., N.S. Eash, and P.W. Unger. 1992. Soil and crop management effects on soil quality indicators. Am. J. Altern. Agric. 7:48–55.

Parkin, T.B., J.W. Doran, and E. Franco-Vizcaíno. 1996. Field and laboratory tests of soil respiration. p. 231–245. *In* J.W. Doran and A.J. Jones. (ed.) Methods for assessing soil quality. SSSA Spec. Publ. 49. SSSA, Madison, WI.

Reganold, J.P., A.S. Palmer, J.C. Lockhart, and A.N. Macgregor. 1993. Soil quality and financial performance of biodynamic and conventional farms in New Zealand. Science (Washington, DC) 260:344–349.

Sarrantonio, M., J.W. Doran, M.A. Liebig, and J.J. Halvorson. 1996. On-farm assessment of soil quality and health. p. 83–105. *In* J.W. Doran and A.J. Jones. (ed.) Methods for assessing soil quality. SSSA Spec. Publ. 49. SSSA, Madison, WI.

20 Impact of Farming Practices on Soil Quality in North Dakota[1]

John C. Gardner and Sharon A. Clancy

North Dakota State University Carrington Research Extension Center
Carrington, North Dakota

An important component of a sustainable future includes the maintenance and well being of the environment directly influenced by agriculture. Though a number of tools are available to evaluate the productivity and economic return among farming systems, there are few that reliably indicate the environmental impact of agriculture upon a given site. Since development of such tool would help guide farmers towards better practices, and lawmakers towards better public policies, we have been searching for a practical and reliable indicator of environmental performance with which to track, evaluate, and compare various alternative farming practices. Both the literature and our experience suggest the most reliable and consistent environmental indicator might be those soil and water characteristics that are influenced by management (National Research Council, 1993). We tested the ability of soil quality indicators to distinguish among the most innovative farming practices in our region of the Northern Great Plains.

Innovative Farming Systems in the Great Plains

The Great Plains region of the USA is a vast agricultural resource, comprising roughly one fifth of the nation's total land area. It is a region where the bulk of the wheat (*Triticum aestivum* L.) and other small grains are produced, and also the majority of the beef cattle (*Bos taurus*). Originally described as the great American desert, farming across this region for the past century has learned to live in delicate balance between productivity and disaster. Especially since the Dust Bowl of the 1930s, farming practices have rapidly evolved around new technology as it has become available.

Haas et al. (1974) and others have focused on the soil and how it has changed over time in response to the evolving farming practices in the Great Plains. Most of the changes have been associated with evolving crop rotation and tillage practices. The dominant crop in the region has been wheat, and most of the changes in rotation have not been with other crops, but with fallow. Leaving the

---

[1] Sponsored in-part through a grant from the Northwest Area Foundation, Grant no. 89-0034, St. Paul, MN.

Copyright © 1996 Soil Science Society of America, 677 S. Segoe Rd., Madison, WI 53711, USA. *Methods for Assessing Soil Quality*, SSSA Special Publication 49.

field unplanted for a growing season has been a widely adopted practice to stabilize wheat grain yield through soil moisture storage, mineralization of soil nutrients, and management of weeds. Unfortunately, this rotational practice also has sped the loss of soil organic matter through the reduction in annual return of C to the soil through plant growth. While tillage practices are always an important management decision, they are of particular importance in the Great Plains because of frequent dry periods and often sparse crop residue cover. The plow has gradually given way to the disk, chisel plow, undercutter, and now use of post-emergence, broad-spectrum herbicides to conserve soil moisture and reduce erodibility. Across the Great Plains today, innovative rotations are being used to purposefully increase the return of organic matter to the soil, and tillage is being reduced to slow organic matter loss.

In the northern part of the Plains, North Dakota has a continental climate of warm summers, cold winters, and moderate precipitation. In comparison to other wheat growing states of the Great Plains, it's northerly latitude has encouraged the production of spring-sown wheat since the threat of winter-kill is usually greater than drought. Though spring-sown cereals don't mature as early, thus escaping drought, they are well adapted to the relatively cool nights and lower evaporative demand common to North Dakota summers.

One of the limitations of spring wheat is the short duration of growth between late April–May and August, leaving 8 mo of the year void of a crop. Coupling the short season with fallow, a period of 20 continuous months without a growing crop is common. The lack of soil cover has increased the susceptibility of soil to erosion and been a primary motivation for many farmers to experiment with limited or no-tillage farming practices. These alternative practices began first in the southwest and northeast parts of the state in the mid-1970s and have spread across the region. Recent herbicide and planting equipment innovations, such as low-cost glyphosate and air-seeders, have made limited and no-till farming both practical and economical.

Another attribute of North Dakota's climate is the relatively low level of crop insect–disease pests, and a growing season adapted to a wide range of short-season, spring-sown crops. This combination of factors has contributed to the success of pesticide-free farming systems and made North Dakota the leading state in the production of organic grains. These farming systems rely upon crop rotation to manage weeds (by alternating early, late, and fall-sown crops), insects and diseases (by alternating susceptible crops), and build soil fertility (through the inclusion of a leguminous green manure crop every 3 to 4 yr and/or the regular use of animal manures). Though tillage is sometimes intensive in these systems, the wider diversity of crops results in near continuous crop growth and soil cover. Also, a greater amount of organic matter is returned to the soil since the rotation generally consists of green manure crops and other crops that return higher proportions of residue.

The link between agriculture and environmental quality is important in North Dakota, since a higher proportion of land is cropped than any other state (64%). And, when combined with grazed land, agricultural practices have a direct impact on more than 91% of the state. Past studies by Bauer and Black (1981) confirmed the impact of improved tillage practices on certain soil properties.

FARMING PRACTICES

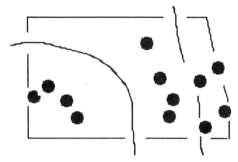

Fig. 20–1. North Dakota's three principal ecoregions and 12 sites samples during 1990 to 1991.

Since that time, further innovation in both tillage and rotation practices have been noticeable across the state.

Like Bauer and Black (1981), our objectives were to compare the impact of these practices on soil properties using actual farm sites, with prairie sites used as a standard of comparison.

METHODS

Though difficult to categorize farms and farmers, especially since farming practices are always evolving, three North Dakota farm management approaches were chosen for study along with perennial grassland sites that served as an ecological (and environmental) benchmark (Clancy et al., 1993):

Site identification	Description
Prairie	Sites stable successionally with native or naturalized vegetation.
Conventional	Farms successfully using practices that are predominant in region.
No-till	Farms making extensive use of conservation (or no) tillage practices as compared with conventional.
Organic	Farms making extensive use of crop rotation to substitute for little (or no) synthetic pesticide or fertilizer inputs as compared with conventional.

Since North Dakota has three dominate eco-regions, sites of each type were carefully chosen in each region that shared similar soils, farming enterprises, labor, and size (Fig. 20–1). Though general data were taken from each whole farm, one specific site (field) per farm was chosen with a soil type that matched other sites within a region. These 12 sites were visited bi-weekly throughout the 1990 and 1991 growing seasons to measure environmental impact as determined by crop yield, fertilizer and pesticide use, energy consumption, water use, diversity and population of insects, extent of vegetation or residue on the soil surface, and soil quality characteristics.

Most of the soil quality parameters were observed using standard laboratory procedures recommended in the North Central region (Dahnke, 1988). Light fraction organic matter was determined by using a procedure similar to that of Spycher (1983). Aggregate size was determined by dry sieving surface soil samples through a series of round-hole sieves ranging from 0.1 to 2 mm in diam. Aggregates retained on the 1-mm sieve were used to determine the percentage of water stable aggregates using the procedure described by Kemper and Rosenau (1986).

RESULTS

All the farms included were carefully selected to represent well-managed and successful operations that had been using similar practices for at least the last decade. Management practices actually observed varied considerably among farms, but did fit the description originally defined. Conventional farms used the most tillage, while no-till farms had the most frequent pesticide applications, especially herbicides. Organic farms made the greatest use of crop rotation in having both the greatest diversity of crops and the highest frequency of legumes in rotation for biological N fixation.

In general, the measurable impact on what are deemed important environmental performance indicators were not as predictable as management among these well-managed farms. For example, in calculating the ratio between harvested crop yield and total inputs in energy units, we found the organic farms most efficient in the east and west regions, with the no-till farm most efficient in the central portion of the state. The no-till systems were generally the least consumptive of water and resulted in the least amount of soil erosion as calculated using current USDA-Natural Resources Conservation Service guidelines. Only two sites exceeded the erosion threshold, or t-value; the conventional site in the east due to wind erosion susceptibility, and the organic site in the west due to water erosion on slopes.

Using soil quality factors as indicators to distinguish among farming systems also had varying results, depending upon the parameter used. Mean values for the soil quality parameters used are found in Table 20–1. Though trends in the data were apparent, few of the sites within a region were statistically different using classical analysis of variance and mean separation techniques. A conscious choice was made in this study, however, to concede statistical power in exchange for the chance to observe actual farm sites. In general, the soil quality factors that proved statistically significant were those associated with structural and C-related properties; all properties that are slow to change over time. Variability among the data (in this case the coefficient of variation was used to normalize variation across means of different magnitudes) is another indicator of the temporal nature of these systems. In general, data gathered from the prairie sites were least variable, while the conventional farms had the greatest variation among the two years of the study.

Among the most variable data gathered were the soil chemical parameters, particularly N, which was highly influenced by time of observation in the context

Table 20–1. Comparative mean values of soil quality data gathered during 1990/1991 across North Dakota from prarie and farm sites identified as using conventional, no-till, or organic farming practices.

Soil quality parameter measured	North Dakota Region											
	East				Central				West			
	Prairie	Conventional	No-till	Organic	Prairie	Conventional	No-till	Organic	Prairie	Conventional	No-till	Organic
Depth A horizon, cm	20.0	47.5	32.5	37.5	20.0	22.5	45.0	42.5	17.5	15.0	15.0	15.0
Bulk density, 0–15 cm g/cm^3	0.62a†	1.04b	0.88ab	1.02b	0.98	1.13	0.90	0.95	0.99	1.17	1.07	1.02
	(6)‡	(16)	(10)	(16)	(5)	(7)	(7)	(8)	(17)	(12)	(12)	(12)
Total N, 0–120 cm kg/ha	1137	1458	2112	1108	1213	1309	1075	2205	824	2202	3394	852
	(12)	(27)	(25)	(20)	(17)	(56)	(35)	(14)	(19)	(8)	(28)	(34)
Inorganic P, 0–15 cm kg/ha	17.3	31.1	19.8	13.8	17.6	39.4	41.6	27.8	15.3a	110.6c	29.4ab	43.6b
	(11)	(9)	(8)	(10)	(7)	(10)	(2)	(7)	(2)	(8)	(28)	(9)
pH, units above 6.0	1.8	2.0	1.6	2.0	1.1	1.2	1.0	1.5	1.5	0.9	1.9	1.4
	(2)	(<1)	(1)	(<1)	(4)	(2)	(2)	(2)	(1)	(2)	(1)	(1)
Organic matter, 0–15 cm(kg/ha) × 1000	101.4a	48.4c	81.0b	71.0bc	87.9	52.4	72.0	59.3	63.7	53.7	55.9	43.1
	(14)	(3)	(10)	(11)	(13)	(6)	(6)	(4)	(12)	(8)	(6)	(9)
Light fraction organic matter, % of total organic matter	14.5	4.7	1.0	2.2	12.7	9.9	9.6	10.1	23.6	7.1	9.4	11.7
	(34)	(39)	(20)	(9)	(87)	(71)	(114)	(77)	(26)	(56)	(11)	(24)
Larger aggregates, % > 1 mm in diam.	90.6a	37.3c	65.9b	64.8b	82.0	49.9	65.1	76.6	61.6a	67.1a	67.6a	52.6b
	(3)	(9)	(6)	(4)	(3)	(5)	(6)	(6)	(8)	(4)	(11)	(5)
Water stable aggregates, %	93a	81b	94a	81b	96a	84b	95a	92a	95a	79b	79b	79b
	(3)	(2)	(1)	(6)	(2)	(1)	(1)	(3)	(3)	(3)	(10)	(2)

† Numbers followed by different letters within the same row and region are significantly different ($P < 0.05$) by Tuckey's mean separation test.
‡ Number is parentheses below the mean is the coefficient of variation in percent.

of the larger whole-farm system. For example, at the east conventional site, a small grain–bean (*Phaseolus* sp.) rotation resulted in a temporal variation of soil N levels that had a 2-yr cycle. This cycle was observed in it's entirety within the 2 yr of data collection. In contrast, cropping systems such as the central organic or the west conventional farm had events that greatly influenced soil N, but occurred only once in 4 or 5 yr. Within the 2 yr of data collection, we observed a sweetclover (*Melilotus alba* Medikus) green manure crop on the central organic farm and the application of livestock manure to fallow at the west conventional farm. Both resulted in relatively large pulses of soil N during the time frame of this study (Table 20–1). While the extremes of dynamic soil characteristics, such as N, may be important determinants of potential environmental impact, they can only be understood if observed throughout a farm's entire rotation cycle. Management of such factors could be improved with the development of rapid and inexpensive means of field measurement, such as those proposed in a soil quality test kit (Liebig et al., 1996).

Other soil quality factors seemed more correlated to the original distinctions among farming practices. Both no-till and organic farms generally resulted in greater quantities of organic matter in the surface soils as compared with conventional practices. The exception was the conventional farm in the west, which despite a wheat–fallow rotation and frequent tillage, had a relatively high organic matter content as a result of regular applications of livestock manure. Beyond the quantity of soil organic matter, measures of organic matter quality had mixed results. The most active, or light fraction (Spycher, 1983) proportion of total organic matter was highly variable, but seemed greatest on organic farms as compared with the no-till, however, the water stability of the soil aggregates tended to be greater among no-till sites.

The soil quality factors that most correlated to farming practice category were the physical properties. Conventional farming practices generally had the highest bulk densities and the smallest proportion of large soil aggregates. Both alternative strategies, either reducing tillage or increasing organic matter return, seemed capable of improving the physical condition of the soil.

Determining the ultimate impact of agricultural practices upon environmental quality is difficult due to the wide variety of values and beliefs about what constitutes a *quality* environment. We included prairie sites in our study to provide an empirical benchmark of performance that most would be able to identify with, and from which to draw their own conclusions. In general, both no-till and organic farming systems were intermediate between the characteristics of conventional farming practices and the prairie. Assuming the prairies were the least environmentally degrading sites observed, both no-till and organic systems exhibited an improvement in environmental performance as compared with conventional practices.

In summary, soil quality factors were helpful in distinguishing among farming practices for their impact on the environment. Among the soil characteristics observed, those associated with the physical properties seemed to best reflect the previous decade of farming practices and be the least sensitive to the time of observation. These also are properties that might be most easily observed through regular visual inspection. Chemical factors, such as soil N content, were sensitive to the time of observation, but provided important information on recent management events. Biological indicators might prove intermediate in temporal variability between the rapid changes in chemical parameters and the long-term changes in physical parameters. Quantification of short, intermediate, and long-term soil quality indicators could be a useful tool to enhance decision-making for the purpose of improving management and reducing the overall environmental impact of agriculture.

REFERENCES

Bauer, A., and A.L. Black. 1981. Soil carbon, nitrogen, and bulk density comparisons in two cropland tillage systems after 25 years in virgin grassland. Soil Sci. Soc. Am. J. 45:1166–1170.

Dahnke, W.C. 1988. Recommended chemical soil test procedures for the North Central Region. North Central Regional Publ. no. 221 (revised). North Dakota Agric. Exp. Bull. 499. North Dakota State Univ., Fargo.

Clancy, S.A., J.C. Gardner, C.E. Grygiel, M.E. Biondini, and G.K. Johnson. 1993. Farming practices for a sustainable agriculture in North Dakota. North Dakota State Univ., Carrington Res. Ext. Ctr., Carrington.

Haas, H.J., W.O. Willis, and J.J. Bond. 1974. Summerfallow in the western United States. USDA-ARS Conserv. Rep. 17. USDA, Washington, DC.

Kemper, W.D., and R.C. Rosenau. 1986. Aggregate stability and size distribution. p. 425–442. *In* A. Klute (ed.) Methods of soil analysis. Part 1. 2nd ed. Agron. Monogr. 9. ASA and SSSA, Madison, WI.

Liebig, M.A., J.W. Doran, and J.C. Gardner. 1996. Evaluation of a field test kit for measuring selected soil quality indicators. Agron. J. 88:683-686.

National Research Council. 1993. Soil and water quality: an agenda for agriculture. Natl. Acad. Press, Washington, DC.

Spycher, G., P. Sollins, and S. Rose. 1983. Carbon and nitrogen in the light fraction of forest soil: Vertical distribution and seasonal patterns. Soil Sci. 135 (2):79–87.

21 Use of Soil Quality Indicators to Evaluate Conservation Reserve Program Sites in Iowa[1]

Douglas L. Karlen and Timothy B. Parkin

USDA-ARS, National Soil Tilth Laboratory
Ames, Iowa

Neal S. Eash

University of Tennessee
Knoxville, Tennessee

Approximately 14.7 million ha (36.4 million acres) of highly erodible land (HEL) in 36 states were enrolled in the Conservation Reserve Program (CRP) following passage of the 1985 Food Security Act. In return for retiring this land from row crop production for a 10-yr period, the U.S. Department of Agriculture paid CRP participants an annual per-acre rent and one-half of the cost of establishing a permanent land cover (Young & Osborn, 1990). The land use or C factor within the Revised Soil Loss Equation (RUSLE) was then adjusted for this change, and the public was assured that for an annual expenditure of approximately $1.8 billion, they were *purchasing* several benefits including reduced soil erosion. Secondary benefits associated with the CRP included protecting the nation's ability to produce food and fiber, improving water quality, reducing sedimentation, fostering wildlife habitat, curbing the production of surplus commodities, and providing income support to farmers.

Unfortunately, there were no plans to quantitatively evaluate whether the CRP really accomplished these goals. The lack of quantitative data documenting changes that can be directly attributed to the public investment in the CRP may now make it more difficult to convince taxpayers that they should continue to invest their tax dollars in our nation's soil and water resources. This situation demonstrates the need for natural resource information and its proper interpretation. It also should encourage soil scientists to use their knowledge of soils, as finite living systems that are vital to life on Earth, to help guide policy decisions.

The use of soil quality assessments was suggested as a method to evaluate the effectiveness of various land use practices and policy decisions by the U.S. National Research Council (NRC) in their publication entitled *Soil and Water*

---

[1] Funded in part by Northwest Area Foundation, Grant no. 92-82, Minneapolis, MN.

Copyright © 1996 Soil Science Society of America, 677 S. Segoe Rd., Madison, WI 53711, USA. *Methods for Assessing Soil Quality,* SSSA Special Publication 49.

Quality: An Agenda for Agriculture (National Research Council, 1993). To accomplish this goal, indicators that are sensitive, reliable, and capable of detecting changes in soil physical, chemical, and biological properties and interactions must be developed. The indicators must be qualitative or semiquantitative for immediate use by land managers, and quantitative for developing long-range strategies that will improve land-use, enhance plant, animal, and human health, and sustain biological diversity within natural and managed ecosystems.

The development of soil quality indicators will require a combination of holistic and reductionist approaches because of the different scales at which soil quality can be assessed. Holistic evaluations must integrate the effects of basic biological, chemical, and physical properties and processes with uncontrollable factors such as seasonal weather patterns, and controllable soil and crop management factors including the choice of tillage practices, how to handle crop residues, and preferences for pest control. A broad understanding of these interactions and their relative importance is needed to develop useful indices of soil quality that integrate this information in meaningful and scientifically valid ways. Reductionist approaches are needed to develop the basic criteria or standards that can be used to evaluate various soil quality indicators within a specific land resource or ecosystem boundary.

In 1993, cooperative research was initiated in four of the top 12 CRP enrollment states (Iowa, Minnesota, North Dakota, and Washington) to evaluate the effects of the CRP by measuring several soil quality indicators. Results from North Dakota studies are described by Gardner and Clancy (1996, this publication). In Iowa, paired comparisons between CRP and adjacent cultivated sites were made in Henry and Butler Counties. The effects of post-CRP management practices on several soil quality indicators were assessed in Taylor County. Our hypothesis for paired comparisons was that enrollment of HEL into the CRP would improve soil quality by increasing soil C (organic matter) and the amount of water-stable soil aggregates. These changes were expected to improve soil quality, because they would create an improved soil structure with reduced surface crusting, increased water infiltration rate, lower runoff, and reduced soil erosion (Lal, 1991; Wilson & Browning, 1945). Our hypothesis for the post-CRP management studies was that if and when these areas were returned to row crop production, use of no-till practices would maintain soil quality benefits achieved as a result of public investment in the CRP.

METHODS AND MATERIALS

Paired Comparisons: Conservation Reserve Program and Cultivated Sites

Soil quality indicators were measured in October 1993 at six Henry County sites located north and west of Mt. Pleasant, IA. Four sites had Pershing (fine-silty, mixed, mesic Typic Argiudolls) silt loam, one Givin (fine, montmorillonitic, mesic Udollic Ochraqualfs) silt loam, and one Ladoga (fine, montmorillonitic, mesic Mollic Hapludalfs) silt loam soil. Three of the Pershing sites had slopes of

0 to 2%, the other Pershing and the Ladoga site had slopes of 5 to 9%, and the Givin site had a slope of 0 to 2%.

Pershing soils (USDA, 1985) are found on convex ridgetops and side-slopes, with individual areas ranging in size from 2 to 8 ha (5 to 20 acres). The surface layer is typically a very dark gray silt loam about 18 cm thick, while the subsurface layer is a dark grayish brown silt loam, about 10 cm thick. Permeability is slow, available water capacity is high, and runoff is medium.

Givin soils are found on convex ridgetops in somewhat poorly drained, irregular areas ranging in size from 2 to 12 ha (5 to 30 acres). The surface layer is a very dark grayish brown silt loam, about 20 cm thick, while the subsurface layer is dark grayish brown silt loam about 7.5 cm thick. The soil has moderately slow permeability, high available water capacity, and slow runoff.

Ladoga silt loam also is found on convex ridgetops and side-slopes in narrow, long or irregular areas ranging from 2 to 12 ha (5 to 30 acres). The surface layer is very dark grayish brown silt loam, about 20 cm thick. Plowing has generally mixed some streaks and pockets of brown silty clay loam subsoil material into the surface layer. Ladoga soil has moderate permeability, high available water capacity, and medium runoff. The Iowa Soil Properties and Interpretation Database (ISPAID) indicates that for areas that have been in cultivation for >20 yr, typical soil organic matter levels in the top 18 cm for all three of these soils are 3.0 ± 0.5 % (T.E. Fenton, 1995, personal communication).

Paired comparisons also were made in October 1993 and October 1994 at four sites in Butler County, about 20 miles west of Cedar Rapids, IA. Three sites had Waukee (fine-loamy over sandy or sandy-skeletal, mixed, mesic Typic Hapludolls) loam and were located on sites with 2 to 5% slopes, while the other site had Bassett (fine-loamy, mixed, mesic Mollic Hapludalfs) loam on a 9 to 14% slope. In general, Waukee loam (USDA, 1982) is found on nearly level stream benches and in upland areas, which was representative of the sampling sites chosen for this study. The surface layer is black loam, about 20 cm thick, while the subsurface layer is a very dark brown loam, about 15 cm thick. The soil has moderate subsoil permeability, moderate available water capacity, and slow surface runoff. Soil organic matter levels according to the ISPAID are expected to be 3.5 ± 0.5%.

Bassett loam (USDA, 1982) is located on a strongly sloping, convex side-slopes. The surface layer is a dark brown loam, about 20 cm thick. It is generally mixed with brown loam from the upper part of the subsoil. The soil is moderately permeable, with high available water capacity, and medium surface runoff. Soil organic matter levels for this soil are expected to be 2.5 ±0.5%.

Sampling sites were located in cooperation with local USDA-Natural Resources Conservation Service (NRCS) personnel to identify adjacent fields where soil and crop management histories prior to enrollment in the CRP were essentially the same. In Henry County, the cultivated sites were being managed in a 2-yr corn (*Zea mays* L.) and soybean [*Glycine max* (L.) Merr.] rotation. A chisel plow was used for primary tillage followed by field cultivation to prepare the final seedbed. When sampled, the CRP sites had been in grass for 2 yr. Predominant species appeared to be fescue (*Festuca* sp.) and smooth bromegrass (*Bromus inermis* L.), but the exact composition was not determined. In Butler

County, the conventional sites were being managed in a corn and soybean rotation using minimum tillage or no-till operations. The predominant grass species in the 6-yr old CRP areas appeared to be smooth bromegrass. Cropping history prior to CRP enrollment was not obtained.

The CRP and cultivated sites within every pair of samples had A horizons of similar thickness, similar slope aspect and steepness, and were from the same soil map unit. Several measurements including bulk density, water-filled pore space, aggregate stability (% greater than 0.25 mm), pH, total C and N, NO_3–N, electrical conductivity (EC), microbial biomass, respiration, hyphal length, and ergosterol were made on samples from each site. The methods used were essentially the same as described by Sarrantonio et al. (1996, this publication). Soil NO_3–N concentrations were measured with a Cardy[2] meter (Spectrum Technologies, Plainfield, IL). Respiration was measured using methods similar to those described by Parkin et al. (1996, this publication). Aluminum cans were inserted into the soil to a depth of 7.5 cm and covered with a lid. The headspace was sampled at three timepoints for determining CO_2 concentrations on a gas chromatograph. Total organic N and C concentrations were measured by dry combustion using a Carlo-Erba NA1500 NCS analyzer (Haake Buchler Instruments, Patterson, NJ). Hyphal length was determined with a grid-line intersect method (Olson, 1950) using Calcofluor white M2R (Sigma Chemical Co., St. Louis, MO) stain. Ergosterol analyses followed the method developed by Grant and West (1986). Microbial biomass was determined using the fumigation–extraction (Vance et al., 1987) with a K-value of 0.39 for conversion from C extracted to biomass. Soil aggregate stability was determined using a nested sieve apparatus (4-, 2-, 1-, 0.5-, and 0.25-mm sieves) similar to that described by Yoder (1936). Data were converted to ecologically relevant units as suggested by Doran and Parkin (1996, this publication), and analyzed for each County using a paired t-test as described under PROC MEANS in the SAS Institute (1985) user guide.

Post-Conservation Reserve Program Evaluations

In Taylor County, several soil quality indicators were monitored to evaluate the effects of various soil and crop management practices that might be used if CRP land is returned to row-crop production. The treatments included soybean planted into (i) CRP that was killed with herbicides in autumn and fertilized with swine manure, (ii) CRP that was killed with herbicide in autumn but not fertilized, (iii) CRP that was killed with herbicides in the spring, and (iv) CRP that was killed by moldboard plowing in either autumn or spring. A fifth treatment was established to evaluate the effects of planting corn into CRP that was killed with herbicide in autumn prior to planting. Soil measurements were made in November 1993, May 1994, and October 1994 to compare the moldboard plowing with no-till production practices.

---

[2] Mention of trademark, proprietary product, or vendor does not constitute a guarantee or warranty of the product by the USDA and does not imply its approval to the exclusion of the other products or vendors that also may be suitable.

The experimental site was chosen by our farmer cooperator who also established the five treatments in strips that were approximately 18 m (60 ft) wide and 91 m (300 ft) long. The two predominant soils at this experimental site are Clarinda (fine, montmorillonitic, mesic Typic Argiaquolls) silty clay loam and Clearfield (fine, montmorillonitic, mesic Typic Haplaquolls) silty clay loam. Each strip contained both of these poorly-drained soils that are normally found on moderately sloping, convex side slopes in long-narrow or irregularly shaped areas of 2 to 15 ha (5 to 35 acres). The surface layer of Clearfield is a black, friable silty clay loam about 23 cm thick. The subsurface layer is very dark gray, mottled, friable silty clay loam about 12 cm thick. Permeability is moderately slow, available water capacity is high, and runoff is medium. The surface layer of Clarinda is typically a black, friable silty clay loam about 15 cm thick. The subsurface layer is very dark gray, friable silty clay loam about 12 cm thick. The permeability of Clarinda soils is very slow and runoff is medium. There is a seasonal high water table at a depth of 30 to 100 cm, and available water capacity is high. Organic matter content in the surface of both soils is typically 3 to 4%.

A moldboard plow treatment was included since many farmers felt that would be the only method they could use to return the CRP land to row crop production, however, four of the five treatments focused on the use of no-tillage practices, since our hypothesis was that no-till would preserve many of the soil quality benefits obtained by having the land enrolled in the CRP.

RESULTS AND DISCUSSION

Paired Comparisons: Conservation Reserve Program and Cultivated Sites

Abnormally high amounts of rainfall throughout Iowa in 1993 caused the soils in both Henry and Butler Counties to be very wet when sampled. This was reflected by the water-filled pore space (WFPS) that averaged 67 to 75% and showed no significant difference between cultivated and CRP areas at any of the sampling sites (Tables 21–1 and 21–2). It is important to know the WFPS when sampling for soil quality assessments because when values are above ~60%, conditions are no longer optimum for aerobic soil microbial activity and anaerobic processes such as denitrification become more predominant (Doran et al., 1990; Linn & Doran, 1984).

Despite the wet conditions, microbial biomass and respiration measurements were significantly higher for CRP than for cultivated areas (Table 21–1) even though the Henry County CRP sites had been planted to grass for only 2 yr. Hyphal length and ergosterol concentrations, which reflect fungal activity, were also significantly higher for CRP than for cultivated areas. Thus, all four indicators suggest that returning these soils to grass improved their quality from a biological perspective. From a physical perspective, bulk density was lower and the percentage of water-stable aggregates higher for the CRP as compared with cultivated sites (Table 21–1). The Pershing silt loam sites (Pairs 1 to 4) had lower bulk density at all four CRP sites, but the percentage of water-stable aggregates

Table 21–1. Selected soil quality indicators measured for the 0- to 7.5-cm depth in paired Conservation Reserve Program (CRP) and tilled sites on Pershing, Givin, and Ladoga silt loams in Henry County, Iowa, during October 1993.

CRP	Cultivated	CRP	Cultivated	CRP	Cultivated
Water-filled pore space (%)		Soil pH		Electrical conductivity (dS m^{-1})	
73	83	5.8	6.9	0.1	0.5
57	73	5.6	5.6	0.2	0.3
75	66	5.9	5.5	0.1	0.2
78	66	5.9	5.9	0.1	0.1
48	50	6.4	6.9	0.3	0.5
88	88	6.6	6.4	0.2	0.1
0.801†		0.489		0.158	
Aggregate stability (%)		Total organic C (kg ha^{-1})		Microbial biomass (kg C ha^{-1})	
19.3	8.3	18383	11556	487	261
29.5	32.5	15610	11989	592	274
16.0	17.3	14062	9448	408	241
18.7	11.1	16273	9882	372	366
25.0	14.7	15793	11107	576	339
39.2	31.4	18933	18314	558	230
0.080		0.004		0.007	
Bulk density (g cm^{-3})		Total organic N (kg ha^{-1})		NO_3–N (kg ha^{-1})	
1.43	1.61	1609	628	11.8	13.3
1.30	1.73	1287	882	12.7	19.5
1.26	1.47	1096	650	8.5	16.5
1.30	1.44	1170	724	10.7	14.0
1.31	1.33	1218	1217	6.9	13.0
1.11	1.07	1099	781	9.7	14.2
0.067		0.015		0.012	
Respiration (kg C ha^{-1} d^{-1})		Hyphal length (m g^{-1} soil)		Ergosterol (g g^{-1} soil)	
0.80	0.33	156	62	2.27	1.67
0.79	0.09	113	147	3.81	1.72
0.72	0.32	141	117	2.62	1.83
0.76	0.31	132	59	2.57	2.54
1.70	1.69	140	42	4.18	1.95
0.85	0.44	126	104	3.31	1.90
0.007		0.079		0.020	

† Probability of a greater *t* value based on paired comparisons of CRP and cultivated sites.

was higher at only two sites. The higher percentage of water-stable aggregates in the cultivated area for the second pair may reflect a prior loss of unstable aggregates due to erosion, since that pair of points was located on a 5 to 9% slope and was classified as being moderately eroded. Particle size distribution was not measured, but erosion could have resulted in a higher clay content in the surface or simply resulted in a sorting that left only the most water-stable or resistant aggregates at that site. Organic C and N contents in CRP soils were significantly higher than in the cultivated sites. Nitrate–N concentrations of CRP soils, however, were significantly lower than NO_3–N at cultivated sites (Table 21–1). This presumably reflected the longer growing season for grass and its ability to rapidly assimilate any available NO_3–N. Measurements of soil NO_3–N may not be the best indicator of soil quality because of the dynamic and temporal changes that can occur within the soil; however, if soil and crop management practices result in an accumulation of excess NO_3–N, the potential for greater leaching and possible degradation of water quality would have to be considered as an indicator of

Table 21–2. Selected soil quality indicators measured for the 0- to 7.5-cm depth in paired Conservation Reserve Program (CRP) and tilled sites on Waukee and Bassett loam in Butler County, Iowa, during October 1993.

CRP	Cultivated	CRP	Cultivated	CRP	Cultivated
Water-filled pore space (%)		Soil pH		Electrical conductivity (dS m^{-1})	
100	65	7.2	6.5	0.2	0.1
70	69	5.7	6.4	0.1	0.1
54	64	6.4	6.0	0.1	0.1
74	68	5.5	6.4	0.1	0.1
0.466†		0.773		0.391	
Aggregate stability (%)		Total organic C (kg ha^{-1})		Microbial biomass (kg C ha^{-1})	
42.0	21.7	18699	13685	662	547
21.1	15.9	12163	13770	418	434
27.4	22.2	19656	18443	598	279
48.1	45.1	14990	17039	750	365
0.125		0.705		0.118	
Bulk density (g cm^{-3})		Total organic N (kg ha^{-1})		NO$_3$–N (kg ha^{-1})	
1.58	1.43	1801	1566	8.3	9.7
1.31	1.44	1287	1469	10.8	14.0
1.44	1.41	1372	1385	11.9	15.9
1.44	1.48	1069	1154	9.7	10.0
0.969		0.907		0.077	
Respiration (kg C ha^{-1} d^{-1})		Hyphal length (m g^{-1} soil)		Ergosterol (g g^{-1} soil)	
1.06	0.92	139	78	0.97	2.29
1.16	0.86	224	63	2.92	2.71
0.37	0.46	350	147	2.15	2.04
0.66	0.68	241	71	2.62	3.85
0.413		0.017		0.272	

† Probability of a greater t value based on paired comparisons of CRP and cultivated sites.

poorer overall soil quality.

Comparisons at four sites in Butler County (Table 21–2) showed statistically significant differences ($P \leq 0.10$) only for hyphal length and NO$_3$–N concentrations even though the CRP had been established for 6 yr. The high percentage of water stable aggregates for both the cultivated and CRP sites on Basset loam (Pair 4) probably reflected the moderately eroded conditions associated with the 9 to 14% slope. Hyphal length was significantly higher in the CRP sites than cultivated sites, reflecting higher potential fungal activity. Concentrations of NO$_3$–N were once again lower in CRP than in adjacent cultivated soils, probably because of the longer season for plant uptake and the presence of active plant roots in the surface 7.5 cm. Single-ring infiltration measurements were attempted in both counties in October 1993 but abandoned because the high water content and slow permeability required an excessive amount of time (>60 min) for <2.5 cm of water to enter the soil.

Several soil quality indicators were measured again in October 1994 at the four Butler County sites. Sampling conditions were nearly ideal since the WFPS for CRP and cultivated sites averaged 64 and 61%. Bulk density averaged 1.30 g cm^{-3} for both CRP and cultivated sites. Organic C averaged 18 430 and 14 600 kg ha^{-1}, while organic N averaged 1700 and 1430 kg ha^{-1}, respectively, for CRP and cultivated sites. Although total organic C and N concentrations tended to be higher in the CRP sites than in adjacent cultivated sites, the differences in these soil quality indicators were not statistically significant. The lack of significant differences at the Butler County sites was somewhat surprising since the CRP had been established for six years compared with only 2 yr in Henry County; however, upon further evaluation it was noted that the cultivated sites in Butler County were being managed using no-till or very reduced tillage practices, whereas in Henry County, the cultivated sites were being chisel plowed and disked several times each year. This demonstrates that the soil quality indicators were indeed reflecting soil management practices, and also supports the hypothesis that use of no-till practices will help preserve the CRP benefits if this land is returned to row-crop production.

Post Conservation Reserve Program Evaluations

Initial measurements (Tables 21–3 and 21–4) were taken in November 1993 when the soil was quite wet. As a result the WFPS values were above the 60% optimum value for aerobic microbial activity. The water content and WFPS were more favorable at the 1994 sampling dates. One year of either corn or soybean production decreased the amount of organic C and N in the surface 7.5 cm in the Clarinda silty clay loam, while NO_3–N concentrations increased by more than three-fold (Table 21–3). Soil pH and salinity measurements (EC) indicated those factors were not a problem and were not affected by management at this site. Soil quality indicators in samples from Clearfield silty clay loam (Table 21–4) showed more fluctuation, with no clear or distinct trends other than having higher NO_3–N levels in October 1994 than in November 1993. Once again, this presumably reflected a loss of plant roots in the surface 7.5 cm to assimilate the NO_3–N as it was mineralized or lost from the crop residues.

In addition to measuring the amount of total C and N in the surface 7.5 cm of soil, particulate grass and weed residues lying on the soil surface also were measured just before planting in 1994. After 7 yr of CRP management, there was an average of 7150 kg ha^{-1} of litter accumulated on the soil surface. For comparison, the average amount of corn residue remaining on the soil surface in an adjacent field, following a rather low-yielding 1993 crop, averaged 5100 kg ha^{-1}.

Table 21–3. Surface (0 to 7.5 cm) soil quality indicators measured in Taylor County, Iowa, on Clarinda soil during transition from 7 yr of Conservation Reserve Program (CRP) into row crop production.

Soil quality indicator	Nov. 1993 7-yr-CRP	Oct. 1994 No-till	Oct. 1994 Plow
Bulk density, g cm^{-3}	1.09	1.07	1.22
WFPS†, %	70	54	68
pH	6.5	6.8	7.0
EC‡, dS m^{-1}	0.15	0.2	0.15
Aggregate stability, %	33	—	—
Total C, kg ha^{-1}	21107	18065	16263
Total N, kg ha^{-1}	1769	1591	1502
NO_3–N, kg ha^{-1}	6.7	21.0	21.2

† WFPS, water-filled pore space.
‡ Electrical conductivity.

Table 21–4. Surface (0 to 7.5 cm) soil quality indicators measured in Taylor County, Iowa, on Clearfield soil during transition from 7 yr of Conservation Reserve Program (CRP) into row crop production.

Soil quality indicator	Nov. 1993 7-yr CRP	May 1994 No-till	May 1994 Plow	Oct. 1994 No-till
Bulk density, g cm^{-3}	1.09	1.31	1.01	1.03
WFPS†, %	78	45	31	54
pH	6.4	7.1	6.5	6.9
EC‡, dS m^{-1}	0.1	0.1	0.2	0.1
Aggregate stability, %	47	--	--	--
Total C, kg ha^{-1}	16980	26167	13849	18579
Total N, kg ha^{-1}	1507	2192	1116	1746
NO$_3$–N, kg ha^{-1}	6.7	3.3	8.2	20.0

† WFPS, water-filled pore space.
‡ Electrical conductivity.

First-year crop production results from these studies were excellent. Initial plant stands in the corn plots were severely reduced by birds and rodents, presumably because the plots were surrounded by CRP grassland. Most of the area was replanted in late May 1994 and the resulting grain yield averaged 10.9 Mg ha^{-1} (174 bu/acre). The portion that was not replanted, however, produced a yield of only 1.3 Mg ha^{-1} (21 bu/acre). Plots planted to soybean averaged 4.59 and 4.42 Mg ha^{-1} (68 and 66 bu/acre) for autumn and spring moldboard plow treatments, respectively, and 4.71 Mg ha^{-1} (70 bu/acre) for fall-killed no-till, 4.66 Mg ha^{-1} (69 bu/acre) for fall-killed no-till plus swine manure, and 4.52 Mg ha^{-1} (67 bu/acre) for spring-killed no-till treatments, respectively.

Water infiltration rates were measured in May and again in October 1994 using single 15 cm diameter metal rings that were inserted 7.5 cm into the soil. Interrow measurements in May, when WFPS averaged 45% for no-till and 31% for plowed treatments, resulted in average values of 0.4 and 0.15 minutes for the first 2.5 cm of water, respectively. Infiltration of the second 2.5 cm of water required 30 min in plowed treatments compared to 4 min for the no-till treatment.

Inter-row infiltration measurements in October were highly variable (Table 21–5), but were generally most rapid for the no-till corn or spring-plowed soybean treatments. Autumn-plowed and no-till soybean treatments had the slowest infiltration rates. The autumn-plowed area had the highest WFPS percentage.

Table 21–5. Time required for intrarow infiltration of two, 2.5-cm increments of water following crop harvest in October 1994.

Treatment	Minutes		Water-filled porosity
	1st increment	2nd increment	%†
No-till corn	2	6	53
Soybean, autumn plowed	16	44	77
Soybean, spring plowed	2	7	59
Soybean, autumn killed	19	42	54
Soybean, spring killed	23	85	56
LSD 0.05	NS	33	--

† Measurements made on composite soil samples from the area before infiltration measurements were made.

This presumably slowed the infiltration rate, but the other treatments had essentially the same percentage WFPS. The slow infiltration rate following soybean was not expected, but was consistent with previous soil erosion measurements and conclusions by Laflen and Moldenhauer (1979) that soybean leaves soil in a more erodible condition than does a prior corn crop.

These preliminary infiltration results need to be verified with more detailed studies, but the semiqualitative assessments demonstrate how relatively simple soil quality indicators can be used to identify potential differences. Assuming that more detailed studies confirm the slower infiltration rates, this information suggests the following protocol for CRP areas that are brought back into corn and soybean production. No-till soybean should be grown the first year because of potential N immobilization, weed control, and pest problems. This should be followed by a cover crop such as spring oats (*Avena sativa* L.) or winter rye (*Secale cereale* L.) to assimilate any residual NO_3–N and then be planted to no-till corn. To minimize potential runoff and erosion losses following the soybean crop, consideration also should be given to using buffer strips, narrow strip intercropping, or other soil and crop management practices that will help reduce slope length. This type of integrated soil and crop management would help ensure that the 10-yr benefits of CRP are not lost in just a few months.

SUMMARY AND CONCLUSIONS

These field evaluations demonstrate that soil quality indicators such as aggregate stability, bulk density, total C, total N, NO_3–N, microbial biomass, respiration, hyphal length, and ergosterol can be measured at various sites and used to detect differences that can be attributed to various soil and crop management practices. By using these indicators, we were able to detect significant differences between CRP and adjacent cultivated sites after just 2 yr of CRP in Henry County while in Butler County there were very few significant differences even though the CRP sites had been in grass for 6 yr. Location differences presumably reflected how the cultivated sites were being managed, since in Henry County they were chisel plowed and disked each year, while in Butler County the cultivated sites were managed with no-till or very reduced tillage practices. These assessments were also able to detect differences in soil quality after CRP land was returned to crop production with various soil and crop management practices. Preliminary results suggest no-till practices can extend the soil quality benefits obtained through public investment in the CRP.

REFERENCES

Doran, J.W., and T.B. Parkin. 1996. Quantitative indicators of soil quality: A minimum data set. p. 25–37. *In* J.W. Doran and A.J. Jones (ed.) Methods for assessing soil quality. SSSA Spec. Publ. 49. SSSA, Madison, WI.

Doran, J.W., L.N. Mielke, and J.F. Power. 1990. Microbial activity as regulated by soil water-filled pore space. p. 94–99. *In* Trans. of the 14th Int. Congr. of Soil Sci. ISSS, Kyoto, Japan.

Gardner, J.C., and S.A. Clancy. 1996. The impact of farming practices on soil quality in North Dakota. p. 337–343. *In* J.W. Doran and A.J. Jones (ed.) Methods for assessing soil quality. SSSA Spec. Publ. 49. SSSA, Madison, WI.

Grant, W.D., and A.W. West. 1986. Measurement of ergosterol, diaminopimelic acid and glucosamine in soil: evaluation as indicators of microbial biomass. J. Microbiol. Meth. 6:47–53.

Laflen, J.M., and W.C. Moldenhaurer. 1979. Soil and water losses from corn–soybean rotations. Soil Sci. Soc. Am. J. 43:1213–1215.

Lal, R. 1991. Soil structure and sustainability. J. Sust. Agric. 1:67–92.

Linn, D.M., and J.W. Doran. 1984. Effect of water-filled pore space on carbon dioxide and nitrous oxide production in tilled and nontilled soils. Soil Sci. Soc. Am. J. 48:1267–1272.

National Research Council. 1993. Soil and water quality: An agenda for agriculture. Natl. Academy of Sci. Press, Washington, DC.

Olson, F.C.W. 1950. Quantitative estimates of filamentous algae. Am. Microscop. Soc. Trans. 69:272–279.

Parkin, T.B., J.W. Doran, and E. Franco-Vizcaíno. 1996. Field and laboratory tests of soil respiration. p. 231–245. *In* J.W. Doran and A.J. Jones (ed.) Methods for assessing soil quality. SSSA Spec. Publ. 49. SSSA, Madison, WI.

Sarrantonio, M., J.W. Doran, M.E. Liebig, and J.J. Halvorson. 1996. On-farm assessment of soil quality and health. p. 83–105. *In* J.W. Doran and A.J. Jones (ed.) Methods for assessing soil quality. SSSA Spec. Publ. 49. SSSA, Madison, WI.

SAS Institute. 1985. SAS user's guide: Statistics. Version. 5 ed. SAS Inst., Cary, NC.

USDA-SCS. 1982. Soil survey of Butler County, Iowa. USDA-NRCS. Washington DC.

USDA-SCS. 1985. Soil survey of Henry County, Iowa. USDA-NRCS. Washington DC.

Vance, E.C., P.C. Brookes, and D.S. Jenkinson. 1987. An extraction method for measuring soil microbial biomass C. Soil Biol. Biochem. 19:703–707.

Wilson, H.A., and G.M. Browning. 1945. Soil aggregation, yields, runoff, and erosion as affected by cropping systems. Soil Sci. Soc. Am. Proc. 10:51–57.

Yoder, R.E. 1936. A direct method of aggregate analysis of soils and a study of the physical nature of erosion losses. J. Am. Soc. Agron. 28:337–351.

Young, C.E., and C.T. Osborn. 1990. Costs and benefits of the conservation reserve program. J. Soil Water Conserv. 45:370–373.

22 Quantifying Soil Condition and Productivity in Nebraska[1]

Gail L. Olson
Lockheed Martin Idaho Technologies
Idaho Falls, Idaho

Betty F. McQuaid
USDA Natural Resources Conservation Service
Raleigh, North Carolina

Karen N. Easterling
Pharmaceutical Product Development
Morrisville, North Carolina

Joyce Mack Scheyer
USDA Natural Resources Conservation Service
Lincoln, Nebraska

In 1993, the Agricultural Lands resource group of the U.S. Environmental Protection Agency's (USEPA) Environmental Monitoring and Assessment Program (EMAP-Ag Lands) conducted a pilot program in Nebraska to evaluate soil quality. EMAP-Ag Land's interest in soil quality arose from a commitment to evaluate the environmental condition of agricultural lands nationwide. EMAP-Ag Lands also evaluates landscape structure, cropland diversity, crop productivity, crop rotations, and N use efficiency in order to further assess agricultural land condition. Most data are collected by the National Agricultural Statistics Service (NASS) at selected sites using a probability-based sampling design (Hellkamp et al., 1995).

Soil quality assessments by EMAP-Ag Lands have taken several forms. A surface soil sampling program was introduced and successfully implemented in North Carolina in 1992. The objectives of the surface soil program were to (i) estimate soil quality by evaluating a suite of indicators from the plow layer (0 to 20 cm) of agricultural fields, and (ii) develop logistics to employ a team of non-soil scientists to collect soil samples from privately-owned farm fields and trans-

---

[1] Research funded in part by the U.S. Environmental Protection Agency under Interagency Agreements with the USDA, USDA-NRCS, and USDA-ARS and supported in part through a Specific Cooperative Agreement between the USDA-ARS and North Carolina State University.

Copyright © 1996 Soil Science Society of America, 677 S. Segoe Rd., Madison, WI 53711, USA.
Methods for Assessing Soil Quality, SSSA Special Publication 49.

port the samples to multiple laboratories in a timely manner. Soil quality indicators included pH, cation-exchange capacity, clay content, organic matter, phosphorus, N, and free-living nematodes.

Soil sampling procedures, from site selection to sample collection, handling, and transportation, were standardized and demonstrated in a training video so that NASS enumerators could reliably implement the field program. Enumerators are typically nonsoil scientists who are trained by NASS to interview farmers. In the North Carolina pilot program, NASS enumerators sampled more than 300 fields in about a month, thereby providing a statistically based sample from all three major land resource regions (Mountain, Piedmont, and Coastal Plain) in a short period of time.

Results from North Carolina surface soil samples revealed significant differences among land resource regions (Campbell et al., 1994a). The surface soils in the Coastal Plain region of North Carolina had less clay, more organic matter, slightly lower pH, and higher cation-exchange capacity than surface soils in the Piedmont or Mountain regions.

If this study was repeated over 20 yr, changes within a region over time also could be determined. Based on a knowledge of crop needs, positive or negative changes in surface soil properties could be identified and an assessment made as to whether or not, in general, soil quality for crop growth was improving or declining. Our immediate challenge, however, was to convey whether soil quality was good or bad based on one year's data. Making value judgments about soil quality was one step beyond our routine efforts of characterizing soil properties.

Our challenge as the project expanded to Nebraska was to graduate from characterizing data to interpreting data using a logical scientific process. If eastern Nebraska soils had higher productivity than western Nebraska soils, did that mean that soil conservation programs were more effective there? Were soils in better condition? In order to answer these questions, we needed to develop a complement of baseline data for each soil and then determine the extent to which results deviated from that baseline.

We also realized the limitation of assessing the surface soil because it represents only a fraction of the root zone. Furthermore, surface soil samples collected in North Carolina were a composite of 20 subsamples across a transect. Therefore, if the transect traversed a hillside, a subsample from the surface of an eroded area might be mixed with a subsample from a depositional area. Because, there were no provisions for documenting site or soil descriptions of the sampling locations, so if the sample was a composite across a hillside with varying degrees of soil erosion and deposition, no one would know except for the original NASS enumerator. How could we properly interpret soil sample data without site and pedological information that soil scientists traditionally consider when evaluating a soil? What did a pH of 5.1 mean if the sample was from the Piedmont vs. if it was from the Coastal Plain? Or from North Carolina versus Nebraska?

After discussions with scientists at the Natural Resources Conservation Service (NRCS) in Lincoln, NE, we added a soil profile component to our sampling program to allow for supporting site and soil data necessary to better estimate soil quality of an agricultural field. The soil profile program was to be implemented in the field by trained NRCS soil scientists. Data from the soil surface and profile programs could then be compared with existing NRCS databases that had been collected over many years.

In 1992, the NRCS presented a model for assessing soil quality from laboratory data and soil description information. The Soil Rating for Plant Growth (SRPG) model was developed by Scheyer et al. (1992) to rate Nebraska soils for their ability to produce corn (*Zea mays* L.). The model, described later, was designed for use with the Natural Resource Conservation Service (NRCS) State Soil Survey (3SD) database.

The SRPG model appeared to be an appropriate tool to quantify soil productivity of agricultural fields. Data could be characterized as well as evaluated within the context of producing corn. Thus, the soil productivity of a specific field could be rated based on field measurements and a similar rating could be calculated for the map unit from which the soil sample came, using the 3SD database. Theoretically, we could estimate if soil was in optimal condition, i.e., if it was measuring up to its expectation.

Given the time and expense of obtaining soil profile data from soil pits, it was not feasible to implement an extensive soil profile sampling program. Thus, we incorporated a detailed soil profile evaluation at 26 sites, and retained an extensive surface soil sampling program across the state. Using this approach, ratings based on soil profile data could be compared with ratings based on surface soil data and on 3SD data to determine if similar results and interpretations could be obtained regardless of the sampling scheme and rigor of analysis.

This chapter briefly outlines the SRPG model, describes how we used it to quantify soil productivity in Nebraska, and discusses results from the soil profile and surface soil sampling programs. Soil condition, within the context of this chapter, refers to the relative degradation of the soil (e.g., erosion, salinization, and others) from its original state. The 3SD database was considered to be the best estimate of the original state of the soil.

SOIL PRODUCTIVITY MODEL DESCRIPTION

Soil productivity at each field site was quantified using the Soil Rating for Plant Growth (SRPG) model developed by a group of scientists in cooperation with the NRCS (Scheyer et al., 1992). The model is specific to Nebraska soils and overall SRPG ratings reflect the suitability of a soil for growing corn. In developing the model, scientists adjusted the soil property ratings so that soils known to have high corn productivity were associated with high SRPGs. The SRPG model (Scheyer et al., 1994) includes seven soil categories considered critical for plant (corn) growth:

$$SRPG = S \times P \times W \times T \times R \times C \times L$$

where
  S = surface soil properties,
  P = soil profile characteristics,
  W = soil water features,

T = soil toxicity,
R = soil reaction,
C = soil climatic factors, and
L = landscape features.

Each category contains several properties (Table 22–1), which are each rated from low to high, with 100 as the maximum. Calculation of a SRPG is a 3-step process:

1. Determine a rating for each soil property within each category based on guidelines developed for the model (Table 22–1).
2. Calculate a rating for each category (e.g., water features) by averaging the property ratings then dividing the average by 100.
3. Obtain an overall SRPG for soil productivity by multiplying the seven category ratings together and then multiplying by 100.

Note that within each category, ratings for different soil properties are averaged to represent the suitability for crop growth with respect to that feature; however, for the overall SRPG, category ratings are multiplied together. Multiplying the category ratings together gives more weight to low values, reflecting the impact of one or more limiting factors on potential soil productivity. For example, a soil may have a category rating of 0.4 for soil water due to low infiltration, and a 1.0 rating for each of the other six categories. The multiplicative product of the seven categories is 0.4 while the average is 0.91. The 0.4 rating more accurately reflects the severe limitation imposed on productivity due to poor infiltration, and thus suboptimal soil water relations.

METHODS

Surface Soil Transect Sampling

Using a probability-based sampling framework, 209 fields were selected for the surface soil sampling program (Fig. 22–1). For each field, NASS enumerators interviewed farmers and then followed a standard protocol for identifying a random starting point within that field, from which a 100-m transect was established. An Oakfield probe was used to collect 20 surface soil subsamples (0 to 20 cm) at regular intervals along the transect. Subsamples were mixed together and a 500 mL portion of the composite sample was sent to the NRCS laboratory in Lincoln, NE, for analysis of soil quality indicators (texture, cation-exchange capacity, and others; Campbell et al., 1994b).

Soil Profile Sampling

At 26 of the 209 fields sampled by NASS for surface soil quality, a NRCS soil scientist determined the soil map unit, dug a soil pit within the transect, described the soil and landscape features, and collected soil samples by horizon. In three fields, the NRCS scientist dug and described two pits because the tran-

Table 22–1. Categories, soil properties and rating system used in the Soil Rating for Plant Growth (SRPG) model. Ratings presented here are generalized in some cases. See Scheyer et al. (1994) for details.

Category	Property	Low	Medium	High
Surface soil	Organic matter	<2% Rate 75	2–5% Rate 85	>5% Rate 100
	Bulk density	> average for textural class	average for textural class	< or average for textural class
	Clay content	Rate 75 <15% or >40% Rate 75	Rate 100 27–40% Rate 90	15–27% Rate 100
	Avalable water capacity	0.01 to 0.1 cm/cm Rate 75	0.11 to 0.15 cm/cm Rate 85	>.15 cm/cm Rate 100
	pH	<6.1 or >7.8 Rate 75		6.1–7.8 Rate 100
	Sodium adsorption ratio	>4		<=4.0 Rate 100
	Carbonates	>2% Rate 75		<2% Rate 100
	Gypsum	>2% Rate 75		<2% Rate 100
	Cation-exchange capacity	<16 cmol/100g Rate 75		>16 cmol/100g Rate 100
	Shrink-swell	More than moderate Rate 75		Low-moderate Rate 100
Soil profile	Depth to restrictive layer†	<50 cm Rate 25–40	50–150 cm Rate 50–85	>150 cm Rate 65–100
	Available water capacity in the root zone	<6 cm available water in top 60 cm Rate 80	6–9 cm available water in top 60 cm Rate 80	>9 cm available in top 60 cm Rate 100
Water above restrictive layer	Water table during growing season	<45 cm	<45 cm and xeric aridic or ustic	>75 cm Rate 100

(continued on next page)

Table 22–1. Continued.

Category	Property	Low	Medium	High
	Permeability	>5 cm/h Rate 75	.25 to 1.5 cm/h Rate 90	1.5 to 5 cm/h Rate 100
	Available water capacity	<.11 cm/cm in most limiting layer below surface	.1–1.5 cm/cm in most limiting layer below surface	>.16 cm/cm in most limiting layer below surface
Toxicity above restrictive layer	SAR	Rate 75 >30	Rate 85 4–30	Rate 100 <4
	EC	Rate 50 >16	Rate 75 to 85 8–16	Rate 100 <8
	CEC	Rate 50 <7 cmol/100g	Rate 75 7–16 cmol/100g	Rate 100 >16 cmol/100g
Reaction above restrictive layer	pH	Rate 75 <4.4	Rate 85 4.5–5.5 or >7.9	Rate 100 5.6–7.8
Soil climate	Soil moisture regime	Rate 50 Aridic to xeric	Rate 90 Xeric to Ustic	Rate 100 Ustic to Udic
	Soil temperature regime	Rate 30 to 60‡ Pergilic, cryic, frigid	Rate 60 to 80	Rate 80 to 100 Others Rate 100
Landscape and slope	Slope	Rate 70 >15% slope Rate 40 to 60	8–14% slope Rate 80	<8% slope Rate 90 to 100
	Erosion	Channelled, gullied, or Class 5 erosion Rate 50	Class 2, 3, or 4 erosion or flooded Rate 60 to 80	Class 1 erosion or non-eroded Rate 90 to 100
	Channelled, gullied or Class 5 erosion	Flooded or Class 2 erosion Rate 50	Class 1 erosion Rate 80	Rate 90
	Class 3 or 4 erosoin No noticeable erosion	Rate 60 to 70		Rate 100

† Restrictive layers include pH < 3.5, bedrock, or available water capacity change greater than 2 classes between adjacent subsurface horizons. Specific ratings are determined by a combination of depth and degreee of profile development. Soils with minimal profile development (e.g., Entisols) generally receive higher ratings than highly developed profiles.
‡ Irrigated fields are rated 100 for moisture regime. Additional adjustments are made to the ratings for moisture-by-temperature interactions.

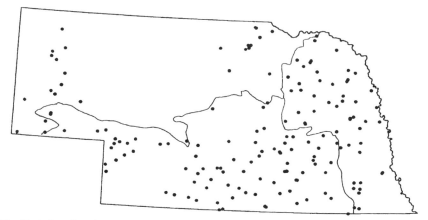

Fig. 22–1. Sampling sites and land resource region boundaries for the surface soil sampling program in Nebraska.

sect traversed two soil types. These soil pit samples were also sent to the NRCS Laboratory for analysis.

Sample Analyses

Composited surface soil samples from the transects and soil profile samples from the pits were analyzed for pH (in calcium chloride), cation-exchange capacity (buffered solution), organic C (Walkley–Black method), and texture (by hydrometer). Carbonates, gypsum, and Na adsorption ratio were measured on a selective basis depending upon the results of the dilute HCl quick test for carbonates.

Soil profile samples also were analyzed for oven dried and 1/3 bar bulk density (clod method), and 15-bar and 1/3-bar water content. Analytical methods for all analyses are described in NRCS methods manual (Soil Survey Laboratory Staff, 1992). In the field, the NRCS soil scientist described the soil profile, including information such as soil color, consistence, structure, and field pH of each horizon to a depth of 60 cm, using standard NRCS soil profile description protocol as outlined in the National Soil Survey Handbook (USDA, 1993).

Determination of Soil Productivity

A Transect SRPG was calculated for each of the 209 surface soil transect samples. Three different SRPGs were calculated for each of the 26 sites and summarized by region. Field SRPG ratings were based on field data from all horizons of the soil profiles. Surface SRPG ratings were based on field data from the surface horizon of each soil profile. Database SRPG ratings were based on tabulated data from all soil horizons of each map unit associated with a field transect. Surface SRPGs were correlated with Field SRPGs and Database SRPGs. All SRPG calculations and statistical analyses were performed using SAS (SAS Institute, 1989). Soil properties of samples evaluated in the laboratory or in the field were assigned ratings as described above (Model Description).

The Database SRPG uses values for soil properties stored in the 3SD soil survey database having the same map unit, soil series, surface texture, and slope class as those described at the 26 field sites. Database SRPG ratings were generated by Iowa State University with a PL1 program written specifically to calculate SRPGs from 3SD data. The 3SD database contains typical ranges for soil properties (e.g., organic matter, cation-exchange capacity by horizon) based on a combination of laboratory analyses performed on typical pedons of the map unit and professional judgment. The PL1 program uses the average of the minimum and maximum values for each 3SD soil property to calculate the SRPG.

Variability of Scale

The impact of spatial scale was determined by collecting surface soil transect samples. Within-sample, within-field and between-field variability were evaluated. Within-sample variability was assessed by splitting a composited sample from a transect, thus, providing two separate 500 mL portions of soil for analysis. Split samples were collected from every twelfth field. At every sixth field, samples for within-field variability were obtained. A second transect was sampled using the same protocol as for the first transect but using a different starting point. A single 500 mL portion of the composited sample was used for analysis. Between-field variability was based on the 209 original transect samples.

Variance component analysis was used to quantify similarities of soil properties and SRPGs at the three spatial scales. Reliability ratios were calculated for each parameter to determine the noise or scatter in the data and indicate the likelihood of detecting trends. The reliability ratio was calculated as

$$\text{Reliability Ratio} = \frac{\sigma^2_f}{\sigma^2_f + (\sigma^2_t/t) + (\sigma^2_d/td)}$$

where
- σ^2_f = between-field variance
- σ^2_t = within-field variance
- σ^2_d = within-sample variance
- t = number of transects, and
- d = number of determinations.

Ratios range from 0 to 1 with 1 representing no noise in the data. A value of 0.5 results when the within-sample variance per sample plus within-field variance per transect equal the between-field variance.

RESULTS

Soil Profile Analysis

Field and Database SRPGs were evaluated individually and as a ratio of Field/Database SRPG for Nebraska Land Resource Regions (USDA, 1981). Field SRPG ratings were significantly lower than Database SRPG ratings in the eastern and southern parts of the state ($P < 0.05$; Fig. 22–2; Table 22–2). Both the

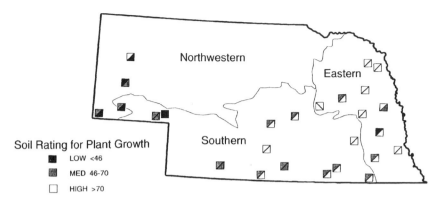

Fig. 22-2. Soil Rating for Plant Growth (SRPG) in Nebraska based on soil profile assessment. Field SRPGs are represented in the upper one-half of each box and Database SRPGs in the bottom one-half of each box.

Field and Database SRPGs were higher in the eastern than the southern and western parts of the state ($P < 0.05$) indicating a regional trend of decreasing soil productivity to the west. Field/Database SRPG ratios (Fig. 22-3; Table 22-2), calculated to estimate soil condition, were not significantly different between the eastern and southern regions, suggesting that although there is a soil productivity gradient across Nebraska (as demonstrated by the Field and Database SRPGs), soils generally are in about the same condition across the state.

Comparison of Surface and Field SRPGs with Database SRPGs revealed that the two sampling approaches provide different results. Surface SRPGs were not correlated with Database SRPGs and were only weakly correlated with Field SRPGs ($r = 0.42$, $P < 0.04$). The $r = 0.42$ indicates that only 16% of the variation between surface and full profile ratings can be accounted for by this correlation. The weak correlation suggests that soil profile analysis provides us with information that cannot be obtained from the surface samples alone. If our intent is to make a best estimate of soil productivity, soil pits must be dug or some alternative method devised to obtain comparable data.

Surface Soil Analysis

Ratings for soil properties of surface samples were not significantly different for eastern ($n = 68$) and southern ($n = 110$) Nebraska soils while both regions had significantly higher ratings than western ($n = 31$) Nebraska soils ($P < 0.02$; data not presented). Western soils had significantly lower soil property ratings

Table 22-2. Mean field and database Soil Rating for Plant Growth (SRPGs) and Field/Database SPRG ratios for each region of Nebraska.

Region	Field SRPG	Database SRPG	Field/Database SRPG
East ($n = 10$)	70	86	0.82
South ($n = 13$)	58	68	0.85
West ($n = 3$)	57	39	1.46

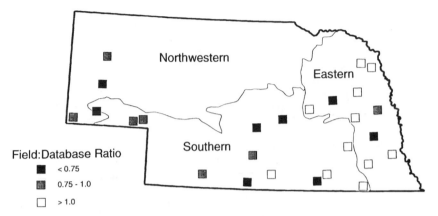

Fig. 22–3. Field/Database Soil Rating for Plant Growth (SRPG) ratios for Nebraska based on soil profile assessment.

than the southern and eastern regions for clay content, organic matter, pH, and cation exchange capacity ($P < 0.07$).

Results from variance component analysis for clay, sand, pH, organic C and the Transect SRPG indicated that variability was within-sample<within-field<between fields (Table 22–3). For cation-exchange capacity, the ranking was within-field, within-sample and between-fields. Reliability ratios ranged from 0.69 to 0.97 for soil properties and was 0.73 for Transect SRPGs (Table 22–3). Organic matter typically considered to be an excellent indicator of soil quality, had the lowest reliability ration at 0.69.

DISCUSSION

In two of the three regions of Nebraska, the Database SRPGs were significantly higher than Field SRPGs. One might conclude that lower Field SRPGs are an indication that the soil has lost productivity and that the soil has been degraded since the database was collected; however, a comparison of methods for calculating SRPGs reveals that these two SRPG ratings are not directly comparable.

Field SRPGs were calculated from actual soil profile data and the Database SRPG ratings were calculated using the average of the minimum and maximum values for the 3SD data. Thus, if data in the 3SD database have a skewed distribution, the average may not be a good representation of the soil property. For example, the 3SD data may indicate that the minimum and maximum soil depth for a given map unit is 20 and 40 in., respectively; however, 20 to 25 in. may be the most frequent soil depth recorded in the 3SD database. The Database SRPG would be calculated as the average for the range of data, i.e., 30 in., while the Field SRPG may use a soil depth of 20 in. Given the different methods of calculating the SRPG ratings, we cannot make a direct comparison between the Field and Database SRPG ratings to arrive at a simple method of estimating soil productivity and potential degradation for a site.

Table 22–3. Statistical summary and variability component analysis of soil properties and Transcect SRPGs for the surface sampling program in Nebraska ($n = 209$).

Property	All samples			Within sample	Within field	Between fields	Reliability ratio
	x	σ^2	CV (%)	σ^2	σ^2	σ^2	
Cation exchange capacity (cmol/100g)	19.3	46.8	35.4	9.93	0	37.0	0.79
Organic C (%)	1.33	0.25	37.2	0.01	0.07	0.17	0.69
pH	5.89	0.83	15.4	0.01	0.17	0.66	0.78
Sand (%)	30.1	671	86.1	3.75	18.13	647.	0.97
Clay (%)	22.5	93.2	43.0	0.34	5.21	88.3	0.94
Transect SRPG	94.3	11.7	3.63	0.06	0.20	0.73	0.73

If we assume that the skewed data and resulting over- or underestimation of soil properties is consistent for all soil types and regions, then Field/Database ratios can be compared across regions. SRPG ratios would not be appropriate to determine if an individual field is in good condition, as we cannot say with certainty that a ratio <1 indicates a declining condition; however, we can compare ratios across regions. In our results, the SRPG ratios were not significantly different between the eastern and southern regions of Nebraska, suggesting similar soil conditions in the two regions.

An alternative method to estimating Database SRPGs would utilize the NRCS Pedon database instead of the 3SD database. The Pedon database contains soils data from actual soil pits that have been evaluated and assigned a map unit. A Pedon Database SRPG could be generated for each pit and a range of Pedon Database SRPGs could be developed for the associated map unit. The Field SRPGs could then be interpreted in light of the variation in Pedon Database SRPGs expected for the map unit. A Field SRPG rating of 63 might be associated with the 78th percentile of the Pedon Database SRPGs for the map unit. Percentiles could be summarized, and relative soil condition (based on percentiles) of each region could be compared. Areas with lower percentiles may be further examined to determine if soil conservation or policy measures could be implemented to facilitate the restoration of soil condition.

Summary statistics, variance component analysis and reliability ratios suggest that cation-exchange capacity, sand, clay, pH and possibly organic matter are fairly well-behaved indicators that can be used to detect regional differences in a monitoring program. When ratings for these parameters are used to calculate Transect SRPGs, the resulting values are also well-behaved, showing increasing variability with scale and a high reliability ratio.

Further testing of the SRPG model and its applications to soil quality and condition are warranted. An independent soil productivity indicator is needed to verify the extent to which SRPGs are correlated with soil productivity. The SRPG model was based upon correlations between SRPGs and corn productivity. Yield alone is not a good soil productivity indicator since it is sensitive to management practices (which have varying effects on soil condition) and to weather patterns within a cropping season and from year to year. Crop productivity indices, using ratios of observed/expected yield, might be more robust.

The SRPG appears to be a promising tool for assessing regional patterns of soil productivity and for identifying possible changes in soil properties that would signal a tendency toward soil degradation. The SRPG does not substitute for biological indicators of ecosystem function. Additional work on soil biological indicators, in conjunction with the SRPG, would add another dimension to our analysis and help determine if degraded soils are still cycling nutrients and otherwise performing necessary ecosystem functions that would sustain productivity.

CONCLUSIONS

The Field and Database SRPG appear to quantify relative productivity of soil and can be used to compare soil productivity across regions. Database SRPGs provide a baseline with which to contrast deviations in Field SRPGs. Assuming that Database SRPGs are a good representation of potential productivity, deviations from that potential (e.g., Field/Database SRPG ratios) provide a rough indication of soil condition.

The 3SD database is not the best source of soil property values for calculating Database SRPGs. Future efforts to develop and apply Field/Database SRPG ratios might be more valuable if the Pedon Database is used to establish Database SRPGs instead of the 3SD database. The Pedon Database contains actual data from soil profile evaluations, rather than ranges of property values as in the 3SD database. Unlike the 3SD Database SRPGs, Pedon Database SRPGs would be directly comparable to Field SRPGs.

Future efforts also should focus on a more thorough characterization of expected values of surface soil properties. Changes in soil condition will probably show up first in the surface soil, where management practices and environmental factors are constantly exerting influence. In contrast, changes in subsoil properties may be very slow and go undetected for years or even decades. A detailed characterization of the expected values of surface soil properties would promote a monitoring program that focuses on the easily accessible surface horizons.

The Transect SRPG for the composited surface samples was a well-behaved indicator, with increasing variability with scale and a reliability ratio of 0.73. If a robust, meaningful baseline can be developed for comparing the Transect SRPGs with Surface SRPGs, the monitoring program could exclusively focus on surface soil sampling and increase sample collection five-fold. With a statistically representative sample and a good basis for comparison, soil condition could likely be estimated from a single year's sampling.

DISCLAIMER

Although the research described in this article has been funded wholly or in part by the U.S. Environmental Protection Agency under Interagency Agreements with the United States Department of Agriculture, Natural Resources Conservation Service and Agricultural Research Service, it has not been subjected to

Agency review and, therefore, does not necessarily reflect the views of the Agency, and no official endorsement should be inferred. This work also was supported in part through a Specific Cooperative Agreement between USDA-ARS and the North Carolina Agricultural Research Service.

REFERENCES

Campbell, C.L., J.M. Bay, C.D. Franks, A.S. Hellkamp, N.P. Helzer, G.R. Hess, J.J. Munster, D.A. Neher, G.L. Olson, S.L. Peck, J.O. Rawlings, B. Schumacher, and M.B. Tooley. 1994b. Environmental monitoring and assessment program: Agroecosystem pilot field program plan, 1992. USEPA/620/R-93/014. U.S. Environ. protection Agency, Washington, DC.

Campbell, C.L., J.M. Bay, A.S. Hellkamp, G.R. Hess, J.J. Munster, K.E. Nauman, D.A. Neher, G.L. Olson, S.L. Peck, B.A. Schumacher, K. Sidik, M.B. Tooley, and D.M. Turner. 1994a. Environmental monitoring and assessment program: Agroecosystem pilot field program report. 1992. USEPA/620/R-94/014. U.S. Environ. Protection Agency, Washington, DC.

Hellkamp, A.S., J.M. Bay, K.N. Easterling, G.R. Hess, B.F. McQuaid, M.J. Munster, D.A. Neher, G.L. Olson, K. Sidik, L.A. Stefanski, M.B. Tooley, and C.L. Campbell. 1995. Environmental monitoring and assessment program: Agricultural lands pilot field program report. 1993. USEPA/620/R-95/004. U.S. Environ. Protection Agency, Washington, DC.

SAS Institute. 1989. SAS user's guide. SAS Inst., Cary, NC.

Scheyer, J.M., R.D. Nielsen, and H.R. Sinclair, Jr. 1992. Soil survey use for plant growth rating. p. 113. *In* Agronomy abstract. ASA, Madison, WI.

Scheyer, J.M., R.D. Nielsen, and H.R. Sinclair, Jr. 1994. Rating for plant growth. Version 1.0. Natl. Soil Surv. Ctr., USDA-Nat. Resour. Conserv. Serv., Lincoln, NE.

Soil Survey Laboratory Staff. 1992. Soil Survey Laboratory Methods Manual. Soil Survey Investigations Rep. 42. Version 2.0. USDA-Soil Conservation Service National Soil Survey Center, Washington, DC.

United States Department of Agriculture. 1981. Land resource regions and major land resource areas of the United States. USDA Agric. Handb. 296. U.S. Gov. Print. Office, Washington, DC.

United States Department of Agriculture. 1993. National Soil Survey Handb. 430-V1-NSSH-1993. USDA-Soil Conservation Service, Washington, DC.

23   Soil Quality Assessment Training for Environmental Educators of Grades 5 through 12[1]

Betty F. McQuaid

USDA-Natural Resources Conservation Service
Raleigh, North Carolina

A teacher training workshop entitled *Project Soil* was sponsored by the Wake Soil and Water Conservation District in Raleigh, NC, in 1994. The workshop involved 13 teachers of Grades 5 through 12 who were interested in expanding their environmental education curricula to include soil science. The purpose of the workshop was to provide a learning experience about soil quality and soil and water conservation practices and to receive hands-on training in the use of a soil quality field test kit for evaluating physical, chemical and biological indicators of soil quality (Fig. 23–1). The workshop activities included field measurements of soil reaction (pH), soil respiration, infiltration, and bulk density in two different soil types. This was followed by an interactive session to discuss teaching students in Grades 5 through 12 about soil quality and the soil quality measurements in relation to sustaining natural resources using a soil *report card*. The workshop was capped off by a tour of the Durham County's District Conservation Farm where teachers were given the opportunity to identify and further learn about best management practices (BMPs).

The USDA-ARS soil quality test kit (discussed in Sarrantonio et al., 1996, this publication) was used in the workshop to (i) estimate ranges of soil properties that are related to soil productivity and environmental quality and (ii) to assist farmers and others in assessing the effects of management practices on soil quality (Cramer, 1994a,b). The Agricultural Lands Resource Group of the U.S. Environmental Protection Agency Environmental Monitoring and Assessment Program (EMAP) has used the test kit to develop and test indicators for regional soil quality assessment. Modifications were made to the USDA-ARS soil quality field test kit for EMAP and it has been tested at more than 48 sites in Maryland, Pennsylvania, and North Carolina. Wake County, North Carolina, Soil and Water Conservation District personnel contacted the EMAP group for assistance in developing soil quality activities for the *Project Soil* Workshop.

---

[1] Agricultural Lands Resource Group (ARG) of U.S. Environmental Protection Agency's (USEPA) Environmental Monitoring and Assessment Program (EMAP), 1509 Varsity Dr., Raleigh, NC 27606.

Copyright © 1996 Soil Science Society of America, 677 S. Segoe Rd., Madison, WI 53711, USA. *Methods for Assessing Soil Quality,* SSSA Special Publication 49.

Fig. 23–1. An education specialist at the Wake County, NC Soil and Water Conservation District demonstrates components of the soil quality field test kit to workshop participants.

A soil quality test kit training guide for the workshop was prepared following Doran (1993, unpublished data). Requirements for modifying the test kit and developing the training guide were (i) teachers must have ready access to the supplies needed to set up a kit in a classroom setting, (ii) field tests must be easy to carry out by most students, (iii) test kit results must be easily applied to natural resource assessment, and (iv) time constraints of the workshop (< 1 h).

Following is a summary of the field procedures and handouts designed for the workshop. As a result of participation in the workshop, teachers of Grades 5 through 12 should have a better understanding of soil quality, ability to perform and teach field assessment of selected soil quality indicators, and relate these indicators to natural resource management and monitoring using a soil quality report card (Table 23–1).

Table 23–1. Soil quality report card used by workshop participants.

Measurements Indicator	Site 1	Site 2
Bulk density		
Soil weight, g		
Can volume, cm^3	344	344
Bulk density = weight/volume		
Infiltration		
Start time (s)		
End time (s)		
Depth of water added (in)	1	1
Infiltration rate = depth of water added/(start time – end time)		
pH		
Meter/color chart reading		
Respiration		
Start time, min		
End time, min		
CO_2 reading, %		

Scoring
Assign points to each indicator as follows:

Bulk density	75 points if 2 to 1.5 g cm^{-3}	
	100 points if < 1.5 g cm^{-3}	
Infiltration	75 points if ≤ to 0.02 in s^{-1}	
	100 points if >0.02 in s^{-1}	
pH	50 points if <5.5 or >8.0	
	75 points if ≥5.5 but <6.5 or >7.5 but ≤8.0	
	100 points if ≥6.5 but ≤7.6	
Respiration	50 points if <0.1% or ≥2.0%	
	75 points if ≥0.1% but <0.5%	
	100 points if ≥0.5% but <2.0%	

	Site 1	Site 2
Bulk density		
Infiltration		
pH		
Respiration		
Total		
Average (total/4)		

GRADES 5 THROUGH 12 TEACHER'S OVERVIEW: WHAT IS SOIL QUALITY?

Soil is more than just dirt! Soil allows us to produce crops and timber. Soil also acts as a filter to protect our environment by holding on to and breaking down wastes, chemicals, and pollutants in the environment (Hellkamp et al., 1994). A simple definition of soil quality is the ability of the soil to perform these functions. Quality soils drain and warm up quickly, soak up water after heavy rains, store moisture for dry spells, are teeming with earthworms and millions of microorganisms and produce healthy crops to eat (Cramer, 1994a). Environmental scientists are presently working to identify indicators of soil quality. Just as your doctor checks a number of indicators of your health, like your temperature, heart rate, and weight, soil scientists use indicators to check soil health.

The soil quality field test kit includes the tools needed to check a number of indicators of a soil's quality or health. Indicators are divided into three different categories: physical, chemical, and biological.

The physical soil quality indicators are bulk density and infiltration. Bulk density is the weight of dry soil per known volume of soil. This measurement is the soil's density, which can tell us how compacted it is. Compacted soil restricts movement of roots and water. Infiltration is the amount of time that it takes a known amount of water to move through the soil surface. Infiltration is a good measure of both the soil's ability to take up water and its potential to erode. As infiltration rate increases, water intake increases and erosion decreases.

The chemical soil quality indicator is soil reaction or pH. pH tells us how acidic or basic the soil is. Plant nutrients such as Ca, P, K, and Mo may be unavailable for plant uptake and growth if the pH is too acidic. If the pH is too basic, plant nutrients such as P, Fe, Mn, Zn, Cu, Co, and B may be unavailable. Most plants grow best at a pH between 6.5 and 7.5 because almost all nutrients are readily available in this range.

The biological soil quality indicator is respiration. Respiration is the measure of how much CO_2 is given off by living organisms in the soil: fungi, bacteria, worms, crickets, and others. Respiration gives us an idea of how many organisms are breathing and working to break down nutrients so that they are available for plant uptake. Higher respiration rates indicate a higher population of soil life.

Soil quality indicators selected for this exercise are included in Soil and Water Quality: An Agenda for Agriculture (National Research Council, 1993). These indicators also have been proposed as key indicators by several Soil Quality Working Groups and are being studied by numerous researchers involved with soil quality. Other physical, chemical, and biological indicators could have been used as well but ease of field implementation and assessment were the overriding criteria in this teacher training workshop.

SOIL QUALITY FIELD EXERCISE

Workshop participants used the soil quality field test kit to measure the physical, chemical, and biological indicators at two farm sites in Durham County, North Carolina. Measurements were recorded on a soil quality report card (Table 23–1) and used to compare the soil quality between the two sites. The soil quality report card developed by Karlen and Stott (1994) provided the framework for the report card used in this exercise. Field sites were located and soil quality indicator data sets were obtained before the workshop. The report card ranges for bulk density, infiltration, pH, and respiration were established based on the initial data sets taken at the two sites. Range breaks and scoring were assigned to achieve site differences and to allow participants the opportunity to associate the differences in scores to management practices, e.g., bare tilled sites had lower respiration than no-till sites with observable surface residue cover.

Participants were introduced to the concept of inherent soil quality because these sites used in the exercises were managed similarly. Thus, many differences in the soil quality measurements were probably due to soil type. Just as children

Table 23–2. Materials, suppliers, and approximate cost for items in the soil quality field test kit.

Item	Cost	Supplier
To determine bulk density you will need:		
Soup can	n/c	recycle
Garden trowel	$5	variety store
Wooden board (8 by 2 by 4)	n/c	recycle
Hammer	$5	variety store
Food scale	$10	variety store
Plastic bags (pint size)	$1	grocery store
To determine infiltration you will need:		
6-inch diameter coffee can (no. 10) or irrigation pipe cut to a height of 5-in	n/c or $1	recycle or agricultural supply
Plastic wrap	$2	grocery store
2-cup measuring cup filled with water	n/c or $1	personal or variety store
Watch with a second hand	n/c	personal
To determine soil reaction (pH) you will need:		
Digital pocket pH meter - or -	$30	scientific supply company
pH paper - or -	$7	scientific supply company
Commercial pH kit	$34	scientifid supply company
Distilled water	$2	grocery store
Coffee scoop or measuring tablespoon	$1	variety/grocery store
To determine respiration you will need:		
6-in coffee can or irrigation pipe (from infiltration test)	n/c	
6-in coffee can (no. 10) lid with sides	n/c	
140 mL syringe with needle	$3	scientific supply company
12 in of 0.25 in diameter platic tubing	$1	pet fish store
Drager tube (0.1% CO_2), nonreusable	$34/10	scientific supply company

are born with the ability to walk or talk sooner than other children, so are soils born or formed with different qualities that largely determine their ability to grow crops or filter chemicals in the environment.

Project Soil was concluded with a farm tour thus providing a chance to learn about the impact of best management practices on soil quality. Pesticide use, erosion control, crop residue additions, and other practices were linked to soil quality. For example, additions of organic matter such as soybean [*Glycine max* (L.) Merr.], wheat (*Triticum aestivum* L.), and grass enhance infiltration and result in a higher infiltration rate score on the report card. Limited use of pesticides may enhance soil organism populations and result in a higher respiration rate (CO_2) score on the report card. By observing soil quality indicators and management practices, participants learned about the influence of humans on soil quality and the vital role that land managers and farmers play as stewards of our valuable, slowly renewable soil resources.

METHODS FOR SOIL QUALITY ASSESSMENT

The soil quality field test kit used for the *Project Soil* Workshop consisted of equipment to measure bulk density, infiltration, pH, and soil respiration. All measurements and calculations were recorded on the soil quality report card.

A list of equipment and supplies for the soil quality field test kit is provided in Table 23–2. A kit can be created for any or all of the indicators, according

Fig. 23–2. A soup can, scale, and hammer are among the tools are used to measure bulk density.

to student ability and availability of supplies. For example, you may wish to omit respiration since it is the most costly procedure to run and materials must be ordered through a science supplier. Also, the teacher should use equipment such as the syringe and needle when younger children are determining the measurements.

Bulk Density

Cut out the top and bottom from a 12 oz. metal food can. A standard can has a diameter of 7.8 cm, height of 7.2 cm, and volume of 344 cm^3. If the can to be used is a different size, measure the height and diameter in centimeters. The volume of the can is then calculated by squaring the radius of the can, multiplying it by the can height, and then multiplying by 3.14 (π).

Bulk density of the top 3 in of soil can be determined by pushing the can into the soil. A wood board may be placed on top of the can and then pressed in by hand or by pounding with a hammer (Fig. 23–2). When the top of the can is even with the soil surface, carefully dig out the can with a trowel or shovel and remove all soil from the outside of the can. Use a knife to cut off any soil that may be sticking out from the top or bottom of the can.

Extract the soil from inside the can and place it in a plastic bag. The soil should be air dried for about 4 d or in a microwave oven using 3- to 4-min cycles. Be sure to keep the bag open and stir the soil several times during the drying process to ensure that all the water is evaporated. Then weigh the soil and plastic bag on a balance. All weights should be recorded in grams. The weight of the plastic bag is negligible as compared with the weight of the soil so it is not important to adjust the soil weight for the bag weight. If a heavier container is used to hold the soil as it dries, the container weight should be subtracted from the total weight. To do this, weigh the soil and container together. Remove the soil and reweigh the container alone. The difference between the soil plus container weight and the container weight is the weight of the soil. Divide the soil weight by the can volume to determine the bulk density of the soil. Bulk densities may range from <1.0 g cm^{-3} to about 1.7 g cm^{-3}. Soils with bulk density <1.0 g cm^{-3} are usually high in organic matter, such as peat soils. Bulk densities greater than about 1.5 g cm^{-3} are usually associated with compaction or have high sand content.

Infiltration

Cut out the top and bottom of a no. 10 coffee can and trim the sides so that it is 5 in. tall or obtain a 5-in. length of 6 in. (15.2 cm) diam (i.d.) irrigation pipe. Push the can into the soil so that 2 in. of the can remains aboveground. As with the bulk density measurement, a wooden board can be used on top of the can to push or pound it in. Line the inside walls of the can with plastic wrap. Be sure to carefully wedge a small potion of the plastic wrap tightly between the can and the soil where the crack was created by pushing the can into the soil. This will reduce the chance for water to drain into the crack and give false measurements.

Pour 2 cups (about 473 mL, mL = cm^3) of water into the can, all at one time. This is equivalent to a depth of 1 in. of water in the 6-in. diameter can or pipe. Using a stop watch or wrist watch with a second hand, record the time when the water was first poured into the can. Watch to see when all the water has infiltrated into the soil and record the time once again. All the water has infiltrated when there is no free water on the soil surface but the surface still glistens. Time for the beginning and end of infiltration should be recorded in seconds. Calculate infiltration rate by dividing the 1 in. depth of water by the number of seconds required for all the water to infiltrate. Infiltration rates may be grouped according to the following classification scheme:

Seconds for 1 in of water to infiltrate	Infiltration class
More than 72 000 (20 h)	Very slow
18 000 to 72 000 (5 to 20 h)	Slow
4500 to 18 000 (75 min to 5 h)	Moderately slow
1450 to 4500 (24 to 75 min)	Moderate
720 to 1450 (12 to 24 min)	Moderately rapid
360 to 720 (6 to 12 min)	Rapid
<360 (6 min)	Very rapid

pH

A coffee scoop or measuring tablespoon, distilled water, and a pH probe or pH paper are needed to assess soil pH. Combine a level scoop of soil with a scoop of

Fig. 23–3. Respiration measurements are recorded on the soil quality report card.

distilled water in a plastic bag. Knead the mixture with your fingers on the outside of the bag. Be sure that all soil is wet up. Allow the mixture to sit for about 10 min. Place the digital pH probe or pH paper into the soil–water mixture. Obtain the pH from the digital readout on the probe or by comparing the color of the pH paper to the pH color chart on the container. If using a pH probe, it should be calibrated before use to ensure its accuracy. pH values near neutral range from 6.5 to 7.5 while values below 6.5 are acidic and values above 7.5 are basic.

Soil Respiration

Soil respiration measurements can be made using the same coffee can as for the infiltration test. A lid is needed for the coffee can and may be made from the top of a no. 10 coffee can. The lid, with the top intact, needs to have sidewalls about 1 in. in height and a one-quarter inch hole drilled in the top. Insert the coffee can into the soil leaving 2 in. aboveground. A rubber stopper should be placed in the hole of the lid and the lid placed over the coffee can in the soil. Leave the covered can in the soil for 30 min. Depress the plunger of a 140 mL syringe as if to push out all the air and then connect it to one end of the plastic tubing. Connect the other end of the tubing to a 0.1% Drager CO_2 tube. To the other end of the Drager tube connect one end of plastic tubing. Connect the other end of the tubing to a needle (Fig. 23–3).

After the 30-min waiting period, insert the needle into the rubber stopper in the lid of the coffee can. Slowly over a period of 20 s, pull 100 mL of air from the coffee can into the Drager tube by pulling out plunger on the syringe until the 100 mL mark is reached. The respiration reading is measured as a volume percentage of CO_2 in the air. This reading can be determined by reading the height of the purple color train on the Drager tube. A CO_2 content of 0.5% represents a high respiration rate and correlates to a gaseous loss of C of about 60 lb per acre per day. Specific details on use of Drager gas detection tubes to measure soil respiration are given by Parkin et al. (1996, this publication).

SUMMARY AND CONCLUSIONS

Training in the use and interpretation of the soil quality field test kit for Grades 5 through 12 teachers can add another dimension to environmental education. Many soil quality indicators typically require expensive laboratory tests; however, some are inexpensive and others can be adapted for in-field measurements by nonscientists, such as students in Grades 5 through 12. Modifications to the USDA-ARS soil quality field test kit proved useful in providing hands-on workshop activities with a scientific basis that can be easily, quickly, and inexpensively used in the classroom. Participants in the workshop were able to assess soil quality indicators of different soils and learn the role of land managers and farmers in sustaining this valuable natural resource. Combining in-field soil quality measures with the best management practices tour provided a good way to relate in-field data to land management.

Drawbacks to the field exercise were the preparation time involved in finding sites and taking indicator measurements prior to the workshop. In addition, sometimes indicator data results, as in research, were difficult to explain and may require an expert to interpret. It is best to use an entire day to conduct the farm tour and to learn about and perform the soil quality tests. A 1-h time period was allocated for the soil quality testing but 1 1/2 h was needed. More than the allotted time also was needed for other workshops using all or part of the soil quality field test kit, such as the Environmental Report Card Teacher Institute (USEPA and North Carolina State University) and the Sci-Link Program (North Carolina State University).

To date, no feedback has been received from teachers who have tried the soil quality field test kit procedures in the classroom. Although this workshop and soil quality field test kit were designed for teachers of Grades 5 through 12, the kit and teaching approach can easily be tailored to Scout troops, FFA, 4-H, college classes, and other groups interested in natural resource education.

ACKNOWLEDGMENTS

The author wishes to thank Shiela Jones, education specialist, Wake County Soil and Water Conservation District, and Anne Hellkamp, USEPA EMAP Agricultural Lands, for editorial assistance; and Debbie Anderson, soil scientist, USDA Natural Resources Conservation Service, and Gail Olson, Lockheed Idaho Technologies, Raleigh, North Carolina for technical assistance.

DISCLAIMER

Although the project described in the chapter has been funded wholly or in part by the US- Environmental Protection Agency under Interagency Agreements with the USDA-Natural Resources Conservation Service and Agricultural Research Service, it has not been subjected to Agency review and, therefore, does not necessarily reflect the views of the Agency, and no official endorsement should be inferred. This work also was supported in part through a Specific Cooperative Agreement between USDA-Agricultural Research Service and the North Carolina Agricultural Research Service.

REFERENCES

Cramer, C. 1994a. Test your soils' health: New kit helps track improvements in soil quality. The New Farm Magazine. 16(1):17–21.

Cramer, C. 1994b. Test your soils' health: more tests to help track soil quality. The New Farm Magazine. 16(2):40:43–45.

Hellkamp, A.S., B.F. McQuaid, D.A. Neher, and G.L. Olson. 1994. Beginner's guide to EMAP-agricultural lands soil quality. USEPA Environ. Monitoring and Assessment Progr., Agric. Lands Resour. Group, Raleigh, NC.

Karlen, D.L., and D.E. Stott. 1994. A framework for evaluating physical and chemical indicators of soil quality. p. 53–72. *In* J.W. Doran et al (ed.) Defining soil quality for a sustainable environment. SSSA Spec. Publ. 35. SSSA, Madison, WI.

National Research Council. 1993. Soil and water quality: An agenda for agriculture. Natl. Acad. of Sci. Press, Washington, DC.

Parkin, T.B., J.W. Doran, and E. Franco-Vizcaíno. 1996. Field and laboratory test of soil respiration. p. 231–245. *In* J.W. Doran and A.J. Jones (ed.) Methods for assessing soil quality. SSSA Spec. Publ. 49. SSSA, Madison, WI.

Sarrantonio, M., J.W. Doran, M.A. Liebig, and J.J. Halvorson. 1996. On-farm assessment of soil quality and health. p. 83–105. *In* J.W. Doran and A.J. Jones (ed.) Methods for assessing soil quality. SSSA Spec. Publ. 49. SSSA, Madison, WI.

24 A Comparative Study of Soil Quality in Two Vineyards Differing in Soil Management Practices[1]

Stamatis Stamatiadis, A. Liopa-Tsakalidi, L. M. Maniati, P. Karageorgou, and E. Natioti

Goulandris Natural History Museum
Kifissia, Greece

Conventional agriculture is characterized by the intensive use of pesticides, chemical fertilizers and tillage in order to attain increased crop yields; however, these management practices have caused economic and environmental problems due to increased cost of synthetic chemicals, environmental pollution, and soil degradation. Consequently, there is an increased interest in crop production systems that optimize yields while conserving soil, water, and energy and protecting the environment. Such systems reduce the use of soil tillage and chemicals while incorporating crop residues or organic conditioners on the soil surface.

Alternative management practices can reestablish the natural physical, chemical, and biological properties of the soil and lead to a sustainable soil environment. Reduced tillage and residue management decrease soil aeration and microbial activity and the associated processes of decomposition and nutrient cycling but create a cooler and wetter soil environment with higher organic matter content (Barber, 1990; Doran, 1980a,b; Doran et al., 1988; Hendrix et al., 1986; Holland & Coleman; 1987 Mielke et al., 1986). These changes are expected to improve soil structure, increase water storage capacity, and reduce soil erosion. Consequently, the adoption of more sustainable soil management practices has increased crop production potential for rainfed agriculture in semiarid and arid climates and has application on the arable lands used for summer crops in Greece.

In this study, the soil quality and plant growth of two adjacent grape (*Vitis* sp.) vineyards having different soil management practices were compared. The research objectives were to: (i) quantify soil properties, plant growth, and productivity; (ii) relate observed differences to soil processes and management prac-

---

[1] This project was cofunded by the World Wide Fund for Nature Greece.

Copyright © 1996 Soil Science Society of America, 677 S. Segoe Rd., Madison, WI 53711, USA. *Methods for Assessing Soil Quality*, SSSA Special Publication 49.

Table 24–1. Perennial growth characteristics of the organic and conventional vines (mean ± SE, $N = 4$ for trunk length, $N = 8$ for main branches, $N = 16$ for terminal branches).

	Cultivation		Ratio (organic/conventional)
	Organic	Conventional	
Trunk length, cm	37.8 ± 10.6	30.8 ± 5.1	1.2
Length of main branches, cm	54.3 ± 5.7	30.5 ± 1.7	1.8*
Length of termilal branches, cm	29.6 ± 2.5	35.3 ± 3.4	0.9

* Significant difference between the two means at $P < 0.05$.

tices; and (iii) identify those practices that maintain soil quality and sustain productivity.

METHODS

Field Selection

The two vineyards selected for study are cultivated for raisins and are located near the village of Mamousia (Aegion, Greece) on the slopes of Mt. Mamousovounion at an altitude of 600 m. These vineyards are located adjacent to one another, have the same soil type and topographical features, and were planted 35 yr ago, but have received different soil management practices during the 4-yr period prior to this study.

The organic vineyard has been included in a program conducted by the Agricultural Union of Aegialia to establish organic farming practices according to regulation 2092/91 of the Council of the European Union. Practices in the organic field include pruning branches and applying 150 to 200 cm^3 of poultry manure around each vine in late January or early February, reduced tillage to a depth of 10 to 15 cm using a rotavator in March and May each year followed by foliar application of S–Cu solution, and incisions prior to vintage.

In the conventional vineyard, Roundup 36AS (containing 36% Glyphosate as the active ingredient) was applied at a rate of 300 g 1000 m^{-2} in May for weed control. No fertilizers were applied and no tillage was performed in this vineyard in the 4 yr prior to study.

Both fields were densely planted with 2.5 vines m^{-2}. The similarity of vine growth between the two fields during the early years of cultivation indicated no differences in trunk length (Table 24–1). Field sampling occurred in early May, early June and late August before vintage in 1994.

Soil Chemical Properties

In each field, three, 1-kg soil samples were taken from the 0- to 15-cm depth and combined for chemical analysis (Peterson & Calvin, 1986). Among the soil chemical properties evaluated were: organic C by the Walkley–Black procedure (% organic matter = 1.724 × Organic C(%)); total N by the Kjeldahl method; ammonium and nitrate by the Granvand-Liazou method; available P by the Olsen method; available K and cation-exchange capacity by flame photometry; $CaCO_3$ by titration with $KMnO_4$. Soil pH was measured by the saturated paste method and electrical conductivity (1:2 soil to water mixture) in dS m^{-1}. Salt concentration was calculated by the following relationship: salts, % = 0.128 × $EC_{1:2}$.

Soil Biological Properties

Microarthropod populations were assessed in May and earthworm populations in May and July. Microarthropods were extracted from four random 0- to 5-cm depth soil samples in a modified Tullgren apparatus (Crossley & Blair, 1991). The animals were fixed on slides in Hoyer's solution and examined under a phase contrast microscope. Chemical extraction with formaldehyde was used for earthworm sampling at the same four random locations in the field. The collected earthworms were killed in 70% ethyl alcohol, transferred to the laboratory in 12% formalin solution, blotted dry on filter paper, and weighed in order to provide an estimate of biomass.

Carbon dioxide evolution was evaluated to reflect microbial activity. Four randomly selected 50 g (dry wt. basis) soil samples from the 0- to 7.6-cm depth were transported to the laboratory where they were packed to their respective field bulk density (see below) at 60% water-filled pore space so that microbial activity would be maximized (Linn & Doran, 1984). After a 3-d equilibration period, soils were incubated with 10 mL of 0.5 M NaOH for 24 h in shielded containers at 26°C. Titration of the NaOH solution with 0.63 M HCl was used to estimate CO_2 evolution.

Soil Physical Properties

Physical properties were measured at three random locations in each field. Metal infiltration rings (9.6 cm in diam.) were hammered into the soil to a depth of 7.6 cm. A 2.5-cm depth of water was poured into the ring and the time needed for all water to infiltrate was recorded. The procedure was repeated and the time of the second infiltration test recorded as an indicator of infiltration rate.

The infiltration rings were covered with plastic for 24 h and then removed with the intact soil, placed in sealable plastic bags and transported to the laboratory for water content, water field capacity, bulk density, soil porosity, and water-filled pore space determinations (Doran, 1995). A particle density of 2.65 kg m^{-3} was used in calculations.

Vine Growth and Productivity

Plant growth indicators were measured in four randomly selected plants from each field in July 1994. Indicators of perennial growth were length of trunk and main branches (Fig. 24–1). Indicators of annual growth and productivity were number and length of terminal branches, leaf dimensions, weight and length of grape bunches, grape diameter, and grape sugar content.

Leaf length and width and stalk length were measured using the third leaf from the beginning of each of three terminal branches of each plant. Length of all

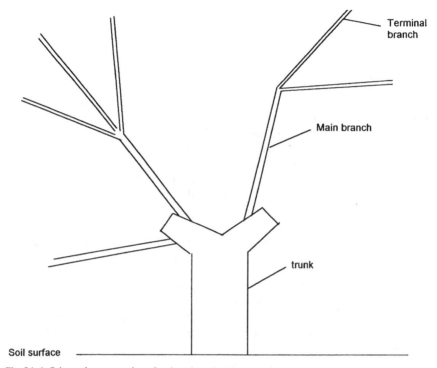

Fig. 24–1. Schematic presentation of a vine plant showing the relative position of the trunk, main, and terminal branches.

bunches of grapes per plant was measured. From each plant, three random bunches of grapes were weighed; the diameter and sugar content was measured from 20 grapes of each bunch using a portable refractometer; grape volume was computed assuming the grapes were spheres.

The weight of bunches, grape diameter, and sugar content were measured again from the same plants in late August prior to harvest. In addition, the number and weight of leaves on one randomly selected terminal branch from each of four plants was measured. Total leaf biomass per field was estimated from leaf weight per terminal branch, the number of terminal branches per plant and the number of plants per field.

Statistical Analysis

Soil physical and biological properties were analyzed using a factorial completely randomized design. Independent variables of the model were cultivation method and time of sampling. Plant growth data was analyzed using a nested design with randomized effects. Independent variables of this model were cultivation method, plants within a cultivation, grape bunches within a plant of each cultivation, and others. The number of model variables was dependent upon the specific independent variable under examination. Means were compared using Bonferronni multicomparison at $P < 0.05$.

Table 24–2. Comparison of soil physical and biological properties (0–7.6 cm) of the organic and conventional cultivation in July (mean ± SE, $N = 3$, except for CO_2–C where $N = 4$).

Property	Cultivation		Ratio (organic/conventional)
	Organic	Conventional	
Bulk density, g cm^{-3}	1.12 ± 0.07	1.42 ± 0.05	0.79*
Water content, %, g cm^{-3}	9.2 ± 0.7	12.0 ± 0.05	0.77*
Pore space, %	57.6 ± 2.7	46.5 ± 1.8	1.24*
Water-filled pore space, %	16.2 ± 2.0	25.9 ± 1.2	0.63*
Water field capacity, %, g cm^{-3}	33.5 ± 2.3	39.0 ± 1.0	0.86
Infiltration rate, min	1.7 ± 0.5	78.3 ± 26.2	0.02*
g CO_2–C m^{-2} 24 h^{-1}	3.6 ± 1.0	5.7 ± 1.5	0.63

* Significant difference between the two means at $P < 0.05$.

RESULTS AND DISCUSSION

Soil Physical Properties

The study fields had similar soil texture with 43% clay, 27% silt, and 30% sand. Physical properties are compared in Table 24–2. Low bulk density of topsoil in the organic field was caused by tillage and was associated with increased porosity. Increased porosity increases soil aeration which, in turn, accelerates aerobic microbial activity and nutrient mineralization (Doran 1980a,b; Hendrix et al., 1986). Therefore during optimal environmental conditions in the spring organic cultivation will have an advantage over conventional cultivation with regard to soil nutrient availability for early plant growth. Alternatively, increased porosity will reduce microbial activity and nutrient cycling more dramatically during periods of water shortage such as during the summer months during the final stages of vine growth. This is portrayed by the greater water-filled pore space, indicative of higher microbial activity and mineralization rates, in the conventional cultivation as compared with the organic cultivation during the droughty period in July (Table 24–2). The difference in water-filled pore space between soils also is reflected in soil water content differences.

Tillage of the organic field appears to be responsible for the 46-fold increase in infiltration rate as compared with the conventional field. Tillage disrupted the surface crust and increased soil porosity. The low infiltration rate of the conventional field poses a risk for surface runoff and erosion, especially where there is high rain intensity and sloping ground. During rainy periods, herbicide runoff also may occur thereby reducing its effectiveness and contaminating adjacent areas.

Soil Chemical Properties

Soil organic C, and available P and K concentrations ranged from moderate to adequate levels (Table 24–3). Cation-exchange capacity also was high because of the high clay content. Electrical conductivity and salt content are low and pose no problem for plant growth. pH was slightly alkaline due to the high $CaCO_3$, but appears optimal for microbial activity. The higher N levels of the

Table 24–3. Comparison of soil chemical properties (0–15-cm depth) of the organic and conventional cultivation.

Property	Cultivation		Ratio (organic/conventional)
	Organic	Conventional	
pH	7.4	7.3	1.0
CEC, cmol kg^{-1}	21.6	18.8	1.1
Conductivity, dS m^{-1}	<3	<3	
Salts, %	<0.11	<0.13	
CaCO$_3$, %	20	8	2.5
Available K, mol kg^{-1}	1.20	0.64	1.9
Organic-C, Mg ha^{-1}	31.2	28.4	1.1
Total N, kg ha^{-1}	2688	2069	1.3
NO$_3$, kg ha^{-1}	63.4	23.7	2.7
NH$_4$, kg ha^{-1}	26.7	18.8	1.4
P Olsen, kg ha^{-1}	58.8	34.1	1.7

organic soil can be explained by manure application. The low levels of NO$_3$ and NH$_4$ in both fields appear normal for late July as these nutrients are generally taken up by the roots earlier in the growing season. The considerably higher CaCO$_3$ in the organic field does not appear to be explained by manure application alone; it is unknown as to what degree earthworm activity (see below) has contributed to Ca enrichment of the top soil (Edwards & Lofty, 1972). High CaCO$_3$ levels pose a risk for leaf chlorosis.

The amount of soil nutrients was always greater in the organic field than the conventional field, especially NO$_3$, P and K. These higher values were associated with the application of organic fertilizer since poultry manure is especially rich in these nutrients.

Soil Biological Properties

Carbon dioxide is a by-product of microbial metabolism. The rate of its evolution from the soil surface is used as an indicator of aerobic microbial activity and decomposition of soil organic matter. Carbon dioxide evolution, measured in the laboratory, was highly variable with no significant difference between the two soils (Table 24–2). The measurement of CO$_2$ evolution of the conventional field was probably underestimated due to the inability to replicate its high bulk density in the laboratory. The bulk density of the conventional soil was brought to ca. 1.3 g cm^{-3}, instead of the field value of 1.42 g cm^{-3}, which corresponds to about 50% WFPS and is below the optimal range of 55 to 61% for microbial activity (Doran et al., 1988).

Earthworms were greater, in both number and size, in the organic field than in the conventional field, which resulted in a difference in total earthworm biomass in the spring (Fig. 24–2). Most earthworms were observed to be anecic that construct tunnels and live deep in the soil. Anecic earthworms are known for their contribution in the improvement of soil structure and fertility by fragmenting organic residues and mixing them with mineral soil, by creating water-resistant organomineral complexes, and by accelerating water infiltration and soil microbial activity (Edwards & Lofty, 1972; Bouche', 1977). For these reasons, earthworms are believed to have contributed to the soil differences between fields in

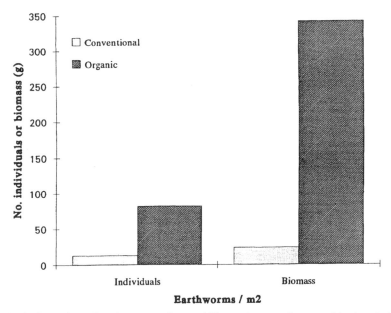

Fig. 24–2. Comparison of earthworm numbers and biomass between the two cultivations in May 1994.

infiltration rate and mineral N levels. No earthworms were found in the soil samples in July due to drought, which normally causes aestivation after vertical migration of earthworms to greater soil depths.

Similarly, springtails (Collembola) and mites (Acari) were almost double in number in the organic field in the spring as compared with the conventional field (Fig. 24–3). The greatest differences between mite suborders in the two fields was in the Mesostigmata or predatory mites (Fig. 24–4). This effect is suspected to result from herbicide application since sharp declines in mesostigmatid mites, and other members of the predatory soil fauna, have been reported after application of a number of pesticides in soil (Wallwork, 1976). Research suggests that microarthropods alter microbial composition and metabolism as well as fragment incorporate organic residues in soil (Werner & Dindal, 1987).

The drastic reduction of both earthworm and microarthropod populations in the conventional field is suspected to be caused by repeated herbicide application since there was no marked difference in organic matter between the two fields. Toxic effects of herbicides on soil biota have been reported in the literature (Doran & Werner, 1990). High bulk density of the conventional field also may have contributed to the reduced populations by restricting access to nonburrowing animals, i.e., microarthropods, to microhabitats within the soil profile (Richards, 1987).

Vine Growth and Productivity

Branch length and leaf dimensions were used as indicators of plant growth (Table 24–4) and were greater for the organic field than the conventional field in July. Terminal branches of the organically grown vines were 2.3 times longer than

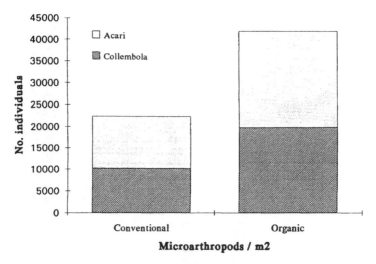

Fig. 24–3. Soil microarthropod populations in the organic and conventional fields in May 1994.

the conventionally grown vines but had only about one-half as many terminal branches per plant by August (Table 24–5). Greater branch length resulted in the production of more leaves with greater leaf dimensions, and thus more leaf biomass for the organic field as compared with the conventional field (Table 24–5).

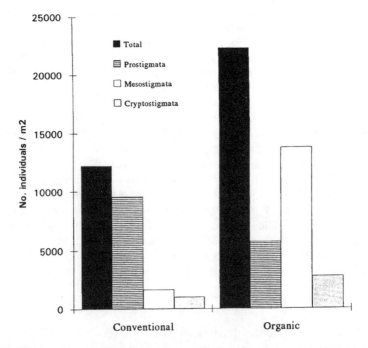

Fig. 24–4. Density of Acari (mite) suborders in the soil of organic and conventional fields in May 1994.

Table 24-4. Comparison of the annual growth of vines between the organic and the conventional cultivation in July 1994 (mean ± SE, $N = 12$, except terminal branches).

Annual growth	Cultivation		Ratio (organic/conventional)
	Organic	Conventional	
Terminal branches			
Branch length, cm†	101.0 ± 5.0	43.0 ± 2.2	2.3*
Leaf dimensions			
Stalk length, cm	12.0 ± 0.7	9.7 ± 0.7	1.2
Leaf length, cm	15.3 ± 0.6	9.4 ± 0.3	1.6
Leaf width, cm	18.5 ± 0.6	11.8 ± 0.5	1.6

* Significant difference between the two means at $P < 0.05$.
† $N = 99$ for the organic, $N = 162$ for the conventional cultivation.

Similar differences and patterns were observed for grape bunches. Organic grape bunches were greater in size and weight but fewer in number than conventional grape bunches in July (Table 24–6). Total weight of grapes produced per plant for the organic field was 1.6 times greater than for the conventional field. Through August organic grapes lost weight but continued to gain volume. The 33% loss of organic grape weight and a slight weight gain for conventional grapes resulted in similar grape weight per plant and sugar content (Table 24–6).

Greater vine development was consistent with higher soil nutrient content in the organic field because of manure application. Soil tillage further accelerated the mineralization of organic fertilizer and plant residues as well as leaching of N.

Carbon Cycling

Leaves constitute the major source of indigenous and readily decomposable organic matter in the vineyard ecosystem. The annual contribution of leaf biomass to the soil was 4.3 and 2.4 kg m^{-2} for organic and conventional fields, respectively. Thus, the organically grown plants provide about 1.75 times more C to the soil environment than does the conventionally grown plants.

Carbon transformation in the soil is primarily due to microbial decomposition. A portion of the C is released as CO_2 to the atmosphere and the remainder is incorporated into microbial biomass or becomes part of the soil humus. Based

Table 24–5. Pattern of growth and estimation of leaf biomass of the two cultivations in August 1994.

Annual growth patterns	Cultivation		Ratio (organic/conventional)
	Organic	Conventional	
Terminal branches			
Length per plant (m)	24.5	17.7	1.38
Number per plant	22.7	40.0	0.57
Number of leaves			
Number per branch	47.2	20.7	2.28
Number per plant	1088	831	1.31
Leaf biomass			
g plant^{-1}	1706	976	1.75
kg m^{-2} of soil	4.27	2.44	1.75

Table 24–6. Comparison of vine productivity between the organic and conventional cultivation during growth and before harvest (mean ± SE).

Month	Variable	N	Cultivation Organic	Cultivation Conventional	Ratio (organic/conventional)
			Bunches of grapes		
July	Length, cm	†	18.0 ± 0.4	15.1 ± 0.3	1.2*
	Weight, g	12	137.0 ± 11.4	52.3 ± 12.3	2.6*
	Number per plant	4	25	40	0.6
	Weight per plant, g		3425	2092	1.6
August	Weight, g	12	91.0 ± 12.5	54.7 ± 9.7	1.7*
	Number per plant	4	25	40	0.6
	Weight per plant, g		2275	2188	1.0
			Grapes		
July	Diameter, mm per grape	280	6.9 ± 0.1	4.0 ± 0.1	1.8*
	Volume, mm^3 per grape	280	197.9 ± 7.4	47.4 ± 2.9	4.2*
	Sugar content, °Be	280	10.1 ± 0.3	11.4 ± 0.2	0.9*
August	Diameter, mm per grape	264	7.4 ± 0.1	5.8 ± 0.1	1.3*
	Volume, mm^3 per grape	264	237.2 ± 10	111.5 ± 3.3	2.1*
	Sugar content, °BE	264	26.1 ± 0.1	26.6 ± 0.1	1.0

* Significant difference between the two means at $P < 0.05$.
† $N = 100$ for the organic, $N = 160$ for the conventional cultivation.

on our laboratory measurements, the amount of CO_2–C released from the conventional field can be greater than the leaf-C entering the soil on an annual basis. The CO_2-C/leaf-C ratio is above unity for the conventional field (Table 24–7) and illustrates that this soil may incur a net loss of soil C over time and lead to soil degradation unless there is a change in soil management. The greater leaf biomass of the organic field contributed to a CO_2-C/leaf-C ratio that approaches unity and indicates a balanced C budget and that the soil management practices being used are sustainable.

Table 24–7. Estimation of the recycling potential of plant detritus in soil based on the annual leaf biomass (August sampling) and CO_2 evolution.

	Cultivation Organic	Cultivation Conventional
Leaves		
Live biomass (kg m^{-2})	4.27	2.44
Dry biomass (kg m^{-2})†	1.28	0.73
Biomass C (kg m^{-2})‡	0.58	0.33
CO_2 Evolution		
CO_2-C (kg m^{-2} mo^{-1})§	0.139	
CO_2-C (kg m^{-2} yr^{-1})¶	1.557	
Ratio (CO_2-C/Leaf-C)	0.96	1.69

† Estimated as 30% of live biomass.
‡ Estimated as 45% of dry biomass (Mengel & Kirkby, 1978).
§ Calculated by taking the mean of CO_2 evolution for both fields (0–7.6 cm depth) under optimized temperature and moisture conditions.
¶ Estimated on the assumption that a 4-mo optimized rate of CO_2 evolution equals the total annual CO_2 production in soil.

SUMMARY

Sustainable agriculture is gaining support worldwide due to high costs, declining produce quality, environmental pollution, and degradation associated with conventional management practices. Practices leading to sustainable soil management include reduction of tillage, elimination of pesticides and fertilizers and the increase in organic matter through mulching and the application of organic conditioners.

In this study, two adjacent vineyards of differing soil management practices were compared in terms of soil quality and plant growth. The sustainable field received reduced tillage for weed control and poultry manure was applied for nutrients. In the conventional field herbicides were applied on the soil surface but tillage and fertilizers had not been used for at least four years.

Differences in soil properties and plant growth were representative of the soil management practices between the two fields. Tillage of the organic field increased soil aeration and water infiltration but reduced surface soil water content during the dry period. Poultry manure application increased soil nutrient content (total N, NO_3, P, K, and Ca) but did not increase organic matter content. Soil invertebrate populations were two times greater in the organic field than the conventional field. Depressed conventional populations were attributed to herbicide application, lessor amounts of crop residue and higher bulk density of the surface soil.

As a consequence of improved soil properties in the organic field, plant growth was much faster during the growing season and grape bunches were larger (but fewer in number) than the conventional field; however, productivity and sugar content among the two fields were similar at harvest.

Organic-C in the form of leaf-C that potentially enters the soil was 1.75 times greater in the organic field than the conventional field. Increased leaf-C resulted in a balanced CO_2-C/leaf-C budget in the organic field and indicated sustainable soil management. Calculations for the C balance have not taken into account the fragmentation and mixing of plant residues with the soil by soil invertebrates, which result in improved soil structure and fertility. Conversely, the estimated annual loss of CO_2-C exceeds the leaf-C produced by the conventional field resulting in a long term loss of organic matter. Therefore, it is expected that soil degradation will continue in the conventional field if the current soil management practices remain unchanged.

DISCLAIMER

Use of trade names does not indicate product endorsement by the authors or organizations involved with this project.

REFERENCES

Barber, S.A. 1990. Soil nutrient bioavailability: A mechanistic approach. John Wiley & Sons, New York.

Bouche', M.B. 1977. Strategies lombriciennes. Biol. Bull. (Stockholm) 25:122–132.

Crossley, D.A., Jr., and J.M. Blair. 1991. A high-efficiency, "low technology" Tullgren-type extractor for soil microarthopods. *In* D.A. Crossley, Jr., et al. (ed.). Modern techniques in soil ecology. Elsevier, Amsterdam, the Netherlands.

Doran, J.W. 1980a. Microbial changes associated with residue management with reduced tillage. Soil Sci. Soc. Am. J. 44:518–524.

Doran, J.W. 1980b. Soil microbial and biochemical changes associated with reduced tillage. Soil Sci. Soc. Am. J. 44:765–771.

Doran, J.W., L.N. Mielke, and S. Stamatiadis. 1988. Microbial activity and N cycling as regulated by soil water-filled pore space. p. 49–56. *In* B.D. Witney et al. (ed.) Tillage and traffic in crop production. 11th Int. Soil Tillage Research Organization Conf., Edinburgh, Scotland.

Doran, J.W., and Werner, M.R. 1990. Management and soil biology. p. 205–230. *In* C.A. Francis et al. (ed.) Sustainable agriculture in temperate zones. John Wiley & Sons, New York.

Doran, J.W. 1995. On-farm measurement of soil quality indices. p. 28–41. *In* A.J. Jones et al. (ed.) After CRP: Soil quality handbook. Univ. of Nebraska and USDA-ARS, Lincoln.

Edwards, C.A., and J.R. Lofty. 1972. Biology of earthworms. Chapman & Hall, London.

Hendrix, P.F., R.W. Parmelee, D.A. Crossley, Jr., D.C. Coleman, E.P. Odum, and P.M. Groffman. 1986. Detritus food webs in conventional and no-tillage agroecosystems. Bioscience 36:374–380.

Holland, E.A., and D.C. Coleman. 1987. Litter placement effects on microbial and organic matter dynamics in an agroecosystem. Ecology 68:425–433.

Linn, D.M., and J.W. Doran. 1984. Effect of water-filled pore space on carbon dioxide and nitrous oxide production in tilled and nontilled soils. Soil Sci. Soc. Am. J. 48:1267–1272.

Mengel, K., and E.A. Kirkby. 1978. Principles of plant nutrition. Int. Potash Inst., Bern.

Mielke, L.N., J.W. Doran, and K.A. Richards. 1986. Physical environment near the surface of plowed and no-tilled soils. Soil Tillage Res. 7:355–366.

Petersen, R.G., and L.D. Calvin. 1986. Sampling. p. 33–51. *In* A. Klute (ed.) Methods of soil analysis. Part 1. 2nd ed. Agron. Monogr. 9. ASA and SSSA, Madison, WI.

Richards, B.N. 1987. The microbiology of terrestrial ecosystems. Longman Scientific & Technical, London.

Wallwork, J.A. 1976. The distribution and diversity of soil fauna. Academic Press, London.

Werner, M.R., and D.L. Dindal. 1987. Nutritional ecology of soil arthropods. p. 815–836. *In* F. Slansky, Jr., and J.G. Rodriguez (ed.) Nutritional ecology of insects, mites, spiders and related invertebrates. Wiley Interscience, New York

25 Soil Quality Information Sheets

Gary B. Muckel
USDA-NRCS, National Soil Survey Center
Lincoln, Nebraska

Maurice J. Mausbach
USDA-NRCS, Soil Quality Institute
Ames, Iowa

BACKGROUND

Interest in soil quality has flourished since publication of the National Research Council report *Soil And Water Quality: An Agenda for Agriculture*; however, technical communications have primarily been restricted to exchanges between scientists and general interest articles in trade magazines. Little technical information has been distributed within the Natural Resources Conservation Service (NRCS), an agency with the capability to implement soil quality operationally through direct assistance to landowners.

A soil quality team of soil scientists, an agronomist, geologist, and public affairs person at the National Soil Survey Center was charged to develop soil quality informational products for the agency. As part of this technology transfer effort, interviews were conducted with field office personnel and other potential users to obtain feedback on their needs. These customers expressed a need for information on soil quality and suggested that it be in the form of a one page information sheet. The team then developed a series of one page (printed front and back) soil quality fact sheets to train and inform field office staffs about different aspects of soil quality. The fact sheets have pictures and/or diagrams and are suitable for use by land managers and others.

Printed copies of the soil quality information sheets are available at State offices of the NRCS. They also are available in digital format on disks or the Internet so that state technical teams can adapt them to local conditions. The Internet url is: http://www.statlab.iastate.edu/survey/sqi/sqiinfo.shtml.

Topics of these initial soil quality information sheets are: (i) Soil quality: Introduction; (ii) Indicators for soil quality evaluation; (iii) Soil quality indicators: Aggregate stability; (iv) Soil quality indicators: Soil crusts; (v) Soil quality indicators: Organic matter; (vi) Soil quality resource concerns: Soil erosion; (vii) Soil quality resource concerns: Compaction; and (viii) Soil quality resource concerns: Sediment deposition on cropland. Sheets 2, 3, and 5 are presented in this

---

Copyright © 1996 Soil Science Society of America, 677 S. Segoe Rd., Madison, WI 53711, USA.
Methods for Assessing Soil Quality, SSSA Special Publication 49.

chapter. They were authored by Gary Muckel, Robert Grossman, and Ellis Knox respectively. Titles and contents of the other information sheets are as follow:

Soil Quality: Introduction

Gary Muckel is the primary author. Contents include: What is soil? What does soil do for us? What is soil quality? How is soil quality important to land owners? How can soil quality be evaluated? How can my awareness of soil quality be applied? What concerns are addressed by soil quality?

Soil Quality Indicators: Soil Crusts

Robert Grossman is the primary author. Contents include: What are soil crusts? Why are soil crusts a concern? How do crusts form? How are soil crusts measured? How can the problem be corrected?

Soil Quality Resource Concerns: Soil Erosion

Gary Muckel is primary author. Contents are: What is erosion? Why should we be concerned? What are some signs of erosion? How can soil erosion be measured? What causes the problem? How can soil erosion be avoided?

Soil Quality Resource Concerns: Compaction

Robert Grossman is author. Contents are: What is compaction? Why is compaction a problem? What causes soil compaction? How long will compaction last? How do organic matter and compaction interact? How can compaction be reduced?

Soil Quality Resource Concerns: Sediment Deposition on Cropland

Lyle Steffen is primary author. Contents are: What is sediment deposition? How is soil quality affected? How is sediment deposition identified? What can be done about sediment deposition?

Arlene Tugel of the Soil Quality Institute and Doug Karlen of the National Soil Tilth Laboratory, Ames, IA, provided excellent review comments.

Suggestions for additional topics are welcomed as is the expertise to develop the fact sheet.

Soil Quality Information Sheet
Indicators for Soil Quality Evaluation

| USDA Natural Resources Conservation Service | April 1996 |

What are indicators?

Soil quality indicators are physical, chemical, and biological properties, processes, and characteristics that can be measured to monitor changes in the soil.

The types of indicators that are the most useful depend on the function of soil for which soil quality is being evaluated. These functions include:
 – providing a physical, chemical, and biological setting for living organisms;
 – regulating and partitioning water flow, storing and cycling nutrients and other elements;
 – supporting biological activity and diversity for plant and animal productivity;
 – filtering, buffering, degrading, immobilizing, and detoxifying organic and inorganic materials; and
 – providing mechanical support for living organisms and their structures.

Why are indicators important?

Soil quality indicators are important to:
 – focus conservation efforts on maintaining and improving the condition of the soil;
 – evaluate soil management practices and techniques;
 – relate soil quality to that of other resources;
 – collect the necessary information to determine trends;
 – determine trends in the health of the Nation's soils;
 – guide land manager decisions.

What are some indicators?

Indicators of soil quality can be categorized into four general groups: visual, physical, chemical, and biological.

Visual indicators may be obtained from observation or photographic interpretation. Exposure of subsoil, change in soil color, ephemeral gullies, ponding, runoff, plant response, weed species, blowing soil, and deposition are only a few examples of potential locally determined indicators. Visual evidence can be a clear indication that soil quality is threatened or changing.

Physical indicators are related to the arrangement of solid particles and pores. Examples include topsoil depth, bulk density, porosity, aggregate stability, texture, crusting, and compaction. Physical indicators primarily reflect limitations to root growth, seedling emergence, infiltration, or movement of water within the soil profile.

Chemical indicators include measurements of pH, salinity, organic matter, phosphorus concentrations, cation-exchange capacity, nutrient cycling, and concentrations of elements that may be potential contaminants (heavy metals, radioactive compounds, etc.) or those that are needed for plant growth and development. The soil's chemical condition affects soil-plant relations, water quality, buffering capacities, availability of nutrients and water to plants and other organisms, mobility of contaminants, and some physical conditions, such as the tendency for crust to form.

Biological indicators include measurements of micro- and macro-organisms, their activity, or byproducts. Earthworm, nematode, or termite populations have been suggested for use in some parts of the country. Respiration rate can be used to detect microbial activity, specifically microbial decomposition of organic matter in the soil. Ergosterol, a fungal byproduct, has been used to measure the activity of organisms that play an important role in the formation and stability of soil aggregates. Measurement of decomposition rates of plant residue in bags or measurements of weed seed numbers, or pathogen populations also can serve as biological indicators of soil quality.

How are indicators selected?

Soil quality is estimated by observing or measuring several different properties or processes. No single property can be used as an index of soil quality.

The selection of indicators should be based on:
- the land use;
- the relationship between an indicator and the soil function being assessed;
- the ease and reliability of the measurement;
- variation between sampling times and variation across the sampling area;
- the sensitivity of the measurement to changes in soil management;
- compatibility with routine sampling and monitoring;
- the skills required for use and interpretation.

When and where to measure?

The optimum time and location for observing or sampling soil quality indicators depends on the function for which the assessment is being made. The frequency of measurement also varies according to climate and land use.

Soil variation across a field, pasture, forest, or rangeland can greatly affect the choice of indicators. Depending on the function, such factors as the landscape unit, soil map unit, or crop growth stage may be critical. Wheel tracks can dramatically affect many properties measured for plant productivity. Management history and current inputs also should be recorded to ensure a valid interpretation of the information.

Monitoring soil quality should be directed primarily toward the detection of trend changes that are measurable over a 1- to 10-year period. The detected changes must be real, but at the same time they must change rapidly enough so that land managers can correct problems before undesired and perhaps irreversible loss of soil quality occurs.

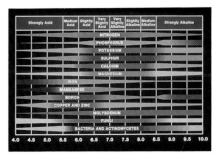

Soil reaction influence on availability of plant nutrients.

What does the value mean?

Interpreting indicator measurements to separate soil quality trends from periodic or random changes is currently providing a major challenge for researchers and soil managers. Soils and their indicator values vary because of differences in parent material, climatic condition, topographic or landscape position, soil organisms, and type of vegetation. For example, cation exchange capacity may relate to organic matter, but it also may relate to the kind and amount of clay.

Establishing acceptable ranges, examining trends and rates of change over time, and including estimates of the variance associated with the measurements are important in interpreting indicators. Changes need to be evaluated as a group, with a change in any one indicator being evaluated only in relation to changes in others. Evaluations before and after, or with and without intervention, also are needed to develop appropriate and meaningful relationships for various kinds of soils and the functions that are expected of them.

The overall goal should be to maintain or improve soil quality without adversely affecting other resources.

(Prepared by the National Soil Survey Center in cooperation with the Soil Quality Institute, NRCS, USDA, and the National Soil Tilth Laboratory, Agricultural Research Service, USDA)

The United States Department of Agriculture (USDA) prohibits discrimination in its programs on the basis of race, color, national origin, sex, religion, age, disability, political beliefs, and marital or familial status. (Not all prohibited bases apply to all programs.) Persons with disabilities who require alternative means for communication of program information (braille, large print, audiotape, etc.) should contact the USDA Office of Communications at (202) 720-2791.

To file a complaint, write the Secretary of Agriculture, U.S. Department of Agriculture, Washington, D.C., 20250, or call (202) 720-7327 (voice) or (202) 720-1127 (TDD). USDA is an equal employment opportunity employer.

Soil Quality Information Sheet
Soil Quality Indicators: Organic Matter

USDA Natural Resources Conservation Service	April 1996

What is soil organic matter?

Soil organic matter is that fraction of the soil composed of anything that once lived. It includes plant and animal remains in various stages of decomposition, cells and tissues of soil organisms, and substances from plant roots and soil microbes. Well-decomposed organic matter forms *humus*, a dark brown, porous, spongy material that has a pleasant, earthy smell. In most soils, the organic matter accounts for less than about 5% of the volume.

What does organic matter do?

Organic matter is an essential component of soils because it:
 - provides a carbon and energy source for soil microbes;
 - stabilizes and holds soil particles together, thus reducing the hazard of erosion;
 - aids the growth of crops by improving the soil's ability to store and transmit air and water;
 - stores and supplies such nutrients as nitrogen, phosphorus, and sulfur, which are needed for the growth of plants and soil organisms;
 - retains nutrients by providing cation-exchange and anion-exchange capacities;
 - maintains soil in an uncompacted condition with lower bulk density;
 - makes soil more friable, less sticky, and easier to work;
 - retains carbon from the atmosphere and other sources;
 - reduces the negative environmental effects of pesticides, heavy metals, and many other pollutants.

Soil organic matter also improves tilth in the surface horizons, reduces crusting, increases the rate of water infiltration, reduces runoff, and facilitates penetration of plant roots.

Where does it come from?

Plants produce organic compounds by using the energy of sunlight to combine carbon dioxide from the atmosphere with water from the soil. Soil organic matter is created by the cycling of these organic compounds in plants, animals, and microorganisms into the soil.

What happens to soil organic matter?

Soil organic matter can be lost through erosion. This process selectively detaches and transports particles on the soil surface that have the highest content of organic matter.

Soil organic matter also is used by soil microorganisms as energy and nutrients to support their own life processes. Some of the material is incorporated into the microbes, but most is released as carbon dioxide and water. Some nitrogen is released in gaseous form, but some is retained, along with most of the phosphorus and sulfur.

When soils are tilled, organic matter is decomposed faster because of changes in water, aeration, and temperature conditions. The amount of organic matter lost after clearing a wooded area or tilling native grassland varies according to the kind of soil, but most organic matter is lost within the first 10 years.

Rates of decomposition are very low at temperatures below 38°F (4°C) but rise steadily with increasing temperature to at least 102°F (40°C) and with water content until air becomes limiting. Losses are higher with aerobic decomposition (with oxygen) than with anaerobic decomposition (in excessively wet soils). Available nitrogen also promotes organic matter decomposition.

What controls the amount?

The amount of soil organic matter is controlled by a balance between additions of plant and animal materials and losses by decomposition. Both additions and losses are very strongly controlled by management activities.

The amount of water available for plant growth is the primary factor controlling the production of plant materials. Other major controls are air temperature and soil fertility. Salinity and chemical toxicities also can limit the production of plant biomass. Other controls are the intensity of sunlight, the content of carbon dioxide in the atmosphere, and relative humidity.

The proportion of the total plant biomass that reaches the soil as a source of organic matter depends largely on the amounts consumed by mammals and insects, destroyed by fire, or produced and harvested for human use.

Practices decreasing soil organic matter include those that:

1. Decrease the production of plant materials by
   - replacing perennial vegetation with short-season vegetation,
   - replacing mixed vegetation with monoculture crops,
   - introducing more aggressive but less productive species,
   - using cultivars with high harvest indices,
   - increasing the use of bare fallow.

2. Decrease the supply of organic materials by
   - burning forest, range, or crop residue,
   - grazing,
   - removing plant products.

3. Increase decomposition by
   - tillage,
   - drainage,
   - fertilization (especially with nitrogen).

Practices increasing soil organic matter include those that:

1. Increase the production of plant materials by
   - irrigation,
   - fertilization to increase plant biomass production,
   - use of cover crops
   - improved vegetative stands,
   - introduction of plants that produce more biomass,
   - reforestation,
   - restoration of grasslands.

2. Increase supply of organic materials by
   - protecting from fire,
   - using forage by grazing rather than by harvesting,
   - controlling insects and rodents,
   - applying animal manure or other carbon-rich wastes,
   - applying plant materials from other areas.

3. Decrease decomposition by
   - reducing or eliminating tillage,
   - keeping the soil saturated with water (although this may cause other problems),
   - keeping the soil cool with vegetative cover.

(Prepared by the National Soil Survey Center in cooperation with the Soil Quality Institute, NRCS, USDA, and the National Soil Tilth Laboratory, Agricultural Research Service, USDA)

The United States Department of Agriculture (USDA) prohibits discrimination in its programs on the basis of race, color, national origin, sex, religion, age, disability, political beliefs, and marital or familial status. (Not all prohibited bases apply to all programs.) Persons with disabilities who require alternative means for communication of program information (braille, large print, audiotape, etc.) should contact the USDA Office of Communications at (202) 720-2791.

To file a complaint, write the Secretary of Agriculture, U.S. Department of Agriculture, Washington, D.C., 20250, or call (202) 720-7327 (voice) or (202) 720-1127 (TDD). USDA is an equal employment opportunity employer.

Soil Quality Information Sheet

Soil Quality Indicators: Aggregate Stability

USDA Natural Resources Conservation Service April 1996

What are soil aggregates?

Soil aggregates are groups of soil particles that bind to each other more strongly than to adjacent particles. The space between the aggregates provide pore space for retention and exchange of air and water.

What is aggregate stability?

Aggregate stability refers to the ability of soil aggregates to resist disruption when outside forces (usually associated with water) are applied.

Aggregate stability is not the same as *dry aggregate stability*, which is used for wind erosion prediction. The latter term is a size evaluation.

Why is aggregate stability important?

Aggregation affects erosion, movement of water, and plant root growth. Desirable aggregates are stable against rainfall and water movement. Aggregates that break down in water or fall apart when struck by raindrops release individual soil particles that can seal the soil surface and clog pores. This breakdown creates crusts that close pores and other pathways for water and air entry into a soil and also restrict emergence of seedlings from a soil.

Optimum conditions have a large range in pore size distribution. This includes large pores between the aggregates and smaller pores within the aggregates. The pore space between aggregates is essential for water and air entry and exchange. This pore space provides zones of weakness through which plant roots can grow. If the soil mass has a low bulk density or large pore spaces, aggregation is less important. For example, sandy soils have low aggregation, but roots and water can move readily.

How is aggregate stability measured?

Numerous methods measure aggregate stability. The standard method of the NRCS Soil Survey Laboratory can be used in a field office or in a simple laboratory. This procedure involves repeated agitation of the aggregates in distilled water.

An alternative procedure described here does not require weighing. The measurements are made on air-dry soil that has passed through a sieve with 2-millimeter mesh and retained by a sieve with a 1-millimeter mesh. A quantity of these 2-1 millimeter aggregates is placed in a small open container with a fine screen at the bottom. This container is placed in distilled water. After a period of time, the container is removed from the water and its contents are allowed to dry. The content is then removed and visually examined for the breakdown from the original aggregate size. Those materials that have the least change from the original aggregates have the greatest aggregate stability.

Soils that have a high percentage of silt often show lower aggregate stability if measured air-dry than the field behavior would suggest, because water entry destroys the aggregate structure.

What influences aggregate stability?

The stability of aggregates is affected by soil texture, the predominant type of clay, extractable iron, and extractable cations, the amount and type of organic matter present, and the type and size of the microbial population.

Some clays expand like an accordion as they absorb water. Expansion and contraction of clay particles can shift and crack the soil mass and create or break apart aggregates.

Calcium ions associated with clay generally promote aggregation, whereas sodium ions promote dispersion.

Soils with over about five percent iron oxides, expressed as elemental iron, tend to have greater aggregate stability.

Soils that have a high content of organic matter have greater aggregate stability. Additions of organic matter increase aggregate stability, primarily after decomposition begins and microorganisms have produced chemical breakdown products or mycelia have formed.

Soil microorganisms produce many different kinds of organic compounds, some of which help to hold the aggregates together. The type and species of microorganisms are important. Fungal mycelial growth binds soil particles together more effectively than smaller organisms, such as bacteria.

Aggregate stability declines rapidly in soil planted to a clean-tilled crop. It increases while the soil is in sod and crops, such as alfalfa.

(Prepared by the National Soil Survey Center in cooperation with the Soil Quality Institute, NRCS, USDA, and the National Soil Tilth Laboratory, Agricultural Research Service, USDA)

The United States Department of Agriculture (USDA) prohibits discrimination in its programs on the basis of race, color, national origin, sex, religion, age, disability, political beliefs, and marital or familial status. (Not all prohibited bases apply to all programs.) Persons with disabilities who require alternative means for communication of program information (braille, large print, audiotape, etc.) should contact the USDA Office of Communications at (202) 720-2791.

To file a complaint, write the Secretary of Agriculture, U.S. Department of Agriculture, Washington, D.C., 20250, or call (202) 720-7327 (voice) or (202) 720-1127 (TDD). USDA is an equal employment opportunity employer.

26 Measuring Sustainability of Agricultural Systems at the Farm Level

A. A. Gomez
University of the Philippines
Los Baños, Philippines

David E. Swete Kelly
Department. of Primary Industries
Maroochy Horticultural Research Station
Queensland Australia

J. K. Syers
Department of Agriculture and Environmental Science
University of Newcastle Upon Tyne
Newcastle Upon Tyne, England

K. J. Coughlan
Australian Centre for International Agricultural Research
Canberra, Australia

OVERVIEW

This chapter bases sustainability evaluation on the multifaceted FESLM (Framework for the Evaluation of Sustainable Land Management) developed by FAO and IBSRAM (Smyth et al., 1993). The work reported evaluates sustainability at the on-farm level. It proposes a preliminary list of field indicators, provides examples from actual measurement and outlines a method to visually, and quantitatively represent results for easy analysis and comparison.

An agricultural system is said to be sustainable at the farm level if it satisfies the farm manager's needs (over time) while conserving the natural resource. Resource conservation is handled separately from farmer satisfaction. Farmer's satisfaction includes issues such as productivity, profitability, stability, and social acceptability.

Copyright © 1996 Soil Science Society of America, 677 S. Segoe Rd., Madison, WI 53711, USA.
Methods for Assessing Soil Quality, SSSA Special Publication 49.

Selection of field indicators is not yet complete. Initial screening for measurable and meaningful surrogates has revealed yield, profit, and frequency of crop failure as field indicators for farmer satisfaction; and soil depth, organic C content, and percentage of ground cover for resource conservation.

An indicator is said to be at a sustainable level if it exceeds a designated trigger or threshold level as given below. The thresholds are tentatively set as improvements on community averages. Those for resource conservation include an absolute minimum. Indicators are expressed as units of their respective threshold levels, where one equals the threshold.

Indication	Threshold level
Yield	20% more than average yield in the community
Profit	20% better than average of the community
Frequency of crop failure	20%, or average frequency for the community whichever is lower
Soil depth	50 cm or average of similar soil types in the community
Organic C	1% or average of the community, whichever is higher
Permanent ground cover	15% or average of the community, whichever is higher

On this basis an agricultural system is not sustainable if the average of indicators for either farmer satisfaction or resource conservation is less than one. The sustainability of the system can be represented in two ways: (i) as the combined average of the ratings for farmer satisfaction and resource conservation, and (ii) as a cobweb polygon for farmer satisfaction and resource conservation.

This procedure for evaluating sustainability was applied to actual data from farms in Guba, Cebu, Philippines. Our experience is that the procedure can be implemented easily. The results also are consistent with our expectations. The process allows the worker to compare farms or farming systems and monitor changes over time. It also allows comparison of different scenarios by altering thresholds or adding other indicators.

Sustainable agriculture has been equated to almost all that is good for the farmer, his farm, and the wider environment. Profitability, stability, productivity, acceptability, and environmental friendliness are some of the qualities now associated with sustainable agriculture. Considering that each of these qualities is complex and can be defined in several ways, it is no surprise that the definition and measurement of sustainable agriculture has been very elusive.

INTRODUCTION

There are two potential approaches for defining and measuring sustainable agriculture. One is based on the principle that the important indicators of sustainability are location specific and change with the situation prevailing on a farm. For example, in the steeplands, soil erosion has a major impact on sustainability, but in the flat lowland rice paddies, soil loss due to erosion is insignificant and may not be a useful indicator. Based on this principle, therefore, the protocol for measuring sustainability starts with a list of potential indicators from which practitioners select a subset of indicators that is felt to be appropriate for the particular farm being evaluated.

The other approach is based on the principle that the definition and consequently the procedure for measuring sustainable agriculture is the same regardless of the diversity of situations that prevails on different farms. Under this principle, sustainability is defined by a set of requirements that must be met by any farm regardless of the wide differences in the prevailing situation. For example, in the steeplands and in the lowland rice paddies, described above, soil erosion is an important indicator of sustainability, accepting that this requirement is more easily met in the latter situation.

There are clear advantages and disadvantages between these two approaches to assessing sustainability. The principle of location specificity avoids the difficulty of selecting and agreeing on a common set of indicators, a task that is always controversial. In addition it allows each practitioner the freedom to choose their own indicators, a feature that is very attractive among workers at the grassroots level. A major drawback with the location specific approach is the difficulty of comparing results from farms where different indicators have been selected. Here lies the strength of the second approach of constant indicators across all farms. All measurements are based on the same indicators and the results are comparable across farms and are easier to analyze for repeatability and replicability.

This chapter assumes that the second principle of a common definition and set of indicators for measuring sustainability is a much more powerful and useful concept for studying sustainable agriculture. It proposes a protocol for measuring sustainability at the farm level by: (i) defining the requirements for sustainability, (ii) selecting the common set of indicators, (iii) specifying the threshold levels, (iv) transforming the indicators into a sustainability index, and (v) testing the procedure using a set of data from selected farms in the Philippines.

This chapter represents the majority of information that was presented at an international symposium on *Advances in Soil Quality for Land Management* held in Ballarat, Victoria, Australia, in April, 1996 (Gomez et al., 1996).

DEFINITION AND REQUIREMENTS FOR SUSTAINABILITY

At the farm level, a farming system is considered sustainable if it conserves the natural resource and continues to satisfy the needs of the farmer, the manager of the system. Any system that fails to satisfy these two requirements is bound to change significantly over the short term and is therefore considered not sustainable.

Farmer satisfaction and resource conservation, the two requirements of sustainability, are not simple characters but are influenced by a host of factors. High yield, low labor requirement, low input cost, high profit, and stability are some of the features that are likely to enhance farmer satisfaction. Natural resource conservation, however, is usually associated with soil depth, water holding capacity, nutrient balance, organic matter content, ground cover, and biological diversity.

This definition is similar to the Framework for Evaluating Sustainable Land Management (FESLM), proposed by FAO and IBSRAM. The first four pillars of FESLM, productivity, stability, viability, and social acceptability are the main components of farmer satisfaction. Social acceptability has more relevance at the community level parameter and is not included at the farm level. The fifth pillar of FESLM, protection–conservation, is handled under resource conservation.

THE INDICATORS OF SUSTAINABILITY

Even with the simplified requirement for sustainability at the farm level, the number of indicators that are commonly mentioned are many. Shown in Appendix 1 is a list of some of these indicators and the procedure for measuring them. It is clear that several indicators are closely related to each other and it is not necessary to measure all of them. Those that should be selected must possess one or more of the following features: (i) be easy to measure, (ii) respond easily to change, (iii) have obvious boundaries (threshold) separating sustainable from unsustainable conditions, and (iv) be directly related to the two requirements for sustainability.

Using the above guidelines, the following indicators were initially selected: yield, net income, and variance of profit as indicators of farmer satisfaction and soil loss, nutrient balance, and organic C as indicators of resource conservation; however, variance of profit, soil loss, and nutrient balance were considered too difficult to measure directly and the following surrogate indicators were used instead: frequency of crop failure, soil depth, and percentage of permanent ground cover.

Of the six indicators selected, only the last one, permanent ground cover poses a problem in terms of universality. For example, in steepland where soil conservation practices are needed, permanent ground cover serves as a useful indicator; however, in the flatlands where soil conservation may be of little importance, ground cover may not be so relevant.

THE TRIGGER OR THRESHOLD LEVEL

The term threshold level is used to denote the boundary between sustainable and unsustainable values. Unless this threshold level is specified for each indicator, it is not possible to distinguish between sustainable and unsustainable conditions.

In this chapter, the primary basis for the threshold level is the average of the community instead of an absolute value for all situations. This seems reasonable since farmers usually judge their state of well being on the basis of their position relative to their neighbors, and since farms that apply good conservation practices are expected to retain their initial resource endowment. With this procedure it is expected that the threshold levels for communities with widely different eco-

Table 26–1. Threshold levels for sustainability indicators.

Indicator	Threshold level	Threshold formulae
Yield (X_1)	20% more than average yield in the community	1.2 (Mean x_1)
Profit (X_2)	20% better than average in the community, whichever is lower	1.2 (Mean x_2)
Frequency of crop failure (X_3)	20%, or average frequency for the community, whichever is lower	0.20 when the mean of $x_3 > 0.20$, mean of x_3 otherwise.
Soil depth (X_4)	50 cm or the average of similar soil types in the community, whichever is greater.	Mean x_4 or 50 cm, whichever is greater.
Organic C (X_5)	1%, or average of community, whichever is higher	0.01 when mean $x_5 < 0.01$, mean x_5 otherwise
Permanent ground cover (X_6)	15%, or average of community, whichever is higher	0.15 when mean $x_6 < 0.15$, mean x_6 otherwise

nomic and biophysical environments also will differ widely. Shown in Table 26–1 are the threshold levels for the indicators used in measuring sustainability.

THE COMPUTATIONAL PROCEDURE

To illustrate the procedure for computing the sustainability index at the farm level, we use data from 10 farms in Guba, Cebu, Philippines (Table 26–2). Guba is a farming community of about 1000 households cultivating the slopes of the mountains surrounding Cebu City. About 15 yr ago, the World Neighbors, a church based organization, decided to introduce contour hedgerow farming into Guba in an effort to conserve the soil and related resources. Today about 60% of the community has adopted the new technology. Of the 10 farms given in Table 26–2, the first six are adaptors of the contour hedgerow technology while the remaining four are not. For data given in the table, yield, profit and frequency of crop failure are survey data while soil depth, organic C and permanent cover are measurement data.

Table 26–2. Sustainability indicators for 10 farms in Guba, Cebu, Philippines.

	Farmer satisfaction			Resource conservation		
Farm no.	Yield	Net income	Frequency of crop failure	Soil depth	Organic C	Permanent ground cover
	T/ha	$/ha	%	cm	%	%
1	1.88	252	15	117	1.15	25
2	1.42	163	20	80	0.52	14
3	1.43	195	20	87	.72	17
4	2.02	247	30	37	.60	14
5	1.75	203	25	86	1.26	16
6	1.62	227	25	70	0.80	14
7	.88	38	20	47	1.61	7
8	.52	30	15	27	0.82	0
9	.98	116	20	100	1.74	0
10	.81	29	15	42	0.82	1
Average	1.33	150	20.5	69.3	1.06	10.8
Threshold	1.60	180	20.0	69.3	1.06	15.0

Table 26–3. Sustainability indices for 10 farms in Guba, Cebu, Philippines.

Farm no.	Satisfaction farmer				Resource conservation				Sustainability index
	Yield	Profit	Crop failure	Index	Depth	Organic C	Ground cover	Index	
1	1.18	1.40	1.33	1.30	1.69	1.65	1.66	1.66	1.48
2	0.89	0.90	1.00	0.93	1.15	0.49	0.93	0.85	NS
3	0.89	1.08	1.00	0.99	1.25	0.68	1.13	1.02	NS
4	1.26	1.37	0.66	1.10	0.54	0.57	0.93	0.68	NS
5	1.09	1.13	0.80	1.01	1.24	1.18	1.07	1.16	1.08
6	1.01	1.26	0.80	1.02	1.01	0.75	0.93	0.89	NS
7	0.55	0.21	1.00	0.59	0.68	1.51	0.47	0.88	NS
8	0.32	0.16	1.33	0.60	0.39	0.77	0.00	0.38	NS
9	0.61	0.64	1.00	0.75	1.44	1.64	0.00	1.02	NS
10	0.51	0.16	1.33	0.67	0.61	0.77	0.07	0.48	NS

The computation of the index of sustainability for each of the ten farms is as follows:

Step 1. Specify the threshold level for each indicator following the formula given in Table 26–1. Convert all measurements into threshold units as shown in Table 26–3.

Step 2. Represent the relative sustainability of farms graphically for visual comparison (Fig. 26–1). Note that the specific components that result in reduced sustainability are easily seen from these graphs.

Step 3. Compute the indices for farmer's satisfaction and resource conservation as the average of their three respective indicators. These two averages must both be ≥1.0 for the system to be judged sustainable. For our example, only farms no. 1 and no. 5 are judged sustainable.

Step 4. For sustainable cases, compute the average of the two indices. This average is the final index of sustainability, which is equal to 1.48 for farm no. 1 and 1.08 for farm no. 5. Note that the sustainability index is computed for sustainable systems only, i.e., no index is computed for farm that are judged nonsustainable. Thus, the sustainability index is always positive and >1.0, the higher the value, the more sustainable.

SOME NOTES ON INDEX

As a consequence of the procedure with which the index is computed, several characteristic features are worth noting. These features are discussed below:

The Requirements for Sustainability

An average rating of more than 1.0 for farmer's satisfaction and resource conservation is necessary for the system to be sustainable. This requirement can be met even if some indicators are below the threshold level (i.e., <1.0). For example, average rating for farmer's satisfaction or resource conservation may exceed 1.0 even if one or more indicator has a rating of <1.0. This means that a deficiency in one indicator can be compensated for by excess capacity in another. For example, in farm no. 5, frequency of crop failure is below threshold but

Sustainability cobweb for two farms in Cebu, Philippines

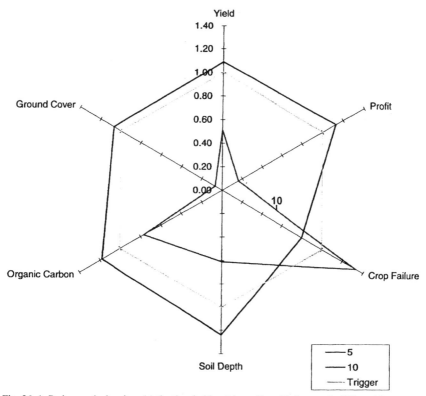

Fig. 26–1. Radar graph showing: (a) the threshold or trigger line, (b) the sustainability of farm number 5 with a bounded area exceeding that of the threshold even as one indicator is below threshold, and (c) the unsustainable situation in farm number ten with five out of six indicators below threshold.

yield and income are high enough to compensate for the deficiency. Note, however, that this ability to compensate is allowable only among indicators of the same index, (i.e., within farmer satisfaction) but not across. Thus excess rating in yield or income cannot compensate for deficiencies in soil depth and organic C.

Sustainability at the Community Level

Changes in the threshold level, over time, is a key indicator of sustainability at the community level. Note that communities that upgrade their management practices should consistently improve their level of productivity and natural resource endowment which then should be reflected in ever improving threshold levels. Thus, improving threshold level, over time, is indicative of sustainability at the community level; and conversely, a decreasing trend indicates nonsustainability.

The Radar Graph

This graph is a good tool to immediately visualize and identify the specific component practices that contribute to reduced sustainability. It helps us understand the differences across farms or over time within the same farm. Hence, overall sustainability is not just reduced to a single analogue derived from a common perspective but becomes a useful visual tool to planning for further action.

Concern has been expressed that the approach gives equal weighting to each of the indices, whereas some workers consider some to be more important than others. The graphical representation goes some of the way to addressing this. Individual workers can see the relative contribution of each index and draw conclusions based on their personal interpretation of the importance of each.

Level of Index

It should be noted that once the sustainability requirement is satisfied, a general index is computed whose value is indicative of the extent to which the threshold level is surpassed. For example, an index of one indicates that the system is at threshold level, an index of two means that the system is two times the threshold, and so on.

Flexibility to Accommodate Additional Indicators

In terms of procedure it should be obvious that there is no difficulty in accommodating additional indicators under each of the two main pillars. Since the indices are averaged across indicators adding more indicators should not unduly complicate the process nor the level of comparability among indices.

Farmer's Satisfaction and Resource Conservation

The net effect of grouping indicators into two requirements for sustainability is to reduce the strictness with which farms can be judged as sustainable. The fact that there is a given level from substitutability among indicators in the same requirement group results from this reduced strictness. Note that if all the six indicators have to exceed the threshold for the system to be sustainable, then fewer farms will pass the requirement for sustainability. This is clearly illustrated by the 10 farms in Guba. Two farms are judged sustainable under the present procedure. Otherwise only one farm (farm no. 1) would pass.

CONCLUSION

The procedure outlined in this chapter has been developed from a definition of sustainability at the farm level that provides two sets of indicators relative to farmer satisfaction and resource conservation. It has been selected for its ease in implementation both in data gathering and in data analysis. Experience in applying this procedure to farms in Guba strongly corroborates this desired simplicity. The data are easy to gather and the analysis is simple. We plan to repeat the process in another community where measurement data will be used for all indicators.

The two approaches to measuring sustainability, i.e., location specific versus constant indicators across farms, is closely related to the principle of substitutability among indicators. The location specific approach does not allow for substitution but requires that all indicators are above their respective threshold level. This is so since the indicators selected for each particular situation are those that are likely to be lower than threshold. This is why soil loss is a good indicator for steeplands where soil erosion can be high, but is not so in the flat lands where erosion is low. If all farms are to be assessed on their weakest points then this is likely to give a very pessimistic picture of sustainability.

In the constant indicator approach, however, a selected indicator is measured for all farms regardless of its likelihood or nonlikelihood of violating the threshold. Indicators are selected for their own merit. Thus this approach is less targeted and more farms are likely to pass the sustainability test.

REFERENCES

Gomez, A.A., D.E. Swete Kelly, and J. K. Syers. 1996. Measuring the sustainability of agricultural systems at the farm level. *In* Proc. of an Int. Symp. on Advances in Soil Quality for Sustainable Land Management: Science, Practice, and Policy, Univ. of Ballarat, Ballarat, Victoria, Australia.

Smyth, A.J., J. Dumanski, G. Spendjian, M.J. Swift, and P.K. Thornton. 1993. FESLM: An international framework for evaluating sustainable land management. A discussion paper. FAO, Rome.

Appendix 1
Summary of the commonly mentioned sustainability parameters.

INDICATORS FOR FARMERS SATISFACTION	MEASUREMENTS
Productivity	
Net return to land	Economic outputs, economic inputs, farm-gate prices (inc. imputed prices), using direct measurement, periodic interviews, market surveys kg/ha, kg/person/year
Net return to labor	
Total factor productivity	
Yield	
Viability	
Cash flow; discounted cash flow	As above, over time (measured or projected; interest rates on farm credit (explicit or implicit); food surveys
Flow of net benefits; net present value	
Net farm income (after farm development)	
Flow of staple food availability	
Stability	
CV of productivity measures	Measurements of inputs and outputs, costs and returns, over time for each test farm; periodic number
CV of net benefits	
Diversity of enterprises	Number and kind of enterprises
Net returns in worst 20% of trials (minimum returns analysis)	Measurement of key elements (e.g., yield, output price) across a sample of farms
Acceptability	
Labor	Person days per year
Membership of community organisations	Number of organisations, type of organisations
Adoption indices	Adoption surveys examining degrees of adoption, farmer opinion, and likely constraints (e.g., tenure status).
Farmer ratings	Opinion poll of farmers, e.g., at a field day

INDICATORS FOR RESOURCE CONSERVATION	MEASUREMENT
Soil loss (gain)	Amount of soil formed - amount of soil loss
Woody perennial population	Area of woody perennials/total farm area
Soil nutrient budget	Added nutrient vs. biomass removed
Turbidity index	Suspended solids in run-off water
Erodability index	Soil loss under controlled rainfall simulation
Ecological diversity	Shannon's index (the total number of species cultivated, collected or used on the farm)
Topography	Slope, slope length
Soil stability	Water dispersable clay
Nutrient cycling	Finn's Cycling Index (Proportion of the nutrients within the system which are recycled within the system)
Bio-resource recycling	The total number of farm generated biological material flow within the farming system.
C:N ratio	Organic Carbon: Mineralisable Nitrogen ratio over time
Soil compaction	Soil resistance to penetration over time
Calico index	Degradation in tensile strength of a calibrated strip of buried calico over time. Surrogate measure of soil biological activity.
Ground cover	Averaged percent of soil surface covered by living or dead mulch during wet weeks (>50 mm rainfall per week)
Water stress	Crop rotation stress days per year